T0136244

# The Colorado Plateau V

# The Colorado Plateau V

RESEARCH, ENVIRONMENTAL PLANNING,
AND MANAGEMENT FOR COLLABORATIVE
CONSERVATION

Edited by Charles van Riper III, Miguel L. Villarreal,
Carena J. van Riper, and Matthew J. Johnson

The University of Arizona Press
Tucson

This volume is based on research presented at the Tenth Biennial Conference of Research on the Colorado Plateau held at Northern Arizona University, Flagstaff, Arizona, and hosted by the U.S. Geological Survey Southwest Biological Science Center Colorado Plateau Research Station, the Miriam-Powell Center for Environment Research, Bureau of Land Management, National Park Service, Diablo Trust, and the Center for Sustainable Environments at Northern Arizona University.

ISBN 978-0-8165-2978-0
Library of Congress Control Number: 2012936876

Manufactured in the United States of America on acid-free archival-quality paper containing a minimum of 30% post-consumer waste and processed chlorine free.

15  14  13  12  11  10   6 5 4 3 2 1

*This book was published from formatted electronic copy that was edited and typeset by the volume editors.*

# CONTENTS

## FOREWORD

The Colorado Plateau covers 140,000 square miles of sparsely vegetated plateaus, mesas, and canyons in the Four Corners region, which includes Arizona, Utah, Colorado, and New Mexico. Despite its arid climate, the Colorado Plateau is one of the most ecologically diverse regions in North America. With elevations ranging from 3,000 to 14,000 feet, the natural systems found across the Colorado Plateau are as varied as the landscape. As a result, ecosystems that represent all of Merriam's life zones, ranging from desert through grasslands and forests to alpine habitats, are found within. The region's diverse and distinctive ecosystems have resulted in high rates of plant endemism and species richness for vertebrates and invertebrates alike.

Human impacts on the Colorado Plateau's natural environment have been extensive and, in some cases, have altered the natural function of the region's ecosystems. For example, the Colorado River, which drains some 90 percent of the plateau, has been harnessed to produce hydropower and diverted to provide water to Las Vegas, Los Angeles, Phoenix, and Tucson. Extraction of the region's extensive energy resources—coal, oil, natural gas, uranium, and oil shale—has resulted in impacts that may be localized but can be severe. It is not extractive human activities alone that impact the region, things as seemingly innocuous as ecotourism, the impacts of which are more diffuse and transparent, can nonetheless lead to a "tragedy of the commons." For example, as more and more people put on running shoes, tie up hiking boots, and jump on their mountain bikes to explore the plateau, alternate routes are created, soil is eroded and compacted, and wildlife feeding and nesting sites may be disrupted. As the magnitude of human impacts on the natural systems of the Colorado Plateau increases, it is imperative that the relationships at work among people, plants, animals, and environment, particularly in the face of ongoing climate change, are better understood.

Developing a better understanding of the living things and natural systems of the Colorado Plateau was the genesis for the first Biennial Conference for Research on the Colorado Plateau, which took place in 1999 on the campus of Northern Arizona University in Flagstaff, Arizona. Over the past two decades, the Colorado Plateau Biennial Conference has continued to be made possible by a partnership between Northern Arizona University and the U.S. Geological Survey, along with many federal, state, and private partners. The series of proceedings produced as a result of these conferences, including this edition, focuses on providing information to Colorado Plateau scientists and land managers. Today, some 20 years, 10 conferences, 1,000-plus presentations, and ten books later, this research continues to provide resource managers with the information they need to improve management decisions to maintain the vitality, diversity, and natural beauty of the Plateau eco-region.

This tenth book presents scientific findings that were shared during the 10th Colorado Plateau Biennial Conference, which took place in 2009, and, as a result, the chapters that follow address the conference's focus on "Collaborative Conservation in Rapidly Changing Landscapes." The book contains 17 chapters of papers contributed by Federal, State, and private sector scientists investigating the cultural, biological, and physical resources of the Colorado Plateau. We are all indebted to the many thoughtful and capable scientists who come together with land managers every other year to share scientific information to support better land management activities on the Colorado Plateau. The value of creating a forum for scientists and managers to share information about current research and resource management issues cannot be overstated.

As climate change, land use, and increasing human populations alter physical, cultural, and biological resources on the lands of the Colorado Plateau, science can—and will—serve to guide decision-making in the face of change. Science can help resource managers, elected officials, and the people of the Colorado Plateau chart a path to a future that can sustain our society and those who follow us—a future that meets not only our physical needs but also feeds our minds and imaginations with the bounty of the Colorado Plateau's natural beauty and cultural heritage. There remains much to be done.

JOHN P. HOFFMANN
U.S. Geological Survey
Acting Center Director
Southwest Biological Science Center
Flagstaff, AZ

# DEDICATION

The Colorado Plateau Biennial conference has been largely sustained by a partnership between Northern Arizona University and the U.S. Geological Survey. The editors would like to dedicate this 10th volume of The Colorado Plateau conference proceedings to two individuals who have committed themselves to this partnership, and also for promoting education and conservation on the Colorado Plateau: Dr. John D. Haeger, President of Northern Arizona University and Dr. Marcia McNutt, Director of the U.S. Geological Survey.

John D. Haeger

## John D. Haeger

Dr. John D. Haeger has been Northern Arizona University (NAU) president since November 2001, after joining the university to serve as NAU provost in June 2000. Over the last decade, Dr. Haeger has endorsed and supported five Colorado Plateau Biennial conferences, with his commitment to education and facilitating learning into management actions—a commitment enhanced by the university's ongoing efforts in research, graduate education and distance learning. Dr. Haeger ushered the university into its largest building boom since the 22-year term of former NAU President J. Lawrence Walkup, and one of the buildings is the High Country Conference center where the Colorado Plateau Biennial conferences are presently held.

As the nation grapples with an increasing need for scientists, resource managers and technical experts, Dr. Haeger is re-emphasizing the university's efforts to attract and retain qualified teachers who will recruit and train students into these fields. These efforts will ultimately benefit resources and their management on the Colorado Plateau.

Further, he has fully endorsed the cooperative national effort with USGS of Colorado Plateau Biennial conferences, such that these events are now recognized nationally as a signature event where managers and research scientists can share information in a collegial environment. This information-sharing has resulted in numerous presentations and products and has created an environment in which Northern Arizona University is contributing on a level highly beneficial to both students and to society.

## Marcia K. McNutt

Director Marcia K. McNutt is a distinguished scientist and administrator and the first woman to act as USGS director in the 130-year history of the agency.

Dr. McNutt previously served as president and chief executive officer of the Monterey Bay Aquarium Research Institute (MBARI), in Moss Landing, CA. She is a native of Minneapolis, MN, where she graduated class valedictorian from Northrop Collegiate School in 1970. She then received a BA degree in Physics, summa cum laude, Phi Beta Kappa, from Colorado College in Colorado Springs, and thus is familiar with the Colorado Plateau ecosystem and supports agency research over this ecosystem. As a National Science Foundation Graduate Fellow, she studied geophysics at Scripps Institution of Oceanography in La Jolla, California, where she earned a PhD in Earth Sciences in 1978. She then spent three years with the USGS in Menlo Park, CA, working on earthquake prediction. Presently, as Director of USGS, Dr. McNutt has continued the cooperative arrangement and relationship between NAU and her agency in support of the Colorado Plateau Biennial conferences.

As a scientist, Dr. McNutt has participated in 15 major oceanographic expeditions and served as chief scientist on more than half of those voyages. She has published 90 peer-reviewed scientific articles. Her research has ranged from studies of ocean island volcanism in French Polynesia, to continental break-up in the Western United States, to uplift of the Tibet Plateau. She served as President of the American Geophysical Union from 2000-2002, was Chair of the Board of Governors for Joint Oceanographic Institutions, and helped to bring about its merger with the Consortium for Ocean Research and Education to become the Consortium for Ocean Leadership, for which she served as Trustee.

Dr. McNutt's honors and awards include membership in the National Academy of Sciences, the American Philosophical Society, and the American Academy of Arts and Sciences. She also holds honorary doctoral degrees from the University of Minnesota and from Colorado College. She was awarded by the American Geophysical Union the Macelwane Medal in 1988 for research accomplishments by a young

Marcia K. McNutt

scientist and the Maurice Ewing Medal in 2007 for her significant contributions to deep-sea exploration. She has served on numerous evaluation and advisory boards for institutions such as the Monterey Bay Aquarium, Stanford University, Harvard University, Science Magazine, and Schlumberger.

## INTRODUCTION AND ACKNOWLEDGMENTS

This is the tenth in a series of books that focuses on the integration of research with resource management issues across the Colorado Plateau. Of the previous nine volumes, the last four have been published by the University of Arizona Press (van Riper and Cole 2004, van Riper and Mattson 2005, van Riper and Sogge 2008, van Riper et al. 2010), while the first five volumes were published by the U.S. Government Printing Office (Rowlands et al. 1993, van Riper 1995, van Riper and Deshler 1997, van Riper and Stuart 1999, van Riper et al. 2001). All 10 books have focused on and highlighted efforts to integrate scientific research findings into management of natural, cultural, and physical resources within the biogeographic province of the Colorado Plateau.

This volume, like the nine before, also highlights aspects of biological, cultural, and physical research. This is combined with a series of chapters explaining assessments of management tools and concepts (e.g. rapid biological inventories aka "bio blitzes," ecosystem valuation, and landscape connectivity) that have been and can be utilized by land managers to conserve, maintain or restore ecosystem health within changing Colorado Plateau landscapes. These tools and other decision-making processes explained in this book can be used to better manage resources on a larger scale, and over broader geographic regions of the Plateau. The mix of chapters covering many diverse research and resource management subjects addresses how scientists and land managers can better interface when dealing with management issues and provides information for improving stewardship of resources in a time of rapidly changing social and ecological conditions.

The 17 chapters that constitute this book were selected from 192 research papers, panel sessions, and posters presented at the 10th Biennial Conference of Research on the Colorado Plateau. The conference was held collaboratively with the Society for Conservation Biology at Northern Arizona University's High Country Conference Center in Flagstaff, Arizona, 5 October through 7 October 2009. The hosts of this conference were the U.S. Geological Survey Southwest Biological Science Center, Northern Arizona University (NAU) and their College of Engineering, Forestry and Natural Sciences, Grand Canyon Trust, Merriam-Powell Center for Environmental Research, the Ecological Monitoring & Assessment Program, the National Park Service, and the Colorado Plateau Cooperative Ecosystem Studies Unit. This conference continues, as an outstanding example among a range of partners, to bring together scientists and managers to address regional conservation issues.

Any scientific work is never a single effort, but a direct result of assistance by many individuals. This book is no exception. We would especially like to thank the following scientific peer reviewers; Ken Bagstad, Patrick Bixler, Jenny Briggs, Brookings, Steve Buckley, Nina Burkhart, David Busch, Dave Cos, Ed deSteiger, Charles Drost, Scott Durst, Peter Fflolliott, Pete Fule, Jose Iniguez, Roy Jamison, Matt Johnson, Yeon-Su Kim, Barbara Kus, Adam Landon, Kirsten Leong, Ellis Margolis, Ted Mellis, David Ostergren, Steve Rosenstock, Michael Scott, Scott Shafer, Carl Shapiro, Tom Sisk, Robert Steidl, Thomas Stohlgren, Dennis Stone, John Vankat, Cynthia Wallace, Robert Webb, and Mary Whitfield, all who unselfishly devoted their time and effort to improving each chapter that they reviewed.

This series of books has received continued financial support from the US Geological Survey and the Southwest Biological Science Center.

We especially appreciate the support that Kate Kitchell, former Center Director of the U.S.G.S. Southwest Biological Science Center (SBSC), provided in the early stages of the development of this book. Before joining the U.S.G.S in 2005, Kitchell was the Bureau of Land Management's Deputy State Director for Resources in Utah, serving 16 years with the Bureau and the previous 10 years in Moab, UT, with the National Park Service. Throughout her career, Kitchell has held a passion for the Colorado Plateau. She has built partnerships for public land use and stewardship making her a valuable partner in our efforts at integrating science into resource management. She is now with the BLM in Montana and will be missed on the Plateau.

Kate Kitchell

We also thank Mark Sogge and the new SBSC Center Director, David Lytle, for providing encouragement and financial assistance for this publication. We are very grateful for the support of Northern Arizona University, and especially Laura Heueneke and John Haeger, for providing a location to hold the conference and financial assistance for special conference sessions. We thank the dedicated U.S.G.S Colorado Plateau Research Station staff (T. Arundel, K. Cole, C. Drost, C. Holden, J. Holmes, M. Johnson, D. Mattson, E. Nowak, M. Saul), and particularly volunteers and staff from NAU and the conference co-sponsors. The audio-visual staff of Wynne Geikenjoyner, Jason Westfall, and Chris Taesali provided much needed assistance during this 10th Biennial Conference. We appreciate the assistance of Neil Cobb, Alison Jones, Ed Singleton, John Hall, Ted Neff, Kathy Tonnessen, Carolyn Dunmire, Ron Heibert, Matthew Johnson, Eli Bernstein, Erika Nowak, Carlos Carroll, David Fulton, Frank Matero, Mark E. Miller, Judith Bischoff, Mike Kearsley, Wayne Padgett, Jeff Foster, Matt Skroch, Martin Main, and Jessica Pratt who organized numerous special sessions at the conference. The work of Caroline Gardner as our technical editor is very much appreciated. Shane Selleck, in addition to her editorial duties, helped in many ways, and without her attention to detail, the mock-up of this book would have never been a reality or finished in time for a roll-out at the 11th Biennial Conference. We would also particularly like to thank Kimberly A. van Riper for the art work that is found throughout this book. Finally, we express deep appreciation to each of our families for their support and understanding during the time that this book was in production.

# LITERATURE CITED

Rowlands, P.G., C. van Riper III, and M. K. Sogge. 1993. Proceedings of the First Biennial Conference on Research in Colorado Plateau National Parks. Trans. and Proc. Ser. No. NPS/NRNAU/NRTP-93/10. U.S. Department of the Interior, Washington, D.C.

van Riper, C., III. 1995. Proceedings of the Second Biennial Conference on Research in Colorado Plateau National Parks. NPS Transaction and Proceedings Series NPS/NRNAU/NRTP-95/11, 305 pp.

van Riper, C., III., and K. A. Cole. 2004. The Colorado Plateau: Cultural, Biological, and Physical Research. University of Arizona Press, Tucson, AZ. 279 pp.

van Riper, C., III, and E. T. Deshler. 1997. Proceedings of the Third Biennial Conference of Research on the Colorado Plateau. NPS Transactions and Proceedings Series NPS/NRNAU/NRTP-97/12, 256 pp.

van Riper, C., III., and D. J. Mattson. 2005. The Colorado Plateau II: Biophysical, Socioeconomic and Cultural Research. University of Arizona Press, Tucson, AZ. 448 pp

van Riper, C., III., and M. K. Sogge. 2008. The Colorado Plateau III: Integrating research and resources management for effective conservation. The University of Arizona Press, Tucson, AZ. i-xiv; 393 pp.

van Riper, C., III., and M. A. Stuart. 1999. Proceedings of the Fourth Biennial Conference of Research on the Colorado Plateau. USGS Forest and Rangeland Ecosystem Science Center CPFS Rep. Ser., 99/16, Flagstaff, AZ . 217 pp.

van Riper, C., III, K. A. Thomas, and M. A. Stuart. 2001. Proceedings of the Fifth Conference of Research on the Colorado Plateau. U. S. Geological Survey/ Forest and Rangeland Ecosystem Science Center USGSFRESC/COPL/2001/21 Rep. Ser., Flagstaff, AZ. 209 pp.

van Riper, C., III, B. W. Wakeling, and T. K. Sisk. 2010. The Colorado Plateau IV: Integrating research and resources management for effective conservation. The University of Arizona Press, Tucson, AZ. i-xii; 335 pp.

# REACHING TOWARDS INTEGRATION ON THE COLORADO PLATEAU

# ADMINISTRATIVE BOUNDARIES AND ECOLOGICAL DIVERGENCE: THE DIVIDED HISTORY AND COORDINATED FUTURE OF LAND MANAGEMENT ON THE KAIBAB PLATEAU, ARIZONA, USA

*Christopher M. Holcomb, Thomas D. Sisk, Brett G. Dickson, Steven E. Sesnie and Ethan N. Aumack*

## ABSTRACT

We applied a mixed-methods approach to research interrelated questions concerning ecological divergence and policy coordination across the administrative boundary between Grand Canyon National Park (GCNP) and Kaibab National Forest (KNF) on Arizona's Kaibab Plateau. To assess cumulative ecological changes between jurisdictions, we used field data and spatial models to compare aspen dominance, fir dominance, predicted fire hazard, and predicted fire behavior across these two jurisdictions within a contiguous study area spanning the boundary. Our analyses document abrupt changes in forest composition and structure at the boundary and indicate that aspen dominance is higher on KNF, while fir dominance and the potential for active crown fire is higher in GCNP.

To assess current and potential future approaches to management, we conducted interviews with agency staff, scientists with active research on the Kaibab Plateau, representatives from environmental advocacy organizations, and the principal grazing permitee on the National Forest portion of the Kaibab Plateau. Pressing issues affecting both agencies, including fire management, have stimulated interest in landscape-level planning and management, requiring coordination across the administrative boundary. Interviews captured several recommendations for improving cooperation, in part by identifying and eliminating bureaucratic roadblocks. Most of those interviewed expressed the view that the goals of the two agencies are complementary, and that historical differences in management objectives and approaches have lessened in recent years. We highlight four important recommendations from these interviews that could facilitate increased cooperation at broader spatial scales: (1) ensure consistency in regulations regarding the management of Mexican spotted owls across the administrative boundary; (2) develop and implement a common fire plan that aligns fire management objectives and allows beneficial fires to burn across the administrative boundary; (3) work toward development of a single natural resource plan that covers the entire Kaibab Plateau and directly addresses issues that demand a long-term, broad-scale approach; (4) develop a budget process that will allow the two agencies to jointly fund projects of mutual interest. Implementing these recommendations will require effort, investment, and creativity, but the resulting ecological benefits and budgetary efficiencies could be considerable.

## INTRODUCTION

Administrative boundaries separating national parks and national forest lands rarely coincide with ecological gradients, the boundaries of biotic communities, or even major topographic features (Sellars, 1997; Newmark, 1985, 1987; Reynolds and Schonewald, 1997). Instead, these boundaries reflect public values and political decisions that often revolve around the use of land and natural resources, including timber and grazing, historical and cultural sites, and unique landscape features and scenic areas (Hays 1999; Sellars 1997; Bade 1923; Van Name 1923). In addition, administrative boundaries that separate national forests from national parks often follow straight lines and turn at right angles because of reliance of the U.S. Cadastral Survey and reference to the township and range grid system (Thomas, 2003).

Once boundaries have been established, differences in land-use practices in each jurisdiction may cause a divergence in ecological conditions on either side of the jurisdictional boundary, resulting in an ecological edge between distinct ecological communities or ecosystem types (Schonewald-Cox and Bayless 1986). In some places, such as the South Rim of the Grand Canyon, straight boundaries have led to changes in forest structure and composition (Landres et al. 1998), while the construction of fences to excluded cattle from Park Service lands has influenced other resources and conditions, further affecting management practices (Anderson, 2000).

In contrast to most ecological gradients and ecotones, an ecological edge is an area where conditions change abruptly, either due to steep gradients in ecological conditions, or as an artifact of human land use patterns (Ries et al. 2004). The creation of ecological edges through contrasting land use practices has been termed the "boundary effect" (Schonewald-Cox and Bayless 1986). Boundary effects can lead to undesirable habitat fragmentation, often manifest as "edge effects" (Ries et al. 2004), and they can alter ecological processes, such as fire, dispersal, and watershed dynamics.

While these effects may be ecologically undesirable, cross-boundary management may be driven by other objectives and its outcomes evaluated by economic and aesthetic considerations, in addition to or in exclusion of the consideration of ecological integrity and function.

Walters and Holling (1990) argue that management policies may be viewed as long-term ecological experiments with recognizable ecological outcomes that should inform future policies in an adaptive management framework. However, Landres et al. (1998) believe that the "current understanding of administrative boundaries and their ecological effects is insufficient for effective landscape-scale forest stewardship" (p. 54). They assert that empirical literature on the effects of administrative boundaries is lacking, and that quantitative comparisons are needed to measure cumulative and long-term effects of administrative boundaries on important ecological conditions, including species composition and microclimate, as well as process such as fire, seed dispersal, and nutrient flows. Studies documenting boundary effects across national park boundaries are available (see, e.g., Schonewald-Cox et al. 1992), but we are unaware of any on the Colorado Plateau, despite its having the highest density of national parks in the United States.

The Kaibab Plateau in northern Arizona presents a compelling opportunity for studying the cumulative ecological effects of nine decades of fragmented management on a once continuous sky-island forest. This large, uplifted and isolated region is split in jurisdiction between the U.S. Forest Service (USFS), which manages Kaibab National Forest, and the National Park Service (NPS), which manages Grand Canyon National Park (Figure 1.1). Since 1919, the two agencies have carried out their different but related missions and mandates adjacent portions of the Kaibab Plateau. Tourism, recreation, and land preservation have been high

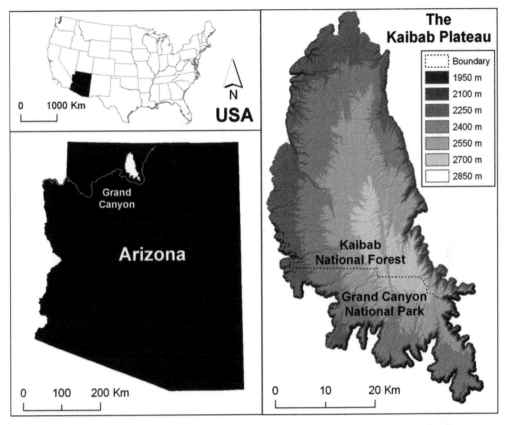

Figure 1.1    Map of the Kaibab Plateau in northern Arizona (lower left), and the administrative boundary between the Kaibab National Forest and Grand Canyon National Park in the shaded relief map of the Kaibab Plateau (lower right). The study area spans this boundary, which is shown here as a dotted line.

priorities for the NPS, while management of timber harvest, livestock grazing, and wildlife resources have been high priorities for the USFS. Now, after nearly 90 years, a distinct ecological edge, representing changes in forest structure and composition across the NPS-USFS boundary, is evident on the ground, and prominent in remotely sensed imagery dating back to the 1970s (Figure 1.2). This ecological boundary runs east-west, bisecting the predominantly north-south trend of the Plateau's valleys and ridge lines, and demarcating divergent composition and structure of what was once a continuous forest landscape.

Across the arid West, where extensive public lands, administered by different state and federal agencies, adjoin large tribal and private holdings, there is increasing interest in large-scale analysis and planning. This collaborative approach is motivated by the observation that major ecological processes – such as wildfire, forest insect outbreaks, and wildlife movements and migration movement – are rarely confined to a single jurisdiction. Landscape approaches that cross multiple ownerships and jurisdictions are needed to address common challenges in a meaningful manner. According to Thomas (2003), interdependence, in some form, is a prerequisite to interagency cooperation. Fire has historically promoted a sense of interdependence among the federal land management agencies and, therefore, encouraged interagency cooperation (Pyne 1982; Pyne 1989). For example, in the

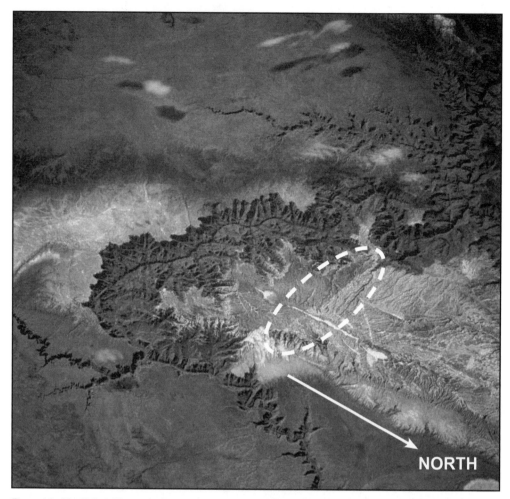

Figure 1.2   The Kaibab Plateau (middle right, with snow) and Grand Canyon from space, winter 1994. After about 90 years of divided jurisdiction between the U.S. Forest Service and National Park Service, a sharp boundary between the two agencies is visible. Note the line between the darker area on the left and the lighter area on the right, within the dashed circle. More recent changes in management practices continue to influence boundary conditions, and greater interagency cooperation may eventually eliminate it altogether (image from Space Shuttle Mission STS060).

working relationship between Glacier National Park and the adjoining national forests, Sax and Keiter (2006) found that "a common concern over the need to coordinate fire management has strengthened the bond between the two agencies."

On the Kaibab Plateau and elsewhere, the occurrence of uncharacteristic large and destructive crown fires in recent years has highlighted the interdependence of the KNF and GCNP, creating a strong incentive to jointly address fire management issues, as well as other issues that span the boundary between jurisdictions and affect entire Kaibab Plateau. Past fires and the likelihood of similar events in the future represent a major concern for agencies in terms of public safety, risk management, conservation of ecological values and wildlife habitat, and protection of the aesthetic values

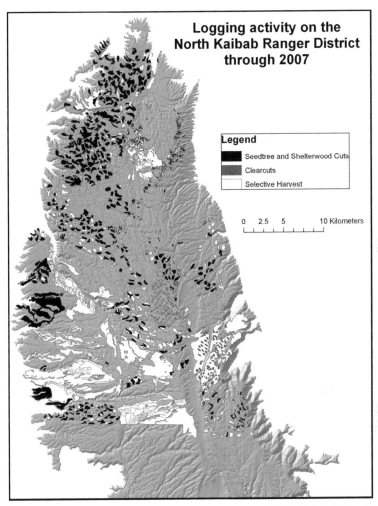

Figure 1.3   Map of logging activity on the Kaibab Plateau, within the Kaibab National Forest (spatial data used with permission of the U.S. Forest Service).

associated with the world-famous Grand Canyon landscape. In 2009, KNF sought to develop consensus on priority areas and treatment guidance across all three ranger districts through a collaborative, science-based process called the Kaibab Forest Health Focus. The outcome of that effort revealed a shared interest in the restoration of cross-boundary fire regimes, and the coordinated use of mechanical tree thinning and prescribed fire to achieve management objectives across forested portions of the Kaibab Plateau (Sisk et al. 2009). Other

resource issues addressed in that and other assessment and planning efforts indicate a similar move toward efforts to manage natural resources from an integrated, landscape perspective, an approach that requires cooperation and a willingness to coordinate management objectives and actions across jurisdictional boundaries

This purpose of this study was to quantify the cumulative ecological changes that have emerged since the forests of the Kaibab Plateau were divided by their separation in GCNP and KNF jurisdictions. We used a mixed-methods approach to examine interrelated questions concerning ecological divergence and policy coordination across the administrative boundary. This approach combined field study and ecological modeling of ecological conditions with interview techniques to assess the degree of compatibility in agency objectives and management styles, and to identify opportunities for increased coordination and cross-boundary management in the future.

Regarding the ecological portion of this study, we hypothesized that aspen dominance (i.e. percentage of total stems) would be higher on KNF than in the GCNP, due to the extensive disturbance footprint of logging (Figure 1.3) and the absence of elk (Cervus canadensis), which often restrict aspen post-disturbance recruitment in southwestern forests, due to heavy herbivory (see Johnston 2001). In contrast, the absence of logging, in combination with an effective fire suppression program in GCNP, has provided little opportunity for widespread aspen regeneration (however, recent increases in the use of prescribed fire and wildfire activity in the park may change this dynamic, see Figure 1.4). Similarly, we expected the dominance of the shade-tolerant white fir (*Abies concolor*), subalpine fir (*Abies lasiocarpa*), and Douglas fir (*Pseudotsuga menziesii*) to be greater in GCNP than on KNF lands, due to this difference in disturbance regimes. Furthermore, we hypothesized that the potential for large crown fire would be greater in the park, due in part to higher densities of shade-tolerant trees and ladder fuels in otherwise similar forest stands.

With respect to the policy portion of this study, our research was largely exploratory. The Kaibab Plateau provided a model for studying the challenges and opportunities related to landscape-scale interagency cooperation across administrative boundaries. We conducted policy analysis based on interviews with key land

management personnel and stakeholders, focusing on agency mission and institutional culture, and identifying current barriers to cross-boundary management and steps that could be taken to enhance cooperation.

## MATERIALS AND METHODS

### Study area

The Kaibab Plateau is a sky island comprising 1730 km$^2$ in north-central Arizona (Reynolds et al. 2005). It extends from the North Rim of the Grand Canyon to just north of the Utah state line, rising approximately 2000 m above the surrounding landscape at its highest reaches. It is bounded on its southern flanks by a complex of steep slopes and cliffs that drop into the Grand Canyon and its tributaries. To the north, gentler slopes descend to large areas of arid sagebrush steppe and grassland. The Plateau is geographically isolated, with the nearest forested area lying 80 km to the west (Reynolds et al. 2005). The interior of the Kaibab Plateau is characterized by north-south trending forested ridges separated by shallow (30 to 70 m) valleys, with deeper canyons incised along its rim.

Vegetation on the Kaibab Plateau changes with increasing elevation from sagebrush (*Artemisia spp.*) steppe or semi-arid grasslands in the surrounding deserts, to pinyon pine (*Pinus edulus*) and juniper (*Juniperus spp.*) woodlands on the lower slopes. The forested area of the Kaibab Plateau includes large stands of ponderosa pine forest in the lower elevations (2200-2500 m), open montane meadows and mixed conifer forest (*Pinus ponderosae, A. concolor, A. lasiocarpa, P. menziesii*) in the middle elevations (2500-2650 m), and subalpine species (*A. concolor, A. lasiocarpa, Picea pungens, Picea engelmannii*) encroaching upon ponderosa pine in the upper elevations (2650-2800 m) (White and Vankat 1993; Mast and Wolf 2004; Mast and Wolf 2006). Aspen (*Populus tremuloides*) occurs in all

Figure 1.4   Map of fire activity on the Kaibab Plateau, 1989-2009. Prior fire events were limited in number and extent, due to fire suppression practices.

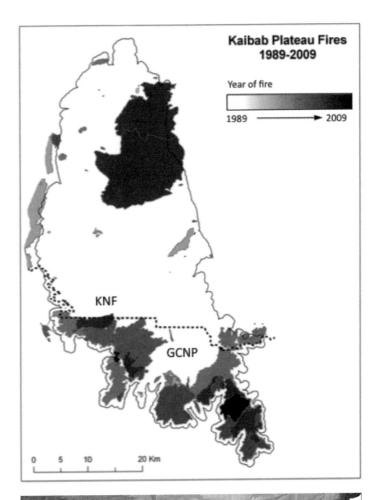

Figure 1.5   Ecological study area, showing the location of field plots sampled on the Kaibab National Forest (open circles) and in Grand Canyon National Park (closed circles).

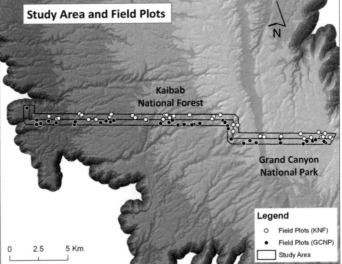

three forest zones, both in monotypic stands and mixed with other conifer species.

The Kaibab Plateau has a montane climate, with a mean summer high of 31° C from June through August, a mean winter low of -3.2° C from November to April, and a mean annual temperature of 12.4° C (USFS Remote Automated Weather Station, Pleasant Valley, Arizona, 36° 30' 19" N, 112° 08' 39" W, 2560 m, 1966-2008). The precipitation regime (mean annual total of 49.8 cm) is bimodal, with frequent monsoonal rainstorms during July and August (mean accumulation of 15.2 cm) and winter snows that persist from November to May (mean depth of 29.2 cm). The Kaibab limestone that covers most of the Kaibab Plateau is heavily karstified, which causes runoff to drain into sinkhole ponds and subterranean fissures. Only a few kilometers of surface permanently flowing streams occur across the entire Kaibab Plateau.

The study area for our ecological research (27.67 km²) was confined to a ~30 km length of the east-west boundary separating KNF and GCNP (Figure 1.5). The study area covers a 460 m elevational gradient that traverses all three forested vegetation zones and extends 500 m into each administrative jurisdiction. Field observations suggest that this distance adequately captures many aspects of the contrast in ecological conditions across the boundary, which is quite abrupt. The land base considered in the interview-based research on interagency policies, management practices, and cooperation included the entire forested portion of the Kaibab Plateau.

## Ecological methods

The study area boundary was delineated in a geographic information system, or GIS (ArcGIS Version 9.3, Environmental Systems Research Institute, Redlands, CA), based on the parameters described above for the ecological research. Within the study area polygon, we randomly generated 100 plot centers throughout the study area, with 50 located in each jurisdiction, using Hawth's Analysis Tools for ArcGIS (http://www.spatialecology.com). Plot centers were located in the field using a Garmin Etrex Legend geographic positioning system (GPS; Garmin Limited, Olathe, KS), and unsuitable locations, characterized by excessive slopes, inaccessibility, or non-forest cover types, were eliminated. Of the remaining plots, we sampled as many plots as feasible over a 3-month field season, resulting in samples from 38 plots on each side of the boundary between KNF and GCNP. No stratification was used in plot selection, but because the boundary cuts across an east-west elevation gradient, and crosses through areas with variable disturbance histories, a range of forest types was sampled, from ponderosa pine stands at the lower elevations on west side of the Plateau, to mixed conifer stands at higher elevations. Most plots were of mixed species composition, and in all cases plots in one jurisdiction were paired with similar nearby sites in the adjoining jurisdiction. With a relatively high number of plots (n=76) sampled randomly across a carefully circumscribed study area, we were able to compare forest composition and structure across the administrative boundary.

Sample plots were circular, with a 10-m radius (area = 314 m²). During May-July, 2008, we navigated to each plot center using a handheld GPS unit. We recorded tree species and diameter at breast height (dbh, 1.37 m from base) for all established trees ≥ 3-m tall. In analyzing these field data, we used a two-sample, one-tailed t-test, assuming unequal variances, in JMP (Version 8, SAS Institute Inc., Cary, NC) to compare tree species composition and forest structural attributes between the two jurisdictions. Specifically, we tested for hypothesized differences in aspen dominance (proportion of the total stems on the plot recorded as aspen) and fir dominance. We pooled true firs with Douglas fir because both have contributed substantially to sub-canopy infilling due to fire suppression, and because timber harvest on KNF within our study area has been very limited in the mixed conifer zone. We employed an alpha of 0.10 because of the high degree of heterogeneity in forest conditions, the large elevation gradient, and the complex mosaic of management treatments.

To explore how differing management has influenced the likelihood of crown fire, we employed spatial models of predicted fire hazard (expressed as heat generated per unit area burned, kJ/m$^2$) and predicted fire behavior (three classes: surface fire, passive crown fire and active crown fire). These models were developed by the Forest Ecosystem Restoration Analysis (ForestERA) Project (see Sisk et al. 2006), using the GIS-based fire simulator FlamMap (Stratton 2004). These models were implemented across the entire Kaibab Plateau and adjoining lands as part of a larger research effort related to the Kaibab Forest Health Focus. These models were generated using seven input layers derived from the U.S. Forest Service's LANDFIRE program (http://www.landfire.gov), based on 1999 and 2000 data. Models were implemented under the assumption of 97th percentile drought weather conditions, low fuel moistures, and sustained 18 mph wind speeds out of the southwest (the prevailing wind vector during the fire season on the Kaibab Plateau). FlamMap is not a dynamic fire simulator, but the wind inputs for these models were generated dynamically using the wind modeling program WindNinja v. 1.0 (U.S. Forest Service Fire Sciences Laboratory, Missoula MT). We clipped spatially explicit model results to the extent of our study area, co-registering them with our study plots and other data layers. These model outputs provided a representation of relative fire hazard and predicted fire behavior across our large study area, as of the year 2000, however they do not capture more recent management actions, wildfires, and other forest disturbances.

For both fire hazard and predicted fire behavior, we compared the two jurisdictions (KNF and GCNP) using every raster cell within the study area. We used a one-tailed z-test to compare the two jurisdictions in terms of mean fire hazard ($\alpha=0.05$) and hypothesized that mean fire hazard would be greater in GCNP. Using the predicted fire behavior model, we determined the total area within each jurisdiction for each of the three fire behavior classes (i.e. surface fire, passive crown fire and active crown fire) and compared the proportion of the study area falling in each class.

## Interviews

We developed a set of open-ended interview questions to help identify policy and management dimensions of cross-jurisdictional coordination between KNF and GCNP on the Kaibab Plateau (Table 1.1). We then compiled a list of potential respondents based on professional expertise and job responsibilities, including program staff and managers from GCNP and KNF, representatives from environmental advocacy organizations that are active on the Kaibab Plateau, the grazing permitee that holds leases for most of the Kaibab Plateau, including the KNF portion of our study site, and ecologists with active research programs on the Kaibab Plateau. Appendix 1-A provides a list of those interviewed, including their affiliation and job title. We sent e-mails to these individuals inviting them to participate in voluntary interviews and attached an overview of the study and the interview questions. We conducted interviews with 18 individuals between May and September, 2009. Interviews lasted approximately 1 hr, beginning with a general overview of the ecological and policy portions of the research, during which respondents were shown images of the boundary phenomenon, including the satellite

| INTERVIEW QUESTIONS |
|---|
| 1. Numerous observers have made reference to sharp changes in forest conditions and a lack of cohesive management between the two agencies in the lands along the jurisdictional boundary between Kaibab National Forest and Grand Canyon National Park. How did it get there, and where did it come from? How long do you think the sharp contrast in forest conditions along the boundary has been there? |
| 2. In what ways do the two agencies already co-manage these forests? In what ways do the two agencies fail to adequately co-manage across the boundary? |
| 3. What are the obstacles or roadblocks – whether financial, political, bureaucratic, ecological, or otherwise – to doing better co-management? |
| 4. Would you expect fire behavior and intensity to change upon crossing from one jurisdiction to another (i.e. KNF to GCNP or visa versa) in any particular areas along the boundary? |
| 5. What would a fully seamless and integrated management regime translate into on the ground? What changes would need to be made, or hurdles overcome, for this to occur? What feasible actions could be taken over the next 5 years by the two agencies to manage these forests more seamlessly? Over the next 15 years? |
| 6. In what ways is it sensible for the two agencies continue to manage separately? |
| 7. What issues, problems or benefits do you think are a result of this sharp change in forest conditions with respect to fire and fuels management, wildlife management, or protected species? Others? |

Table 1.1 Interview questions used for the policy analysis portion of this study.

image included as Figure 1.2. We presented general observations and discussed the major ecological trends that we had hypothesized, based on a review of the scientific literature and our initial field reconnaissance. We then asked each respondent to answer each one of the interview questions in the order presented in Table 1.1. Respondents also contributed general remarks, and often answered additional follow-up questions that were specific to their expertise. Extensive notes were taken throughout the interviews, including verbatim quotations and paraphrased responses. No audio or video recording was made.

We analyzed these data qualitatively and synthesized the information into two major themes – barriers to cross-boundary management, and recommended goals and next steps – with multiple sub-themes

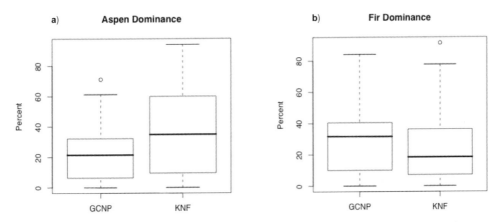

Figure 1.6   Box plots comparing data from field plots located within Grand Canyon National Park (GCNP) and on the adjacent Kaibab National Forest (KNF) in terms of (a) aspen dominance, and (b) fir dominance (n = 76).

### Fire Hazard (kJ/m²)

| Jurisdiction | Mean | S.D. | z-value | p-value |
|---|---|---|---|---|
| GCNP | 29,575 | 28,246 | 15.3 | <0.001 |
| NNF | 24,554 | 25,890 | | |

Table 1.2   Comparison of fire hazard between Grand Canyon National Park and Kaibab National Forest, modeled as a continuous variable over the entire study area. Reported values are derived from spatial models (i.e., rasters) produced by the ForestERA Project at Northern Arizona University (see text).

nested under each. We assessed the relative importance of these themes, based on the number of respondents who voluntarily emphasized them, and also by how important the respondents believed the particular issue was. We identified the most important management recommendations, based on the number of times each was suggested by a respondent, and by how emphatic they were on the issue. For example, 56% of the respondents commented on the differences in mandates as a barrier to coordination, but few thought that those differences were a serious barrier; while 22% of the respondents believed that differences in budgetary

processes were a barrier, and all of them were emphatic about the difficulties that this created. We identified both as important barriers to cooperation among agencies, but considered the former to be more significant.

The results and discussion below are a thematic synthesis of points of consensus that emerged from the interviews. We have attempted to be as transparent as possible in conveying and amalgamating these views, and to preserve differing views as much as possible. We believe that this analytical approach, albeit subjective, provides an informative synthesis of perspectives on complex management issues and

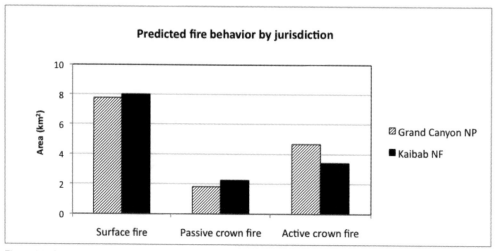

Figure 1.7  Comparison of predicted fire behavior in Grand Canyon National Park and on the Kaibab National Forest. Classification of 90-m pixels into three fire behavior classes (surface fire, passive crown fire, active crown fire) was based on FlamMAP outputs as modeled by the ForestERA Project at Northern Arizona University (see text).

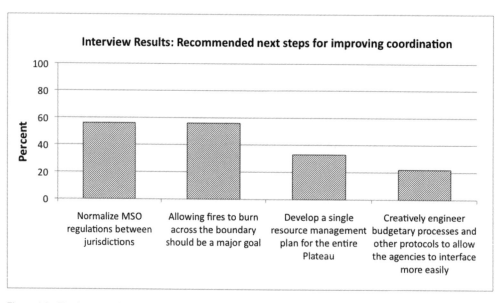

Figure 1.8  The four most frequently identified recommendations for improving interagency coordination and cross-boundary management, as derived from interviews of land management agency personnel, representatives of environmental advocacy organizations active on the Kaibab Plateau, and the primary Forest Service grazing permitee (n = 18).

recommendations, from a diverse array of independent individuals with expertise in these issues.

# RESULTS

### Ecological divergence

Mean aspen dominance was 54% higher on KNF than on GCNP (p=0.012, n=76), the upper quantile of KNF was twice that of GCNP, and the maximum value was 40% higher on KNF (Figure 1.6a). Mean fir dominance was 34% higher on GCNP (p=0.056, n=76) (Figure 1.6b). Fire hazard was 20% higher on GCNP than KNF (p<0.001, z=15.3) (Table 1.2). GCNP had 25% more area than KNF that was predicted to show active crown fire behavior, should a fire occur under the modeled conditions; while KNF had proportionally greater areas predicted to burn as surface fires and passive crown fires. However these differences were not statistically significant (Figure 1.7).

### Cross-boundary management

The views and recommendations of the interviewees were largely independent of job responsibilities and agency affiliation. Interviews revealed that half of the respondents believe that "hard" habitat edges or ecological boundaries between administrative jurisdictions are an undesirable product of fragmented management practices. Most of these believed that forest conditions along the boundary will always reflect a gradient in management goals, but that differences should be minimized, and in the future the boundary studied here should be less apparent – certainly not be visible to the casual observer or from aerial imagery. Instead, they favored a gradual change in management intensity, with changing conditions corresponding to natural patterns generated of fire and insect outbreaks. There was consensus that cross-boundary fires

should be an important means of softening the hard ecological edge that currently conforms to the administrative boundary.

Four recommendations were independently identified and strongly endorsed by multiple respondents (Figure 1.8). Ten individuals (56%) believed that normalizing the differing regulations for the Mexican spotted owl (or MSO, Strix occidentalis lucida) between the two jurisdictions was a very high priority. A different group of ten individuals (56%) believed that a major goal should be allowing fires to burn across the administrative boundary. Six individuals (33%) independently stated the need to develop a single natural resource management plan for the entire Kaibab Plateau, engaging staff from both agencies. Four individuals (22%) emphasized the need for a budgeting system or procedure that would allow the two agencies to jointly fund and staff projects, which they felt was unnecessarily difficult at present.

# DISCUSSION

### Ecological divergence

Our results were consistent with the hypothesis that aspen dominance is higher on KNF than on GCNP. We postulated that disturbances, primarily logging and road-building, have promoted aspen regeneration on KNF, while the relative lack of disturbance (prior to recent wildfires and increased use of prescribed fire) in GCNP, over several decades, has led to a relative decline in aspen. Binkley et al. (2006) found that aspen stems increased by an order of magnitude on KNF during the highest levels of timber harvest (1963-1992), whereas aspen on adjacent GCNP forests increased by only 75% during the same period. Johnston (2001) and Frey (Frey et al 2003) noted that aspen regenerate at very high density (up to 30,000 sprouts per acre) after disturbances, including logging, in areas with an established aspen clone. High sprouting density after logging occurs in areas where the soil remains well aerated and well drained (i.e. not compacted by heavy equipment), undamaged dominant aspen stems are left in place, and browsing from elk is minimal (Johnston 2001; Stone et al. 2001; Stone 2001). Because of isolation from established elk populations and the limited water sources between those populations and potential habitat on the Kaibab Plateau, elk have not colonized the Plateau and vertebrate herbivory is not a significant factor in aspen recruitment. Therefore, the widespread logging on KNF, under conditions that favor aspen sprouting, juxtaposed with over 70 years of fire suppression and regulations prohibiting timber harvest on much GCNP, likely explains the higher level of aspen dominance documented here (see Figures 1.3 and 1.4 for maps of logging and fire activity, respectively, on the two jurisdictions).

Results also show higher fir dominance in GCNP, as expected, and this too is likely due to the absence of fire or other major disturbances within the study area, over a prolonged period, favoring late-successional, shade-tolerant fir species. However, the differences in fir dominance between GCNP and KNF are not great. We expect that the reason for this is that the majority of the mixed conifer forests on both GCNP and KNF have not burned within the study area for several decades. In addition, there was considerably less logging activity in the highest elevations of KNF, which would favor the establishment and persistence of older fir stands. Because the mixed conifer forests in KNF have experienced less disturbance, due to fire and logging, than have the lower elevation forests characterized by ponderosa pine, fir dominance in the two jurisdictions (KNF and GCNP) is more similar than we had expected. Also, prescribed natural fires over the last two decades in the middle elevations of the study area in GCNP have caused disproportionate mortality in

younger fir trees (Fulé et al. 2004a; Fulé and Laughlin 2007), which has likely decreased the magnitude of differences in fir dominance that may have developed between jurisdictions over prior decades. Earlier studies have documented increases in white fir and Douglas fir in the mixed conifer zones of GCNP at the North Rim (Mast and Wolf 2006; White and Vankat 1993), but no studies have compared those findings with the adjacent mixed conifer forests on KNF, until recently (see Sesnie et al., this volume). This fir encroachment is one of the factors contributing to the increased likelihood of active crown fire on GCNP.

Model results support the hypothesis that forests on GCNP have a greater potential for active crown fire, compared to those on KNF. This is not surprising, given the structural differences in forests in the two jurisdictions, with denser, fir-dominated stands in GCNP creating heavier fuel loads, denser ladder fuels (crowns often begin near ground level in these dense stands), and higher estimates of fire hazard. However, it is important to note that these models draw on spatial data that are now a decade old, and recent work with prescribed fire is not well represented by the 2008 field data. Thus, differences in predicted fire behavior may, in some locations, be overstated.

Fulé et al. (2004b) compared reconstructed stand conditions from 1880 to projected conditions in 2040, for lower elevations at the North Rim of GCNP, and found that the crowning index, or the wind speed necessary to generate crown fire given the fuel conditions, would be reduced by 23-80% over this timeframe, indicative of a strong departure from pre-settlement fire severity. Fulé and Laughlin (2007) found that at the middle and high elevations of GCNP, which historically had a mixed-severity fire regime, recent fires have caused disproportionate mortality in the smaller shade-tolerant fir and spruce trees that have encroached since the establishment of effective fire suppression effort and, therefore, structural conditions have trended back toward historical reference conditions. However, even given the mitigating effects of recent prescribed fires in GCNP, our model results suggest that potential fire behavior is still greater in GCNP, when compared to adjacent lands in KNF.

## Interviews and Policy Recommendations

There was an overall consensus throughout the interviews that tighter interagency coordination on the Kaibab Plateau, and a softening of the boundary effects, is desirable. Many respondents noted that the differences in management practices across the boundary are intentional and that they reflect the distinct mandates given the two agencies by Congress, and that this will continue to be the case unless Congress alters those missions in some way. However, only two individuals felt that the differences in mandates and agency culture (emphasis on extractive uses on KNF vs. preservation on GCNP) are so great that it precludes meaningful coordination. In fact, there was general agreement that disjointed policies were a product of rivalries and pronounced conflicts in agency missions in the past, but that rivalry between agencies is greatly diminished and their objectives are generally complimentary, despite remaining differences with respect to their relative emphasis on preservation vs. use. One respondent stated that, "the [ecological] edge is a scar from the past where mandates and cultures truly clashed," whereas territoriality is no longer an issue and the objective of restoring historical fire regimes is a good fit for both agencies now. One individual said, "the turf wars are over, and I wouldn't have wanted to work here then. Interagency coordination is the way of the future." Another offered that "conversations [about increased coordination] would not have been possible twenty years ago, and even ten years ago there was far less interest." One

agency staff member was more emphatic:

"In the past, the Forest Service symbolized something very different from the NPS. This view pervades the discussions, but it's more exaggerated than it should be. The USFS and NPS are closer than they were in the past, and the old battle lines are out of date. We squander incredible opportunities by remaining entrenched."

The policies now pursued by the NPS and USFS on the Kaibab Plateau have changed dramatically in recent years, and the objectives of the two agencies have, to a large extent, converged. Timber harvest on KNF is generally pursued with the goal of maintaining mixed-age stands that move forest structure toward conditions that predominated prior to intensive logging and fire suppression. To some extent this is mandated by wildlife habitat requirements, most notably those of the northern goshawk and its prey species (Reynolds et al. 1992). But this approach is also driven by the objective of restoring fire-adapted forests and conserving biological diversity, in general. Within GCNP, forest management is also driven by the principles of comprehensive ecological restoration, and prescribed fire is used with increasing frequency to achieve desired changes in forest structure. Interestingly, the NPS is also carrying out experimental tree thinning to reduce fuel loads in areas that cannot be safely treated with prescribed fire. Thus, management practices are similar in some areas of the two jurisdictions, and approximately 50% of the respondents believed that management goals are sufficiently complimentary for cooperative, landscape-level planning to be viable.

Several significant barriers to cross-boundary coordination were identified by multiple respondents. Most significant, perhaps, are contradictory policies for wildlife management. In particular rules for managing the Mexican spotted owl (MSO) differ between jurisdictions. On KNF, each individual management requires an environmental impact statement assessing potential impacts to this species, as called for in the biological opinion issued by the U.S. Fish and Wildlife Service (USFWS) for this federally listed threatened species. The opinion also limits thinning and other forest management in the mixed conifer zone, in order to protect foraging habitat. For GCNP, however, the current biological opinion allows certain treatments in the mixed conifer type and the park's entire fire program has been NEPA-cleared, thus NPS does not have to complete NEPA analysis on a project-by-project basis, as does KNF. Interviewees felt that this inconsistency in policy poses a serious roadblock to coordinated fire planning and cross-boundary management, and several noted instances when beneficial fires were extinguished at the boundary because of these and other concerns. Resolving this problem ranked at the top of the list of recommendations emerging from the interview process (Figure 1.8).

Another barrier identified was the difficulty in coordinating prescribed burns and the beneficial use of natural fires to achieve management objectives. Current fire plans and management practices make it difficult for mangers to allow beneficial fires to cross the boundary between GCNP to KNF. Several respondents, including individuals from both agencies, noted recent progress in this area and recommended that a single fire plan be developed for both jurisdictions. Such a plan would, ideally, be considered for program-wide NEPA clearance, with a single policy for assessing impacts to MSO. An integrated fire plan, along with a unified fire management infrastructure, would help both agencies avoid situations where fires are allowed to burn to the edge of the boundary, then are extinguished, thereby exacerbating the divergence in forest conditions because of policy incoherencies that belie the similarity in management objectives.

Moving beyond the focus on fire planning, several interviewees suggested that there should be a single long-term natural resource management plan for the entire Kaibab Plateau, written by an interdisciplinary team drawn from both agencies. They argued that such a plan should consider forested lands on the Kaibab Plateau as a single planning unit demarcated by ecological, rather than administrative boundaries, thereby minimizing and eventually eliminating the contrasting planning and management practices that have resulted in the ecological boundary that exists today. Interview results indicate that the converging mandates of the two agencies make such a document both feasible and desirable. An integrated resource management plan for the entire Kaibab Plateau would provide a formal avenue for enhancing communication and sustaining cooperation between the two agencies, and it could result in increased efficiencies in planning and management. Such a plan would need to clearly recognize the remaining differences in agency mandates and identify ways to meet those mandates while moving toward a more integrated landscape approach.

Significant bureaucratic hurdles will need to be overcome to ease cooperation generally and to address the recommendations described above. Many of the interviewees emphasized the bureaucratic nature of federal land management agencies as a roadblock to coordination, noting the difficulties associated with sharing staff and budgets on joint work projects. Current budgeting and fiscal policies make it difficult for staff or projects to be funded jointly by the NPS and USFS, limiting cross-boundary work. One manager recommended the development of a budget mechanism that would establish "a common source of funds that both agencies contribute to for joint projects" and suggested this would have a positive impact on collaboration and management coordination. Yaffee and Wondolleck

(1997) studied 200 instances where USFS staff engaged in cross-boundary "bridging activities," and found that lack of time, money and personnel, as well as a "lack of flexibility in administrative procedures and budgetary categories," were among the most consistent limiting factors. They also found that:

"... Adequate funding may be present, but narrow budget categories preclude crosscutting activities, and bridging just does not fit comfortably in a budget based on narrow programmatic or line items. Findings ways to bridge budget categories could help provide needed flexibility in funding... Creating a discretionary fund for small expenses connected with bridging activities and a start-up fund for new bridging projects may be worthwhile in many cases (p. 394).

Yaffee (1998) discusses the common problem of securing funding for multi-year, multi-agency projects when budget processes typically work on one-year cycles and allow little latitude for allocating funds beyond traditional line items. Such uncertain funding contributes considerably to the lack of follow-through that has derailed numerous cooperative attempts in the past. The recommendation for developing budgeting practices that facilitate interagency funding of cross-boundary planning and management efforts would address this common bureaucratic hurdle.

## Management Implications

The cumulative effects of a half-century of divided management have caused ecological divergence across the administrative boundary that divides the KNF and GCNP portions of the Kaibab Plateau. The differences measured in this study include greater dominance by shade-tolerant fir species on GCNP and far greater dominance of aspen on KNF. These contrasting conditions contribute to very sharp ecological transitions across the administrative boundary, transitions that are clearly visible from space during the winter months. These conditions arose due to the historical contrast in agency missions, the development of contrasting institutional cultures and management practices, and limited communication and coordination during a prior era of public lands management.

Despite converging agency missions and increased cooperation, various regulatory and bureaucratic roadblocks – including differing approaches and procedures for analyzing effects on sensitive wildlife species, incompatible budgetary processes, and separate planning procedures that interface poorly across jurisdictions – continue to make cooperation difficult. Sustained efforts are required to address these limitations and capture the ecological benefits and operational efficiencies that are likely to derive from greater coordination. Chief among the recommendations identified in this study are policy coordination regarding threatened, endangered and sensitive species; integrated fire plans and management; a comprehensive resource management plan that includes all land ownerships and administrative units; and a budgeting process that facilitates cooperative efforts that are staffed and funded by two or more partnering agencies. These changes would allow the restoration not only of forest structure, but of fire and other ecological processes that operate at landscape scales and require integration, or at least coordination of management efforts across administrative boundaries.

## ACKNOWLEDGEMENTS

This project was made possible by funding and other forms of support from Grand Canyon Trust, Wyss Scholars Program for the Conservation of the American West, Grand Canyon National Park, Kaibab National Forest, Grand Canyon Historical Society, and Garden Clubs of America. We also would like to thank the numerous volunteers who assisted with field work, and those who participated in the interview process during the summer of 2009.

## LITERATURE CITED

Anderson, M.F. 2000. Polishing the Jewel: An Administrative History of Grand Canyon National Park. Grand Canyon Association, Monograph No. 11.

Bade, W.F. 1923. Further Comment on the Proposed Roosevelt-Sequoia National Park and the Barbour Bill. Ecology 4(2): 217-219.

Binkley, D., M. Moore, W. Romme, and P. Brown. 2006. Was Aldo Leopold Right about the Kaibab Deer Herd? Ecosystems, 9: 227-241.

Frey, B., V. Lieffers, S. Landhausser, P. Comeau, and K. Greenway. 2003. An analysis of sucker regeneration of trembling aspen. Canadian Journal of Forest Resources, 33: 1169-1179.

Fulé, P.Z., A.E. Cocke, T.A. Heinlein, and W.W. Covington. 2004a. Effects of an intense prescribed fire: is it ecological restoration? Restoration Ecology 12(2): 220-230.

Fulé, P.Z., J.E. Crouse, A.E. Cocke, M.M. Moore, and W.W. Covington. 2004b. Changes in canopy fuels and potential fire behavior 1880-2040: Grand Canyon, Arizona. Ecological Modelling 175: 231-248.

Fulé, P.Z. and D.C. Laughlin. 2007. Wildland fire effects on forest structure over an altitudinal gradient, Grand Canyon National Park, USA. Journal of Applied Ecology, 44:136-146.

"Grand Canyon from Space Shuttle Mission STS060". http://www.hkhinc.com/arizona/sts060-gc.htm

Hays, S.P. 1999. Conservation and the Gospel of Efficiency: The Progressive Conservation Movement, 1890-1920. University of Pittsburgh Press: Pittsburgh, PA.

Johnston, B.C. 2001. Multiple Factors Affect Aspen Regeneration on the Uncompahgre Plateau, West-Central Colorado, in Sustaining aspen in western landscapes, in Shepperd WD, Binkley D, Bartos D, Stohlgren TJ, Eskew LG, Eds. 2001. USDA Forest Service, Fort Collins, Colorado. General Technical Report RMRS-P-18. 460 p.

Kay, C.E. 2001. Evaluation of Burned Aspen Communities in Jackson Hole, Wyoming, in Sustaining aspen in western landscapes, in Shepperd WD, Binkley D, Bartos D, Stohlgren TJ, Eskew LG, Eds. 2001. USDA Forest Service, Fort Collins, Colorado. General Technical Report RMRS-P-18. 460 p.

Landres, P.B., R.L. Knight, S.A. Pickett, and M.L. Cadenasso. 1998. Ecological Effects of Administration Boundaries, in "Stewardship Across Boundaries," Richard L. Knight and Peter B. Landres, Eds., Island Press, Washington D.C.

Mast, J.N., and J.J. Wolf. 2004. Ecotonal changes and altered tree spatial patterns in lower mixed-conifer forests, Grand Canyon National Park, Arizona, USA. Landscape Ecology, 19: 167-180.

Mast, J.N. and J.J. Wolf. 2006. Spatial patch patterns and altered forest structure in middle elevation versus upper ecotone mixed-conifer forests, Grand Canyon National Park, Arizona, USA. Forest Ecology and Management, 236: 241-250.

National Parks for the 21st Century, The Vail Agenda. 1992. National Park Foundation, Chelsea Green Publishing Company, Post Mills, Vermont.

Newmark, W.D. 1985. Legal and biotic boundaries of western North American national parks: a problem of congruence. Biological Conservation 33: 197-208.

Newmark, W. D. 1987. A land-bridge island perspective on mammalian extinctions in western North American parks. Nature (325)29: 430-432.

Pyne, S. 1982. Fire in America: A Cultural History of Wildland and Rural Fire. University of Washington Press, Seattle, WA.

Pyne, S. 1989. Fire on the Rim: A Firefighter's Season at the Grand Canyon. Weidenfeld and Nicolson, New York.

Reynolds, R.T., J.D. Wiens, S.M. Joy, and S.R. Salafsky. 2005. Sampling considerations for demographic and habitat studies of northern goshawks. J. Raptor Res. 39(3):274–285.

Reynolds, J.J., and C. Schonewald. 1997. Protected Areas, Science, and the 21st Century. The George Wright Forum, (14)3: 5-11.

Reynolds, R.T., R.T. Graham, and H.M. Reiser. 1992. Management recommendations for the northern goshawk in the southwestern United States. USDA Forest Service, Rocky Mountain Forest and Range Experiment Station, Ft. Collins, CO. General Technical Report GTR-RM-217. 90 p.

Ries, L., R.J. Fletcher Jr., J. Battin, and T.D. Sisk. 2004. Ecological responses to habitat edges: mechanisms, models and variability explained. Annual Review of Ecology, Evolution, and Systematics, 35:491-522.

Sax, J. L. and R.B. Keiter. 2006. The realities of regional resource management: Glacier National Park and its neighbors revisited. Ecology Law Quarterly 33: 233-311.

Schonewald-Cox, C., and J. Bayless. 1986. The Boundary Model: A Geographical Analysis of Design and Conservation of Nature Reserves. Biological Conservation, 38: 305-322.

Schonewald-Cox, C., M. Buechner, R. Sauvajot, and B. Wilcox. 1992. Cross-boundary management between national parks and surrounding lands: A review and discussion. Environmental Management, 16(2):273–282.

Sellars, R.W. 1997. Preserving Nature in the National Parks: A History. Yale University Press: Newhaven, CT.

Sesnie, S.E., B.G. Dickson, J.M. Rundall, and T.D. Sisk. In press. Assessment of mixed conifer forest conditions, North Kaibab Ranger District, Kaibab National Forest, Arizona, USA. In van Riper, C., III., and M. Villarreal, editors. The Colorado Plateau V. University of Arizona Press, Tucson.

Sisk, T.D., J.W. Prather, H.M. Hampton, E.N. Aumack, Y. Xu, and B.G. Dickson. 2006. Participatory landscape analysis to guide restoration of ponderosa pine ecosystems in the American Southwest. Landscape and Urban Planning 78: 300–310.

Sisk, T.D., J.M Rundall, E. Nielsen, B.G. Dickson, S.E. Sesnie. 2009. The Kaibab Forest Health Focus: Collaborative prioritization of landscapes and restoration treatments on the Kaibab National Forest. The Forest Ecosystem Restoration Analysis Project, Northern Arizona University. www.fs.usda.gov/Internet/FSE_DOCUMENTS/stelprdb5120031.pdf (last accessed 01/12/2011).123 pp.

Stone, D.M. 2001. Sustaining Aspen Productivity in the Lake States in Sustaining aspen in western landscapes, in Shepperd WD, Binkley D, Bartos D, Stohlgren TJ, Eskew LG, Eds. 2001. USDA Forest Service, Fort Collins, Colorado. General Technical Report RMRS-P-18. 460 p.

Stone, D.M., J.D. Elioff, D.V. Potter, D.B. Peterson, and R. Wagner. 2001. Restoration of Aspen-Dominated Ecosystems in the Lake States in Sustaining aspen in western landscapes, in Shepperd WD, Binkley D, Bartos D, Stohlgren TJ, Eskew LG, Eds. 2001. USDA Forest Service, Fort Collins, Colorado. General Technical Report RMRS-P-18. 460 p.

Stratton, R.D. 2004. Assessing the effectiveness of landscape treatments on fire growth and behavior. Journal of Forestry 102(7): 32-40.

Thomas, C.W. 2003. Bureaucratic Landscapes: Interagency cooperation and the preservation of biodiversity. MIT Press, Cambridge, MA.

Van Name, W.C. 1923. The Barbour Roosevelt-Sequoia Park Bill. Ecology, 4(2): 214-217.

Walters, C.J., and Holling, C.S. 1990. Large-scale management experiments and learning by doing. Ecology, 71(6): 2060-2068.

White, M.A. and J. L. Vankat. 1993. Middle and high elevation coniferous forest communities of the North Rim region of Grand Canyon National Park, Arizona. Vegatatio 109: 161-174.

Yaffee, S.L. 1998. Cooperation: A Strategy for Achieving Stewardship Across Boundaries, in "Stewardship Across Boundaries", Richard L. Knight and Peter B.Landres, Eds., Island Press, Washington D.C.

Yaffee, S.L, and J.M. Wondolleck. 1997. Building Bridges Across Agency Boundaries, in "Creating a Forestry for the 21st Century", Kathryn A. Kohm and Jerry F. Franklin, Eds., Island Press, Washington D.C.

APPENDIX 1-A:
INTERVIEW PARTICIPANTS

Aumack, Ethan. Director of Restoration Programs, Grand Canyon Trust

Crumbo, Kim. Conservation Director, Grand Canyon Wildlands Council

Dickson, Brett. Spatial ecologist, Forest Ecosystem Restoration Analysis Project

Fulé, Peter. Fire Ecologist, Ecological Restoration Institute

Gatto, Angela. Wildlife Biologist, KNF

Hahn, Martha. Science Center Director, GCNP

Higgins, Bruce. Forest Planner from 1994-2007, KNF

Leonard, Ariel. Assistant Forest Planner, KNF

Martin, Steve. Superintendent, GCNP

McKinnon, Taylor. Public Lands Director, Center for Biological Diversity

Mertz, Dave. Joint Fire Management Officer for GCNP and KNF

Noble, Bill. Forest Biologist, KNF

Reynolds, Richard. Wildlife ecologist and goshawk researcher, KNF

Robinson, David S. North Zone Fire Management Officer, KNF and GCNP

Sesnie, Steve. Remote Sensing Specialist, Forest Ecosystem Restoration Analysis Project

Short, Timothy. District Ranger, North Kaibab Ranger District, KNF

Ward, R.V. Wildlife Biologist, GCNP

Williams, Mike. Forest Supervisor, KNF

# ASSESSMENT OF MIXED CONIFER FOREST CONDITIONS, NORTH KAIBAB RANGER DISTRICT, KAIBAB NATIONAL FOREST, ARIZONA, USA

*Steven E. Sesnie, Brett G. Dickson, Jill M. Rundall, and Thomas D. Sisk*

## ABSTRACT

Southwest mixed conifer forest types maintain a diversity of tree species and structural conditions that contribute to desirable ecosystem services (e.g., higher biodiversity, watershed protection, forest carbon pools and aesthetic values). Less is known about mixed conifer forest and historical changes in composition and structure than for other Southwest forest types such as ponderosa pine. The U.S. Forest Service 2009 Kaibab Forest Health Focus initiative identified mixed conifer forest as a priority vegetation type requiring active forest restoration and hazardous fuel reduction in light of recent and severe fire activity. We evaluated contemporary changes in mixed conifer forest conditions on the North Kaibab Ranger District (NKRD) of Kaibab National Forest north of Grand Canyon National Park (GCNP), where other recent studies and historical forest inventories provided an excellent opportunity for comparative analyses. Inventory data from 1909, 1955 and the 1990s on the NKRD showed that average basal area had doubled by 1955. Basal area in 1990 was also double that of 1909, but had decreased by 28%, for trees $\geq$30 cm in diameter since 1955. Tree density for shade-tolerant species such as spruce and true fir showed a >600% increase between 1909 and 1990, whereas ponderosa pine showed little increase. Inventory data indicated a pattern of high basal area accretion prior to 1955 as fire was excluded from MC forest and increased tree recruitment of shade tolerant species following selective logging and insect caused tree mortality during the 1970s and 80s. Separate analyses of elevation and annual solar radiation gradients indicated that tree species composition was significantly different from low to high elevation sites in 1990, as was average canopy height. Densities of shade-tolerant trees were high on all sites. Forest structural attributes associated with fire behavior did not differ significantly across gradients with the exception of relatively mesic sites at high elevations or in shaded areas (i.e., drainages) that showed 18% greater canopy bulk density and 11% lower canopy base height. Restoration activities in mixed conifer forest should be focused on reducing a historically high density of shade-tolerant understory trees while providing opportunities for the regeneration of fire-adapted species, such as ponderosa pine and Douglas fir. Tree thinning and burning activities should seek to restore mixed severity fire regimes that historically maintained tree species and structural diversity, while reducing overall hazardous fuel accumulations that have developed for most site conditions in the absence of fire.

## INTRODUCTION

Mixed conifer (MC) forest on the Kaibab Plateau in northern Arizona is typically characterized by tree species that include ponderosa pine (*Pinus ponderosa*), white fir (*Abies concolor*), Douglas fir (*Pseudotsuga menziesii*), and aspen (*Populus tremuloides*), but also intermixes with spruce-fir forest on more mesic sites with blue spruce (*Picea pungens*), Engelmann spruce (*Picea engelmannii*), and sub-alpine fir (*Abies lasiocarpa*) (White and Vankat 1993). This mixed composition facilitates structural complexity and species diversity, which in turn provide a variety of ecosystem services, such as maintaining watershed values, forest carbon pools and wildlife species habitat (North et al. 2009). The 2009 U.S. Forest Service (USFS) Kaibab Forest Health Focus (Sisk et al. 2009) was a stakeholder-driven initiative and process that identified ponderosa pine and MC forest as priority forest types in need of restoration treatments critical to the maintenance of values and conditions resilient to ecosystem stressors. Stakeholders included local, state and federal land management agencies and non-government environmental organizations.

Studies focused on the condition of western MC forests suggest that many areas are susceptible to increased tree mortality and insect outbreaks as a result of past fire suppression, climate-induced drought and water stress (Guarín and Taylor 2005, Vankat et al. 2005, North et al. 2009, Fulé et al. 2009, van Mantgem et al. 2009). Historically, fires burned with low to mixed severity in MC forest on the Kaibab Plateau (Fulé et al. 2003). Recent wildfires in ponderosa pine and MC forest on USFS land north of Grand Canyon National Park (GCNP), such as the Outlet (2000) and Warm (2006) fires, have shown uncharacteristically extensive and severe fire behavior, particularly in areas of dense and contiguous tree canopy (Wimberly et al. 2009).

While forest conditions in ponderosa pine-dominated sites are well studied, quantitative characterizations of past and present mixed conifer forest composition and structure are relatively few in number (but see White and Vankat 1993, Fulé et al. 2003; Heinlein et al. 2005; Cocke et al. 2005; Fulé et al. 2009; Vankat 2011). Forest reconstruction

and dendrochronology studies for GCNP have demonstrated striking changes in forest composition and structure since historical fire suppression activities began c. 1879 (Fulé et al. 2003; Fulé et al. 2004). However, a loss of historical evidence on more mesic MC forest sites can increase uncertainty about forest conditions and change because of loss of evidence, particularly for small diameter trees, which can be consumed by fire or decay rapidly after mortality (Allen et al. 2002). Forest inventories can provide an added source of information about forest conditions for identifying historical forest structure, composition and dynamics over time to enhance reconstruction data (Sesnie and Bailey 2003; Fulé et al. 2003; Fulé et al. 2004; Vankat 2011).

Less is known about MC forest conditions occurring in forests to the north of GCNP on USFS land, which has been subject to numerous forest management regimes and policy changes over the last century. Our analysis of MC forest structure and composition on the North Kaibab Range District (NKRD) of the Kaibab National Forest (KNF) was undertaken to inform land-use planning and stakeholder discussions about historical changes in MC forest conditions, and to develop restoration treatment recommendations for broad spatial extents. In order to establish an improved understanding of existing MC forest conditions on the NKRD, our study objectives were to: 1) compare historical and contemporary mixed conifer forest structure and composition; 2) characterize environmental conditions that contain MC forest types and determine current forest structure and composition differences across elevation and solar radiation gradients associated with site moisture regimes; and 3) provide a quantitative framework for developing restoration criteria and a desired future condition for MC forest types.

Despite a century of active forest management on USFS lands, which included fire suppression, tree thinning and selective logging, shade-tolerant species have become an increasingly dominant component of MC forest types in the absence of fire (Stein 1988; Fulé et al. 2009; White and Vankat 1993). In addition, forest composition and structure within the MC forest types potentially differ along elevation and topographic gradients associated with site moisture regimes (White and Vankat 1993; Fulé et al. 2003). Kaibab National Forest staff distinguishes between dry and wet site MC forest types, and have developed a set of desired conditions for each type to guide forest management planning. However, site-scale biophysical conditions have not been quantitatively assessed to contrast forest conditions across moisture gradients. We hypothesized that site biophysical variables influence niche factors and disturbance patterns that can drive MC forest structural differences known to influence fire behavior, such as canopy bulk density and canopy base height. Canopy bulk density ($kg/m^3$) and canopy base height (m) are two important forest structural parameters commonly used to assess forest fuel conditions and model wildland fire behavior (Chuvieco et al. 2003). Historical and contemporary forest inventory data can help to characterize potential differences in MC forest structure and composition prior to developing restoration and hazardous fuels mitigation recommendations within MC forest types on the NKRD.

## METHODS AND MATERIALS

### Study Area

The study area includes MC forest on the NKRD in northern Arizona (Figure 2.1). A 65,700-ha area was initially defined as containing MC forest based on the extent of the LANDFIRE program's (http://www.landfire.gov/) existing vegetation map and MC forest categorizations. This area ranges in elevation from 1,700 to 2,800 m. However, only MC forest at elevations above ≥2,550 m (total area = 49,300 ha)

was examined due to the 2006 Warm Fire that burned some MC forest below this elevation. In addition, MC and spruce-fir forest principally exist on sites above 2,500 m (White and Vankat 1993). Kaibab National Forest staff typically considers forests above 2,900 m as the spruce-fir forest type, although MC forest on mesic sites below this elevation may closely resemble spruce-fir species composition. Average annual precipitation and temperature can vary considerably across the Kaibab Plateau according to meteorological data recorded from years 1925 to 2009 (http://www.wrcc.dri.edu/). The Jacob Lake (2,414 m) weather station that is within the ponderosa pine forest type, at the northern end of the Kaibab Plateau, has an annual average of 53 cm of rainfall and 268 cm of snowfall, whereas the Bright Angel Ranger Station (2,560 m) to the south in GCNP shows an annual average of 64 cm of rainfall and 347 cm of snowfall. Temperatures associated with MC forest are generally cooler in the upper elevations and range between a maximum of 25° C in July to a minimum of -9° C in January at the Bright Angel Ranger Station.

Topography in the study area is comprised of low, forested plateaus and ridge tops divided by valley bottoms and more continuous areas of even terrain toward the center of the Kaibab Plateau. Soils underlying MC forest generally are Utric Glossoboralfs and Typic Cryoboralfs and Paleboralfs with a mixture of fine sandy and gravely loam textures (Brewer et al. 1991).

### Available Datasets

No new data were collected with this study as historical and contemporary forest inventory data were available for the NKRD (Sesnie and Bailey 2003). Three principal datasets were consolidated and used to perform comparative analyses:

• Historical inventory data from 1909 and a stand table for the MC forest type on the

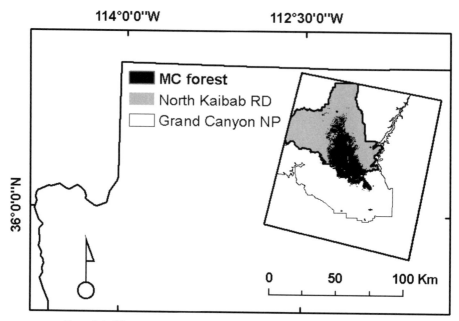

Figure 2.1   Mixed conifer study area on the Kaibab National Forest, North Kaibab
Ranger District in northern Arizona.

Table 2.1  The number and area of mixed conifer forest stand polygons analyzed according to elevation and solar radiation categories.

| Elevation Categories | Descriptor | No. Stands | Area (ha) | % of Area Sampled[a] |
|---|---|---|---|---|
| 1 (2,550 – 2,627 m) | Dry | 50 | 930 | 10.2 |
| 2 (2,627 – 2,693) | Moist | 118 | 4,100 | 52.5 |
| 3 (2,693 – 2,807) | Mesic | 134 | 4,692 | 81.5 |
| **Solar radiation categories[b]** <br> 1 (268,695 – 1,727,604 w/m²/yr) | Mesic | 42 | 1,113 | 49.8 |
| 2 (1,727,604 – 1,810,183) | Moist | 124 | 4,495 | 40.1 |
| 3 (1,810,183 – 1,865,237) | Dry | 77 | 2,803 | 41.0 |
| 4 (1,865,237- 2,023,515) | Very Dry | 40 | 1,208 | 46.6 |

[a] Percent of area sampled indicates the proportion of area inventoried that was in a mixed conifer forest type rather than aspen, ponderosa pine or other hardwood forest types.
[b] Solar radiation categories are associated with slopes of increased exposure to sunlight throughout the year.

NKRD summarizing tree density (trees/ha) for conifer species were compared to the most recent forest inventory data described below. The 1909 inventory data are from a 52-ha "strip cruise" which measured diameter at breast height (d.b.h, 1.37-m above the ground) for all conifer trees ≥15.2 cm d.b.h, in addition to seedlings ranging in height from 0 to 1 m and saplings that ranged from 1m in height to <15.2 cm d.b.h (Lang and Steward 1910). Basal area (m²/ha) was calculated for each tree species within 2.5 cm diameter classes for trees >15.2 cm and <50.8 cm d.b.h and 5 cm diameter classes for trees ≥50.8 cm d.b.h.using the equation for English units ($d^2*0.005454$)* the number of trees per acre in a class, where $d^2$ is the squared d.b.h class. An assumed mid-point d.b.h of 7.6 cm was used to estimate sapling basal area for all tally trees in this size class. Basal area was then converted to m²/ha. In previous studies, Lang and Stewart (1910) stand table data have compared favorably to tables developed Southwest forest types in the early 1900s (Woolsey 1911, Sesnie and Bailey 2003) and forest reconstruction data (Fulé et al. 2003, Fule et al. 2004). These data represent early MC forest structure and composition prior to more extensive USFS timber management and active fire suppression, although fire regime disruption likely occurred prior to 1909 because of heavy livestock grazing beginning c. 1885 (Russo 1964). A second pair of stand tables for open and dense forest conditions developed from 1955 Continuous Forest Inventory (CFI) 10th ha plots (n = 54) in unlogged MC forest was also compared with the 1909 inventory. The 1955 stand tables tallied saplings in diameter categories between 9.1 cm and 19.2 cm d.b.h and 19.3 cm and 29.4 cm d.b.h and no seedlings were accounted for in tables. The mid-point for each diameter class was used to estimate basal area for saplings in addition to 2.5 cm diameter classes for trees ≥29.4 cm.

• Recent inventory data for MC forest types were collected between 1980 and 2000 (USFS Stand Exam inventories) and used for comparison with the 1909 MC stand table. This recent dataset is referred to as the "1990 inventory" (average year measured) and cover site biophysical conditions representative of the NKRD MC forest stands. Tree measurements were collected using variable-radius plots for trees ≥12.7 cm d.b.h and nested fixed-radius plots (typically 1/300 ac or 2 m radius plots size) for smaller trees <12.7 cm d.b.h. A diameter was recorded for all trees greater than 2.5 cm d.b.h (1" d.b.h) and all other seedlings and saplings were counted. Approximately one plot per hectare was established within stand polygons, which were drawn over orthorectified aerial photographs to discriminate different forest types and site conditions. Small tree counts and diameter measurements differed between 1909, 1955 and 1990 inventories as diameters were measured with greater precision in 1990. Basal area calculations for 1909 and 1955 were more uncertain, particularly for saplings, since trees were tallied within a diameter range that can over or underestimate these values. Nevertheless, saplings typically constituted a small proportion of the overall tree basal area on a site.

• Environmental data were evaluated in a geographic information system (GIS), along with a 30-m digital elevation models (DEM) from the National Elevation Dataset (NED, http://seamless.usgs.gov/). The DEM was used to estimate average elevation and annual solar radiation values within 1990 forest inventory units (mapped stand polygons). Terrain features were used to develop biophysical categories for stratifying and comparing MC forest inventories according to low-, mid- and high-elevation sites across the NKRD.

## Analyses

To compare contemporary and historical forest conditions, MC 1990 inventory data (n = 528 stands) were summarized using average tree species density and basal area regardless of geographic location or biophysical setting. Tree density and basal area for conifers were also summarized according five d.b.h classes (<10cm, 10-29.9cm, 30-45.9cm, 46-60.9cm, >61cm) for comparison between inventory dates and tree species. In the KNF inventory database, stand polygons were classified by dominant tree species. For this analysis, only MC forest types dominated by a combination of ponderosa pine, spruce, true fir or Douglas fir trees were summarized and compared with 1909 MC inventory data. 1990 MC stands had a maximum of 50 percent of the basal area comprised by ponderosa pine. Average tree density and basal area for each coniferous tree species were used to compare the two forest inventories since only a summarized stand table for MC forest was available from 1909.

We also evaluated average tree species density and basal area from 1990 MC forest inventories within elevation categories above 2,550 m. These categories are related to precipitation gradients at a macro-scale, according to interpolated mean annual precipitation data (not shown, http://www.worldclim.org/). Elevation categories were selected using three quantile groups, with elevation values ranging from the 25th to 50th percentile (Dry), 50th to 75th percentile (Moist) and values >75th percentile (Mesic) for existing MC forest mapped by LANDFIRE (Table 2.1; Figure 2.2). MC forest polygons below the 25th percentile were <2,550 m, which excluded some lower elevation MC forests that transition from ponderosa pine. MC forest structure and composition within stand polygons for each elevation category were assumed to follow a gradient of increasing annual precipitation and temperature based on weather station records from the Kaibab Plateau and interpolated climate data. The elevation category of an MC stand inventory polygon was determined using the average value calculated with the Spatial Analyst extension to ArcGIS (v.9.3; Environmental Systems Research Institute, Redlands, CA).

To determine if MC forest composition differed between elevation categories, multivariate comparisons were made using Multi-Response Permutation Procedures (MRPP; McCune and Grace 2002) implemented in PC-ORD (v.5; McCune and Medford 1999). MRPP provided a non-parametric multivariate statistical comparison capable of using non-Euclidean distance measures to compare tree species data within elevation categories. MRPP also is a robust means of analyzing data of unequal and small sample sizes (McCune and Grace 2002). Comparisons among elevation categories were made using Sorensen (Bray-Curtis) similarity values derived from the average basal area for each tree species. Average stand tree basal area, density, canopy bulk density, canopy base height and canopy height in each elevation category were compared using two-sample t-tests (test statistic = t, $\alpha = 0.05$).

We also analyzed total annual solar radiation categories to compare forest composition and structure along moisture gradients associated with topography. Solar radiation is a function of local-scale topography and likely interacts with elevation. However, for this study, observations were made without accounting for differences in elevation to avoid limitations associated with a large number of categories and a small number of representative samples. Global solar radiation values (w/m²/year) were derived from the DEM using the Solar Analyst v. 1.0 extension to ArcView (v.3.3; Environmental Systems Research Institute, Redlands, CA). As with elevation, annual solar radiation was divided into four quantile categories (Table 2.1; Figure 2.3). Radiation categories represent an increasing amount of annual sunlight incident upon land surfaces as a function of slope conditions, surrounding terrain, and aspect. The amount of solar radiation on a site drives important ecosystem processes, such as photosynthesis, transpiration and evaporation, plant growth, and species composition in addition to soil and plant moisture conditions (Pocewicz et al. 2004). Only MC forest inventories above 2,550 m were used to evaluate changes in tree density, basal area, canopy bulk density, canopy base height and canopy height according to radiation categories. The multivariate (MRPP) and univariate (t-tests) analyses detailed above were also used for these comparisons. The radiation category of each MC stand inventory polygon was determined using the average value calculated in ArcGIS.

All 1990 forest inventory data were summarized using the Forest Vegetation Simulator (FVS) Central Rockies variant for Southwest tree species (http://www.fs.fed.us/fmsc/fvs/). Canopy bulk density and canopy base height estimates were calculated with the FVS Fire and Fuels Extension (Reinhardt and Crookston 2003), which uses empirical methods described in Scott and Reinhardt (2001).

Most MC forest on the NKRD has undergone selective harvesting in the past three decades, according to post hoc forest change comparisons made using spectral data Landsat Multispectral Spectral Scanner (MSS) imagery from 1973, 1983, and 1993 (data not shown). However, forest disturbances were not accounted for in our comparative analyses of stands, with the exception of relatively small areas that have experienced clear cutting or other severe disturbances (e.g., wind-throw or fire). These stand polygons were typically <6 ha in size and were removed from all analyses because they represented a small proportion of MC stands in the study area with principally early successional forest characteristics.

## RESULTS AND DISCUSSION

### Mixed Conifer Inventories

The 1909 stand table and 1990 forest inventory data indicated that large changes have occurred in MC forest composition and structure over the 20th century (Table 2.2). Tree density for shade-tolerant species, such as spruce and true fir, increased >600% between years 1909 and 1990. Density and basal area increases were primarily from recruitment of shade-tolerant saplings and trees 10 cm to 46 cm d.b.h (Table 2.3). A relatively small increase in tree density (53%) and basal area (24%) was observed for ponderosa pine. Ponderosa pine and Douglas fir were favored for timber harvest on the NKRD, which is a potential explanation for minor increases in basal area that have been observed for these species in other parts of the Southwest (Fulé et al. 2009). For all conifer species combined, tree density was over three times higher in 1990 than was documented in the 1909 inventory, and basal area increased by 83%. High tree density and basal area from 1990 inventories were similar to values reported in previous studies excluding aspen (White and Vankat 1993, Fulé et al. 2003, Cocke et al. 2005). Further analyses of 1990 MC forest conditions within elevation and solar radiation categories below included aspen for comparison with other studies.

Comparisons between 1909 and 1990 inventories should be considered with some discretion, as sampling protocols were different and only a stand table summary is available from 1909. In addition, the specific geographic location of the 52 ha plot reported in the 1909 inventory is unknown. Fulé et al. (2003, 2004) found that the 1909 stand table summary data matched well with reconstructed forest composition and structure data derived from studies in GCNP. The 1909 inventory and contemporary data likely provide a useful reference for identifying general changes in MC forest

Table 2.2  Differences in mean tree density and basal area for all conifer tree species including seedling and saplings between 1909 and 1990. Species are ponderosa pine (PP), Douglas fir (DF), true-fir (TF, sub-alpine fir and white fir) and spruce (SP, Engelmann and blue spruce) to matching 1990 inventory data with the 1909 MC table.

| Species | Average density (trees/ha) | | | Average basal area (m²/ha) | | |
|---|---|---|---|---|---|---|
| | 1909 | 1990 | Increase (%) | 1909 | 1990 | Increase (%) |
| PP | 17.4 | 26.7 | 53 | 5.1 | 6.3 | 24 |
| DF | 6.9 | 36.9 | 435 | 2.0 | 3.1 | 55 |
| TF | 37.8 | 304.2 | 705 | 3.6 | 7.7 | 114 |
| SP | 12.6 | 88.3 | 601 | 2.1 | 6.2 | 195 |
| SUM | 74.7 | 456.2 | - | 12.8 | 23.4 | - |

Table 2.3  Differences in mean tree density and basal area for all conifer tree species between 1909 and 1990 within five diameter classes.

| 1909 average density (trees/ha) | | | | | | 1909 average basal area (m²/ha) | | | | |
|---|---|---|---|---|---|---|---|---|---|---|
| DBH Class | PP | DF | TF | SP | Sum | PP | DF | TF | SP | Sum |
| Saplings | 9.9 | 4.1 | 32.8 | 9.0 | 55.9 | 0.3 | 0.1 | 0.8 | 0.3 | 1.5 |
| 15.2-29.9[1] cm | 3.7 | 1.4 | 2.8 | 1.9 | 9.8 | 0.8 | 0.3 | 0.6 | 0.4 | 2.1 |
| 30-45.9 | 2.2 | 0.5 | 1.3 | 1.5 | 5.2 | 1.4 | 0.3 | 0.8 | 0.7 | 3.2 |
| 46-60.9 | 0.9 | 0.6 | 0.6 | 0.5 | 2.6 | 1.1 | 0.7 | 0.7 | 0.5 | 3.1 |
| >=61 | 0.6 | 0.3 | 0.3 | 0.1 | 1.3 | 1.5 | 0.6 | 0.7 | 0.2 | 3.0 |
| | | | | Sum | 74.7 | | | | Sum | 12.8 |

| 1990 average density (trees/ha) | | | | | | 1990 average basal area (m²/ha) | | | | |
|---|---|---|---|---|---|---|---|---|---|---|
| DBH Class | PP | DF | TF | SP | Sum | PP | DF | TF | SP | Sum |
| Saplings | 16.0 | 29.2 | 280.0 | 69.7 | 394.9 | 0.1 | 0.1 | 0.3 | 0.2 | 0.7 |
| 10-29.9 cm | 6.5 | 5.7 | 19.5 | 14.3 | 45.9 | 1.3 | 1.1 | 3.5 | 2.6 | 8.5 |
| 30-45.9 | 2.2 | 1.5 | 3.6 | 3.5 | 10.8 | 1.5 | 1.0 | 2.3 | 2.2 | 7.0 |
| 46-60.9 | 1.3 | 0.5 | 0.8 | 0.7 | 3.3 | 1.7 | 0.6 | 1.1 | 0.9 | 4.3 |
| >=61 | 0.7 | 0.2 | 0.2 | 0.1 | 1.2 | 1.7 | 0.4 | 0.5 | 0.2 | 2.9 |
| | | | | Sum | 456.2 | | | | Sum | 23.4 |

[1]The breakpoint between sapling and larger trees is 15.2 cm for the 1909 inventory, as these data are derived from a stand table summary.

composition and structure over the last century on the NKRD. In addition, CFI stand tables indicate that tree density and basal had nearly doubled by 1955 on the NKRD. A 1955 stand table derived from 49, $10^{th}$ ha plots in unlogged MC forest showed an average of 34 trees/ha and basal area of 23.2 $m^2$/ha, in comparison with 35 trees/ha and 12.8 $m^2$/ha in 1909 for all conifer trees with a measurable d.b.h. A second stand table from 1955 also showed an average basal area and tree density of 29.4 $m^2$/ha and 69 trees/ha, respectively, but were derived from only five plots in unlogged MC forest. The 1955 stand tables were comparable to total basal area in 1990 (23.4 $m^2$/ha), however, 91% of the basal area in 1955 was from trees $\geq$30 cm d.b.h., compared to 63% in 1990. Only a single CFI plot in MC forest was reported as logged in the 1955 inventory.

These inventory data reflect changes in forest structure and composition that likely occurred as a result of fire suppression between 1909 and 1955 and increased selective logging in MC forest prior to 1990. A doubling of tree basal area between 1909 and 1955 was consistent with simulated biomass accretion reported for the north rim of GCNP by Fulé et al. (2004). Tree biomass, which is highly correlated with basal area, increased by an average of 122% when simulated at 20-year time intervals between 1880 and 2040, increasing as much as 279% on higher elevation sites (Fulé et al. 2004). Reduced basal area for trees $\geq$30 cm between 1955 and 1990 is likely the result of selective logging of large trees in MC forest during this period. Historical tree harvest records from the NKRD and CFI plots show that selective logging in MC forest began shortly after 1955 and increased in subsequent years due to an increase in spruce budworm (*Choristoneura occidentalis* Freeman) activity and tree mortality during the mid-1970s and 80s (Wahlfeld 1993; Sesnie and Bailey 2003).

Changes in forest structure over time

are consistent with those reported by Vankat (2011) in GCNP, which showed extraordinarily high average tree basal area (65 $m^2$/ha) for MC forest in 1935, followed by much lower basal (35 $m^2$/ha) recorded at these sites in 2004. Increased tree mortality was recorded in MC forest on the NKRD during the 1970s and 1980s because of high tree density and spruce budworm defoliation (Wahlfeld 1993). However, average values for basal area reported for MC forest in 1935 are questionable (Vankat 2011), as these values greatly exceeded those estimated in stand tables developed from unlogged CFI plots in 1955 for the NKRD. An average basal area of 65 $m^3$/ha in 1935 would closely match those of old-growth MC forest recorded on highly productive sites in the southern Sierra Nevada of California (68.5 $m^2$/ha), where fire was excluded for 135 years (North et al. 2004). Nevertheless, patterns of high biomass accretion, inflection and recession between 1935 and 2004 in GCNP reported by Vankat et al. (2005) and Vankat (2011) likely reflect forest dynamics in areas of high tree density and basal area where no logging had occurred.

Elevation Gradient

Inventory data and tree measurements summarized over stand polygons were not suitable for discerning fine-scale site differences and relationships between spatial heterogeneity in forest composition and structure. However, these polygons were delineated to encompass a similar site biophysical condition, forest type and successional stage. As such, these data and analyses are best interpreted as distinguishing vegetation differences and site-scale biophysical conditions at a scale of 6-100 ha.

Differences in MC forest observed within the three elevation categories were principally due to dissimilarities in tree species composition (Figure 2.4a, Figure 2.4b). Multivariate comparisons among the elevation categories using MRPP demonstrated that MC forest in elevation categories 1 (dry) and 2 (moist) were not significantly different in tree species composition (Table 2.4). Inventory data indicated that a transition in tree species composition occurred above 2,700 m between MC forest and increasingly spruce- and true fir-dominated forest. A large decrease in basal area was also observed for ponderosa pine above this elevation in category 3 (mesic), but the species was co-dominant in the other two categories (Figure 2.4a, Figure 2.4b). Ponderosa pine was represented by a lower number of trees/acre in all elevation categories, but was proportionally high in basal area, indicating that fewer large trees were typically present. A higher density and basal area of white fir at lower elevations was compensated for by more abundant subalpine fir at higher elevations (Figure 2.4a, Figure 2.4b). Aspen showed very little difference among elevation categories.

Overall, forest structure data indicated a gradual decrease in tree density with increased elevation, but similar basal area across all three elevation categories (Table 2.5). Nevertheless, univariate comparisons of average tree density, basal area, and canopy bulk density among all three elevation categories were not significantly different (Appendix 2-A). Forest structural conditions were generally similar across categories (Table 2.5). An exception was average canopy height, where MC forest in elevation category 1 (22.3 m, dry) was statistically significantly different from categories 2 (21.0 m, moist; t = 2.045, p = 0.022) and 3 (20.6 m, mesic; t = 2.81, p = 0.003). This result likely is due to a shift in species composition or the successional status of MC forest favoring a greater abundance of shorter-stature trees at higher elevations. Average canopy bulk density was also significantly different between elevation categories 2 (2.06 m, moist) and 3 (1.60 m, mesic; t = 1.65, p = 0.007); however, both average tree height and canopy base height exhibit relatively minor differences among categories.

Our results indicate that a change in tree species composition occurs with increased elevation, as do small changes in canopy height. With respect to overall MC forest structural conditions, density, basal area and canopy bulk density were similar across elevation categories. Shade-tolerant trees, although different in species composition, were a substantial component of MC forest at all elevations. These results suggest that that a large number of understory shade-tolerant trees have developed in the absence of disturbance factors (e.g., fire), regardless of potential site differences among elevation categories. Moreover, average tree density and basal area were similar to MC forest conditions observed in GCNP. Fulé et al. (2003) showed average MC tree density was 873 trees/ha and basal area was 38.8 m²/ha for the Little Park area of GCNP. These data are also comparable to forest structure estimates from Thompson Canyon in GCNP by White and Vankat (1993) that ranged from 720 to 1,394 trees/ha and 29.0 m²/ha to 48.3 m²/ha basal area. These tree densities were quite similar to MC forest on USFS land quantified from 1990 inventory data (Table 2.5). Basal area was also similar, but somewhat lower (≤31 m²/ha, Table 2.5), which was expected given the difference in forest management history and selective timber harvests prior to 1990 inventories.

### Solar radiation gradient

Tree species composition differed at increasing levels of annual solar radiation , which can moderate site moisture conditions according to local hillslope topography (e.g., foot slope to ridge-top; Pocewicz et al. 2004). Notably, subalpine fir and spruce decrease in density and basal area with increased annual solar radiation in a consistent fashion (Figure 2.5a, Figure 2.5b).

Figure 2.2   Map of elevation categories used to select mixed conifer stand polygons and inventory data.

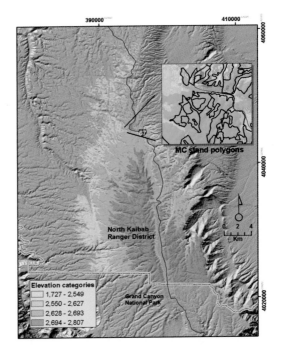

Figure 2.3   Map of annual solar radiation (w/m²/yr) categories used to select mixed conifer stand polygons and inventory data.

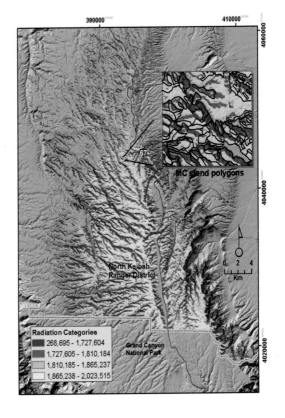

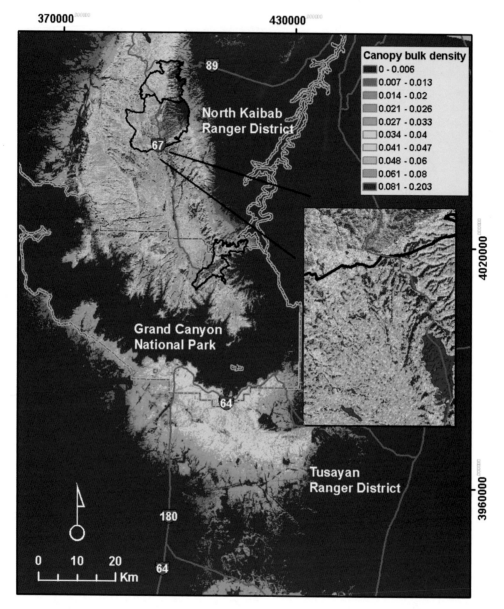

Figure 2.6   Digital map of canopy bulk density (kg/m3) derived for the Kaibab National Forest and GCNP (grey outline) from 2006 Landsat Thematic Mapper (TM) imagery and US Forest Service Inventory and Analysis plots. Large changes in canopy bulk density within map polygons (black) are a result of the Warm (north) and Outlet (south) fires. Validation information for canopy bulk density were derived from 271 validation plots and bootstrapped error estimates (Mean/Median residual error = 0.013/0.006 kg/m$^3$ and 73.4% of the variance was explained by the regression tree model used).

| Elevation comparisons | T | A | *p*-value |
|---|---|---|---|
| 1 vs. 2 | -0.37 | 0.001 | 0.278 |
| 1 vs. 3 | -17.23 | 0.037 | <0.001 |
| 2 vs. 3 | -23.05 | 0.036 | <0.001 |

Table 2.4 Results from MC forest comparisons by tree species basal area and elevation categories using multivariate MRPP statistical tests. T is the test statistic and A is the chance corrected agreement statistic that describes within group homogeneity compared to randomized data.

| Elevation category | TD | BA | CBD | CBH |
|---|---|---|---|---|
| 1 (2,550 – 2,627 m) | 1,006 (115.1) | 31.32 (3.2) | 0.105 (0.008) | 1.76 (0.163) |
| 2 (2,627 – 2,693 | 898 (117.3) | 30.92 (2.0) | 0.104 (0.004) | 2.06 (0.163) |
| 3 (2,693 – 2,807) | 788 (56.5) | 31.45 (1.9) | 0.109 (0.003) | 1.60 (0.004) |

Table 2.5  Information on forest structure along an elevation gradient in the Kaibab Ranger District, Kaibab National Forest, Arizona.  Data are from a 1990 inventory for mixed conifer stands at three elevations.  In the table TD = tree density given as trees/ha.  BA – basal area given as $m^2$/ha.  CBD = canopy bulk density expressed as kg/$m^3$.  CBH = crown-base height given in meters.  All numbers in brackets are standard errors [SE] of the average.

| Solar radiation comparisons | T | A | *p-values* |
|---|---|---|---|
| 1 vs. 2 | -0.40 | 0.0010 | 0.2579 |
| 1 vs. 3 | -2.52 | 0.0087 | 0.0268 |
| 1 vs. 4 | -2.68 | 0.0134 | 0.0226 |
| 2 vs. 3 | -2.07 | 0.0041 | 0.0446 |
| 2 vs. 4 | -2.52 | 0.0061 | 0.0273 |
| 3 vs. 4 | -2.67 | 0.0089 | 0.0216 |

Table 2.6  Results from MC forest basal area comparisons by tree species between each of the radiation categories described as mesic, moist, dry and very dry using multivariate MRPP statistical tests.

| Radiation categories | TD | BA | CBD | CBH |
|---|---|---|---|---|
| 1 (268,695 – 1,727,604 w/$m^2$/yr) | 981(126.6.) | 34.1 (3.5) | 0.127 (0.009) | 1.61 (0.147) |
| 2 (1,727,604 – 1,810,183) | 903 (68.8) | 30.3 (1.9) | 0.104 (0.004) | 1.753 (0.101) |
| 3 (1,810,183 – 1,865,237) | 938 (75.1) | 33.3 (2.6) | 0.106 (0.005) | 1.924 (0.180) |
| 4 (1,865,237- 2,023,515) | 908 (95.5) | 29.2 (3.7) | 0.101 (0.101) | 1.771 (0.169) |

Table 2.7  Average (SE) tree density (TD; trees/ha), basal area (BA; $m^2$/ha), canopy bulk density (CBD; kg/$m^3$), and crown-base height (CBH; m) for mixed conifer stands and summarized by solar radiation category using the 1990 inventory data.

These differences likely explain significant differences in tree species composition observed for all but the first two solar radiation categories (Table 2.6). Other species differed less consistently across radiation categories, although Douglas fir and aspen showed somewhat higher tree densities and basal area on sites receiving greater solar radiation. These results were similar to elevation comparisons that show species specific responses along presumed site moisture gradients related to different amounts of annual solar radiation, annual precipitation and turnover in tree species composition and abundance. Results did show somewhat more complex relationships between solar radiation categories and tree species dominance on a site. For example, ponderosa pine basal area increased and then decreased on sites from low to high solar radiation and tended to be replaced by, or was co-dominant with, Douglas fir on potentially drier sites (Figure 2.5b). These species occurred in low numbers according to tree density data, but were proportionally high in basal area, suggesting that a few large individual ponderosa pine and Douglas fir trees were typically present in MC forest. White and Vankat (1993) have suggested that a large increase in shade-tolerant species density and dominance on these sites poses a potential threat to mixed conifer forest diversity as low numbers of relict overstory ponderosa pine and Douglas fir are replaced by spruce and true fir.

On average, forest structure was more varied among solar radiation categories in terms of basal area than tree density (Table 2.7). Average canopy height also decreased from 22.0 m on mesic sites to 19.5 m on very dry sites. Both basal area and canopy height were significantly different between most solar radiation categories (Appendix 2-A). These data suggest that, in addition to site moisture, other factors such as land use practices may influence forest structural characteristics. For example, sites with low

annual solar radiation (mesic) were generally on steeper north-facing slopes (>25% slope) or within narrow canyon areas (Figure 2.2). It is possible that these sites were less favorable for selective tree harvesting or less susceptible to other disturbances, such as frequent fire. Based on fire-scar data from GCNP, Fulé et al. (2003) observed that north facing slopes exhibited less frequent fires than did south and east facing slopes.

Sites with very low solar radiation (mesic) showed significantly higher canopy bulk density than all other categories, in addition to having the lowest average canopy base height (Table 2.7, Appendix 2-A). Wimberley et al. (2009) showed that un-thinned drainages, associated with the lowest solar radiation values, experienced the greatest burn severity during the 2006 Warm Fire. Canopy bulk density from this study was slightly higher than that estimated in GCNP mixed conifer sites (0.08 kg/m$^3$) by Fule et al. (2004), likely due to different canopy bulk density equations used. Nevertheless, values were consistent with canopy fuels modeled from USFS Forest Inventory and Analysis plots (Figure 2.6, unpublished data). Extensive areas of moderate to high canopy bulk density (0.061 to 0.20 kg/m$^3$) and low canopy base height (<2 m on average) are evidence that contiguous canopy fuels exist for MC and ponderosa pine forest types that dominate the NKRD. When combined with extreme fire weather, these conditions have the potential to promote both passive and active canopy fire activity (Fulé et al. 2001, Wimberly et al 2009).

## CONCLUSIONS

Historically, site biophysical conditions and mixed severity fires created heterogeneous MC forest structure and composition in stands within close proximity to one another (Fulé et al. 2003). Today's MC forests on the NKRD are dramatically different from when forest inventory data were first collected in 1909. On average,

Figure 2.4   Mixed conifer average (A) tree density and (B) basal area and standard error for tree species within three elevation categories.

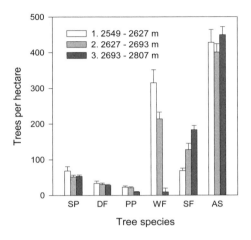
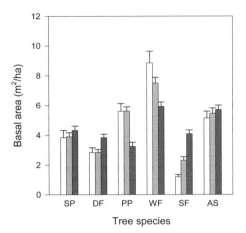

Figure 2.5   Mixed conifer average (A) tree density and (B) basal area and standard error for tree species within four solar radiation categories.

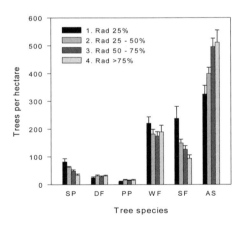
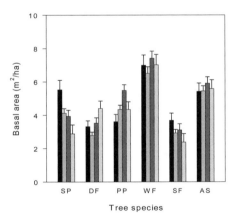

1990 MC tree density is six times greater than shown in 1909 inventory records and basal area has doubled on average over this same period. Perhaps most notable is that current forest inventory data on USFS land showed nearly equal tree densities and only moderately less basal area in comparison with tree data reported from studies in GCNP mixed conifer forest given their different land management histories. Forest management on the NKRD historically resulted in selective logging practices that removed a small proportion of trees and were focused on harvesting primarily overstory ponderosa pine and Douglas fir trees with commercial value (Sesnie and Bailey 2003). In addition, reduced fire frequency has facilitated a large cohort of shade-tolerant tree species resulting in more contiguous vertical and horizontal canopy structure in MC forest. Tree species composition differed within the MC forest type along the elevation and solar radiation gradients studied, but showed little or no significant difference in forest structural conditions. This was contrary to our expectation that overall tree density and basal area would differ significantly along elevation and solar radiation gradients that are linked to site moisture regimes. High tree densities and canopy bulk density have become increasingly contiguous across sites where fire has been mostly absent over the last several decades, increasing the likelihood of landscape-scale canopy fire events (White and Vankat 1993; Fulé et al. 2004; Wimberly et al. 2009).

Prior studies indicate that Southwest MC conditions are less likely to be resistant to tree mortality factors, such as periodic drought, climate change (hotter drier conditions) and other associated environmental stresses (North et al. 2009; Fulé et al. 2009), and suggest important implications for implementing restoration treatments. First, fire undoubtedly played an important role in maintaining lower shade-tolerant tree densities prior to fire

suppression activities, particularly on dry sites. True fir species are sensitive to bole scorch and low intensity fires that would likely have maintained historically lower numbers of established trees (Fulé and Laughlin 2007). Mixed severity fire was a periodic disturbance in MC forest that likely maintained a more spatially heterogeneous distribution of tree species, a diversity of structural conditions, and lower tree densities (Stein 1988; White and Vankat 1993; Fulé et al. 2003; Fulé et al. 2009; North et al. 2009). Returning more frequent and low-intensity fires to contemporary MC forest conditions is essential to achieving forest restoration goals, but could initially increase ground fuel accumulations via understory tree mortality in post-treatment areas, or present fire control risks (Stephens 1998; Agee 2003; but see Fulé and Laughlin 2007). To address uncertainty regarding the role of prescribed fire in MC forest restoration, recent fire monitoring plots established within GCNP can provide information on pre-treatment conditions and post-fire effects on MC forest composition and structure (Fulé and Laughlin 2007). Similarities between USFS land and GCNP forest conditions also provide a potential analog from which to develop methods for achieving desired outcomes using prescribed fire. However, Wimberly et al. (2009) found that forest thinning, followed by prescribed burning, was a significant factor in reducing burn severity in areas experiencing recent and extensive wildfire on the NKRD, in contrast to using these treatments types independent of each other. Greater interagency collaboration may also improve forest stewardship activities across USFS and GCNP lands on the Kaibab Plateau (see also Holcomb et al. this volume). Stakeholder-driven planning efforts, such has the Kaibab Forest Health Focus, can also facilitate socially viable forest restorations recommendations and environmental decision making (Sisk et al. 2009).

Second, silvicultural treatments such as thinning, prescribed burning or other selective tree harvest practices should be aimed at reducing overall tree density and basal area to restore species and size class distributions that support mixed-severity fire regimes. In this context, North et al. (2009) present recommendations for utilizing biophysical gradients associated with differing site moisture and disturbance regimes, which could be used to help guide silvicultural treatments, in conjunction with USFS desired conditions for MC forest types. Additionally, hazardous fuels reduction can likely be accomplished by restoring forest resiliency to natural fire events in areas of high fire risk. Results from Vaillant et al. (2009) indicate that understory thinning treatments prior to applying prescribed fire can be effective when targeted in areas with hazardous canopy fuels conditions for reducing high canopy fire risk in MC forest. Restoration activities should also create opportunities to regenerate fire resistant species, such as ponderosa pine and Douglas fir, and reduce contiguous areas of high canopy bulk density on the NKRD (Figure. 2.6). Historically, dry MC sites were likely dominated by fire resistant tree species, in contrast to mesic or higher elevation sites dominated by spruce and true fir species. Creating canopy-gaps in close proximity to residual ponderosa pine and Douglas fir, and reducing competitive interactions with shade-tolerant species, may promote regeneration for these species. Silvicultural practices that encourage natural stand dynamics and mixed severity fire behavior are needed to restore mixed conifer forest conditions which are less susceptible to extreme disturbance.

Some elements of this study and assessment of MC forest and landscape conditions can be expanded to other southwestern locations where forest inventory data are available. However, historical data of similar quality to the 1909 forest inventory data are relatively rare. USFS Forest Inventory and Analysis plots for MC forest types and other historical forest inventories (e.g., CFI plots) can also be used in combination with multi-temporal remotely sensed data to estimate existing forest structural conditions and changes that have occurred in recent decades. Available forest inventory data, coupled with published studies and historical syntheses provide an important and efficient means of informing collaborative land-use planning efforts, forest restoration goals, and hazardous fuels reduction objectives in southwestern MC forest types.

## ACKNOWLEDGEMENTS

Funds for this project were provided by the Kaibab National Forest. We thank Ariel Leonard and Jared Scott from the US Forest Service for their collaboration and the opportunity to develop information that will contribute to the Kaibab National Forest plan revision. An agreement with the US Forest Service Forest Inventory and Analysis program in Ogden, UT, provided us with the inventory plot data used to derive the digital data layer for canopy bulk density. We also thank two anonymous reviewers who contributed many helpful comments to improve this manuscript.

## LITERATURE CITED

Agee, J.K. 2003. Monitoring post-fire tree mortality in mixed-conifer forest reserves of Crater Lake, OR. Natural Areas Journal 23: 114–120

Allen, C.D., M. Savage, D.A. Falk, K.F. Suckling, T.W. Swetnam, T. Schulke, P.B. Stacey, P. Morgan, M. Hoffman, and J.T. Klingel. 2002. Ecological restoration of southwestern ponderosa pine ecosystems: a broad perspective. Ecological Applications 12: 1418-1433.

Brewer, D.G., R.K. Jorgensen, L.P. Munk, W.A. Robbie and J.L. Travis. 1991. Terrestrial Ecosystem Survey of the Kaibab National Forest: Coconino County and Part of Yavapai County, Arizona. USDA Forest Service Southwestern Region. 319 p.

Chuvieco, E., D. Riaño, J. van Wagtendonk and F. Mordod. 2003. Fule loads and fuel type mapping. In Wildland fire danger estimation and Mapping: the role of remote sensing data. Series in Remote Sensing Vol. 4. Edited by E. Chuvieco. World Scientific Publishing Co. Pte. Ltd. 5 Toh Tuck Link, Singapore. Pp. 119-143.

Cocke, A.E., P.Z. Fulé, J.E. Crouse. 2005. Forest change on a steep mountain gradient after extended fire exclusion: San Francisco Peaks, Arizona, USA. Journal of Applied Ecology 42: 814-823.

Fulé, P.Z and D.C. Laughlin. 2007. Wildland fire effects on forest structure over and altitudinal gradient, Grand Canyon National Park, USA. Journal of Applied Ecology 44: 136-146.

Fulé, P.Z., A.E.M Waltz, W.W. Covington and T.A. Heinlein. 2001. Measuring forest restoration effectiveness in reducing hazardous fuels. Journal of Forestry 99 (11): 24 – 29.

Fulé, P.Z., J.E. Crouse, A.E. Cocke, M.M. Moore, and W.W. Covington. 2004. Changes in canopy fuels and potential fire behavior 1880-2040: Grand Canyon, Arizona. Ecological Modelling 175: 231-248.

Fulé, P.Z., J.E. Crouse, T.A. Heinlein, M.M, Moore, W.W. Covington and G. Verkamp. 2003. Mixed-severity fire regime in a high-elevation forest of Grand Canyon, Arizona, USA. Landscape Ecology 18: 465-486.

Fulé, P.Z., J.E. Korb and R. Wu. 2009. Changes in forest structure of a mixed conifer forest, southwestern Colorado, USA. Forest Ecology and Management 258: 1200-1210.

Guarín, A. and A.H. Taylor. 2005. Drought triggered tree mortality in mixed conifer forest in Yosemite National Park, California, USA. Forest Ecology and Management 281: 229-244.

Heinlein, T.A, Moore, M.M., Fulé, P.Z., Covington, W.W., 2005. Fire history and stand structure of two ponderosa pine-mixed conifer sites: San Francisco Peaks, Arizona, USA. International Journal of Wildland Fire 14, 307-320.

Lang D.M. and S.S Stewart. 1910. Reconnaissance of the Kaibab National Forest. Northern Arizona University, Flagstaff, Arizona, USA, Unpublished report on file.

McCune, B. and J.B. Grace. 2002. Analysis of Ecological Communities. MJM Software Design, Gleneden Beach, Oregon 97388, USA.

McCune, B. and M.J. Mefford. 1999. PC-ORD: Multivariate Analysis of Ecological Data Version 4. MjM Software Design, Gleneden Beach, Oregon 97388, USA.

North, M., J. Chen, B. Oakley, B. Song, M. Rudnicki, A. Gray and J. Innes. 2004. Forest stand structure and pattern of old-growth western hemlock/Douglas-fir and mixed-conifer forests. Forest Science 50: 299-311.

North, M., P. Stine, K. O'Hara, W. Zielinski and S. Stephens. 2009. An ecosystem management strategy for Sierran mixed-conifer forests. Gen. Tech. Rep. PSW-GTR-220. Albany, CA: U.S. Department of Agriculture, Forest Service, Pacific Southwest Research Station. 49 p.

Pocewicz, A. L., P.E. Gessler and A. Robinson. 2004. The relationship between effective plant area index and Landsat spectral response across elevation, solar radiation and spatial scales in a northern Idaho forest. Canadian Journal of Forest Research 34: 465-480.

Reinhardt, E.D and N.L. Crookston. 2003. The Fire and Fuels Extension to the Forest Vegetation Simulator. USDA Forest Service, Rocky Mountain Research Station, Ogden, UT, General Technical Report RMRS-GTR-116.

Russo JP. 1964. The Kaibab deer herd: its history, problems, and management. Wildlife Bulletin No. 7. Phoenix (AZ): Arizona Game and Fish Department.

Scott, J.H and E.D. Reinhardt. 2001. Assessing canopy fire potential by linking models of surface and canopy fire behavior. Res. Pap. RMRS-PR-29. Fort Collins, CO: US Department of Agriculture, Forest Service, Rocky Mountain Research Station. 59 p.

Sesnie, S. and J. Bailey. 2003. Using history to plan the future of old-growth ponderosa pine. Journal of Forestry 101: 40-47.

Sisk, T.D., J.M. Rundall, E. Nielsen, B.G. Dickson, and S.E. Sesnie. 2009. The Kaibab Forest Health Focus: Collaborative Prioritization of Landscapes and Restoration Treatments on the Kaibab National Forest. The Forest Ecosystem Restoration and Analysis Project, Lab of Landscape Ecology and Conservation Biology, School of Earth Sciences and Environmental Sustainability, Northern Arizona University.

Stein, S.J. 1988. Explanations of the imbalanced age structure and scattered distribution within a high-elevation mixed coniferous forest. Forest Ecology and Management 25: 139-153.

Stephens, S.L. 1998. Evaluation of the effects of silvicultural and fuels treatments on potential fire behavior in the Sierra Nevada mixed-conifer forests. Forest Ecology and Management 105: 21–35.

Vaillant, N.M., J.A. Fites-Kaufman and Scott L. Stephens. 2009. Effectiveness of prescribed fire as a fuel treatment in Californian coniferous forests. International Journal of Wildland Fire 18: 165-175.

van Mantgem, P.J., N.L. Stephenson, J.C. Byrne, L.D. Daniels, J.F. Franklin, P.Z. Fulé, M.E. Harmon, A.J. Larson, J.M. Smith, A. H. Taylor, and T.T. Veblen. 2009. Widespread increase in tree mortality rates in the western United States. Science 323: 521-523.

Vankat, J. L., D. C. Crocker-Bedford, D. R. Bertolette, P. Leatherbury, T. McKinnon, and C. L. Sipe. 2005. Indications of large changes in mixed conifer forests of Grand Canyon National Park. p. 121-129 in van Riper, C. III and D. J. Mattson (eds.). The Colorado Plateau II: Biophysical, socioeconomic, and cultural research. Proceedings of the 7th Biennial Conference of Research on the Colorado Plateau. University of Arizona Press, Tucson, Arizona.

Vankat, J.L. 2011. Post-1935 changes in forest vegetation of the Grand Canyon National Park, Arizona, USA: Part 2-Mixed conifer, spruce-fir, and quaking aspen forests. Forest Ecology and Management 261:326-341.

Wahlfeld, D. 1993. North Kaibab logging history. Kaibab National Forest Supervisors Office, Williams Arizona. Unpublished report on file.

White, M.A. and J.L. Vankat. 1993. Middle and high elevation coniferous forest communities of the North Rim region of Grand Canyon National Park, Arizona, USA. Vegetatio 109: 161-174.

Wimberly, M.C., M.A. Cochrane, A.D. Baer and K. Pabst. 2009. Assessing fuel treatment effectiveness using satellite imagery and spatial statistics. Ecological Applications 19: 1377-1384.

Woolsey, T.S. 1911. Western yellow pine in Arizona and New Mexico. USDA Bulletin 101, 64 p.

| Elev. class | BA | | | D | | | HT | | | CBD | | | CBH | | |
|---|---|---|---|---|---|---|---|---|---|---|---|---|---|---|---|
| | df | t | p | df | t | p | df | t | p | df | t | p | df | t | p |
| 1 vs. 2 | 97 | 0.1990 | 0.4213 | 166 | 1.3652 | 0.0870 | 81 | 2.0458 | **0.0220** | 76 | 0.0932 | 0.4630 | 143 | -1.266 | 0.1038 |
| 1 vs. 3 | 73 | -0.079 | 0.4684 | 182 | 1.4384 | 0.0760 | 64 | 2.8105 | **0.0033** | 65 | -0.448 | 0.3276 | 73 | 0.8751 | 0.1922 |
| 2 vs. 3 | 214 | -0.394 | 0.3469 | 250 | -0.106 | 0.4575 | 213 | 0.9780 | 0.1646 | 227 | -0.901 | 0.1842 | 165 | 2.4384 | **0.0079** |

| Rad. class | BA | | | D | | | HT | | | CBD | | | CBH | | |
|---|---|---|---|---|---|---|---|---|---|---|---|---|---|---|---|
| | df | t | p | df | t | p | df | t | p | df | t | p | df | t | p |
| 1 vs. 2 | 164 | 2.2044 | **0.0144** | 164 | 1.0127 | 0.1563 | 109 | 2.4526 | **0.0079** | 59 | 2.31742 | **0.01198** | 78 | -0.745 | 0.2290 |
| 1 vs. 3 | 117 | 0.4259 | 0.3355 | 117 | 0.6368 | 0.2627 | 106 | 1.2000 | 0.1164 | 63 | 2.02850 | **0.02337** | 117 | -1.155 | 0.1251 |
| 1 vs. 4 | 81 | 2.7850 | **0.0033** | 81 | 0.9596 | 0.1700 | 66 | 3.9768 | **0.0001** | 72 | 2.33187 | **0.01125** | 77 | -0.638 | 0.2625 |
| 2 vs. 3 | 199 | -2.141 | **0.0167** | 199 | -0.506 | 0.3064 | 176 | -1.185 | 0.1188 | 176 | -0.3863 | 0.34987 | 111 | -0.705 | 0.2410 |
| 2 vs. 4 | 163 | 1.0349 | 0.1511 | 163 | 0.2185 | 0.4136 | 64 | 2.3509 | **0.0109** | 77 | 0.33897 | 0.36778 | 63 | -0.081 | 0.4677 |
| 3 vs. 4 | 116 | 2.6364 | **0.0048** | 116 | 0.6055 | 0.2730 | 69 | 3.1025 | **0.0014** | 82 | 0.63654 | 0.26310 | 110 | 0.5230 | 0.3010 |

Appendix 2-A    Elevation and solar radiation class comparisons of forest structure parameters using two sample *t*-tests. Degrees of freedom (df) differ for comparisons which assume an unequal variance as a result of F-test comparisons for two sample variances. Significant class differences ($p \leq 0.05$) are in bold. Basal area = BA, tree density = D, canopy height = HT, canopy bulk density = CBD and canopy base height = CBH.

# LIVESTOCK GRAZING FOLLOWING WILDFIRE: UNDERSTORY COMMUNITY RESPONSE IN AN UPLAND PONDEROSA PINE FOREST

*Lauren A. Mork, Thomas D. Sisk and Ethan N. Aumack*

## ABSTRACT

Livestock grazing occurs across extensive tracts of western National Forest lands. Meanwhile, wildfire in western forests is increasing in frequency and severity, changing the ecological context within which grazing occurs. Although disturbance is known to regulate plant community organization, the effects of interactions between fire and ungulate herbivory in herbaceous understory forest communities remain poorly understood. To our knowledge, no prior experiment has addressed the interaction of livestock grazing and fire on ponderosa pine understory communities in the Southwest. We examined the hypotheses that livestock grazing following wildfire 1) alters relative plant species abundance and 2) increases incidence of the invasive exotic *Bromus tectorum*. We conducted a replicated landscape-scale grazing enclosure experiment within an extensive wildfire burn on the Kaibab Plateau north of Grand Canyon, AZ. One significant pattern and several important community trends emerged: 1) Grass cover decreased by 2% in grazed plots, but increased by 19% in ungrazed plots one growing season after grazing, resulting in a significantly greater difference between grazed and ungrazed plots one year after grazing in 2009 than prior to grazing in 2008. 2) Non-native herbaceous cover increased by 13% in grazed plots while decreasing by 6% in ungrazed plots in 2009. 3) Cheatgrass occurrence increased in both grazed and ungrazed plots in 2009, and increased at almost twice the rate in grazed plots as compared to ungrazed plots. While the latter two findings were not statistically significant during the initial period of this experiment, they may nonetheless prove to be biologically meaningful and may have important implications for future understory composition. Results from this ongoing experiment strongly emphasize the need for long-term field experiments that will provide a scientific basis for post-fire grazing management and advance conservation of native plant diversity in western forest ecosystems.

## INTRODUCTION

A century of livestock grazing, fire suppression, and intense logging has altered the structure and composition of ponderosa pine and mixed-conifer western forests, resulting in dense stands highly susceptible to fire (Covington and Moore 1994; Belsky and Blumenthal 1997). Forest wildfire activity has increased dramatically in recent decades, characterized by more frequent large wildfires, longer wildfire duration, and an extended wildfire season (Westerling et al. 2006). Increased temperatures predicted by climate models for the Southwest will intensify this pattern, resulting in an increase in total area burned (McKenzie et al. 2004). Because fire is a key disturbance that shapes vegetation structure and composition, changes in fire regimes can have considerable implications for forest understory communities. Vast areas of these forests, including 103 million acres of National Forest land in 16 western states, are grazed by livestock. Yet, the ecological consequences of livestock grazing are controversial, and specific effects of the interaction between livestock grazing and fire remain poorly understood.

Interactions between grazing by large ungulate herbivores and episodic disturbances such as fire may have important consequences for native species diversity. Disturbance plays an important role in regulating plant communities by influencing community structure and composition. As a result, disturbance may also lead to exotic invasions that alter ecosystem function (Hobbs and Huenneke 1992). For example, D'Antonio and Vitousek (1992) attributed cheatgrass invasion throughout the Southwest to livestock grazing, which subsequently altered fire regimes. Furthermore, interactions between different disturbances are believed to have the greatest effect on species diversity (Hobbs and Huenneke, 1992).

Of particular concern in the debate over post-fire cattle grazing is the potential for livestock to promote invasions by non-native species already favored by post-fire environments. Both substantial increases in exotic species richness and cover (Beaulieu 1975; Crawford et al. 2001; McGlone et al. 2009) and low exotic richness and cover (Foxx 1996; Keeley et al. 2003; Laughlin et al. 2004; Huisinga et al. 2005) have been reported after fire. In extreme examples, populations of exotic species have exploded following fire in ponderosa pine and mixed conifer forests (Crawford et al. 2001; McGlone et al. 2009). Remote location, land-use histories of protection, and low anthropogenic disturbance may explain low incidence and abundance of exotic species following fire, as in two studies from the North Rim of the Grand Canyon (Laughlin et al. 2004; Huisinga et al. 2005). Given the growing acreage of forests burned by wildfire or in prescribed restoration treatments, some researchers advocate the precautionary

principle in introducing livestock grazing after fire. Livestock are often a source of non-native seed that may survive fire and establish a new invasive population, and livestock grazing after fire may also inhibit the recovery of native species, alter plant succession, and exacerbate damage to soil and water resources (Beschta et al. 2004; Keeley 2006).

Belsky and Blumenthal (1997), Milchunas (2006), and Bakker and Moore (2007) provide comprehensive reviews of livestock grazing impacts on ponderosa pine understory communities. Higher herbaceous plant cover has been reported for both grazed areas (Costello and Turner 1941; Bakker and Moore 2007) and for areas protected from grazing (Rummell 1951; Schmutz et al 1967; Potter and Krenetsky 1967; Smith 1967; Krenetsky 1971). Grazing has consistently been found to decrease perennial bunchgrass cover (Arnold 1950; Rummell 1951; Smith 1967; Zimmerman & Neuenschwander 1984; Bakker and Moore 2007) while protection from grazing increases grass cover (Potter and Krenetsky 1967; Krenetsky 1971). Livestock grazing in ponderosa pine forests has also been shown to alter understory composition (Arnold 1955; Clary 1975) and to promote establishment or expansion of annual grasses and weedy and exotic species (Arnold 1950; Smith 1967; Franklin & Dyrness 1973; Johnson et al. 1994). Species-specific responses to grazing or protection from grazing are moderated by grazing intensity and history (Milchunas 2006).

In a review of 82 studies on herbivory by wild and domestic ungulates, Wisdom et al. (2006) found that only 15 (18%) evaluated the interactions between herbivory and episodic disturbances such as fire. Of these, only one study addressed the interactions between domestic ungulate herbivory and episodic disturbances in the West, and none specifically examined these interactions in the arid and semi-arid Southwest, nor in ponderosa pine forests. Federal agencies

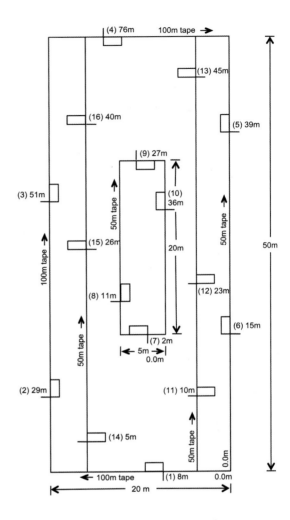

Figure 3.2   The Modified-Whittaker plot (modified from Stohlgren et al. 1995) with sixteen 1m$^2$   (0.5m x 2m) subplots and one 100m$^2$ (5m x 20m) subplot nested within one 1000m$^2$ (20m x 50m) plot. All sampling occurred within the 1000m$^2$ plot. We surveyed species percent cover in each of the sixteen 1m$^2$  plots, and generated a species list for the 100m$^2$ plot. Here we report only results from the species percent cover surveys.

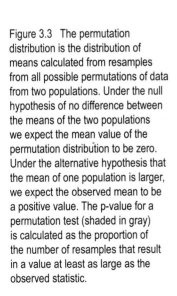

Figure 3.3   The permutation distribution is the distribution of means calculated from resamples from all possible permutations of data from two populations. Under the null hypothesis of no difference between the means of the two populations we expect the mean value of the permutation distribution to be zero. Under the alternative hypothesis that the mean of one population is larger, we expect the observed mean to be a positive value. The p-value for a permutation test (shaded in gray) is calculated as the proportion of the number of resamples that result in a value at least as large as the observed statistic.

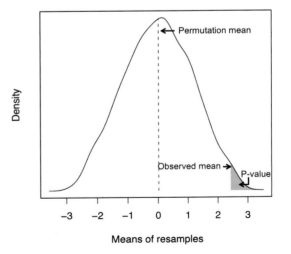

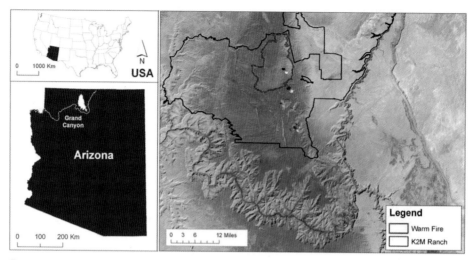

Figure 3.1 a

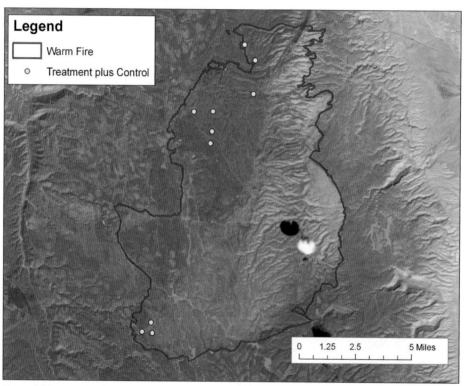

Figure 3.1 b

Figure 3.1   Study area encompassing a) The Kaibab Plateau and Kane and Two-Mile Ranch grazing allotments (USFS and BLM lands), and b) paired treatment (grazed) and control (ungrazed) plots (n=10) established in 2008 within the Warm Fire burn perimeter.

including the US Forest Service (USFS) commonly defer livestock grazing for two years after fire in keeping with an unwritten range management guideline (D. M. Stewart, USFS Region 3, Albuquerque, New Mexico, personal communication) but no research exists to indicate whether this amount of time is adequate for understory community recovery in southwestern ponderosa pine forests. Grazing on National Forest lands is regulated by Annual Operating Instructions (AOIs). AOIs are established each year by the permit holder and the Forest Service. The permitted number of animals allowed to graze on an allotment is increased or decreased, depending on the condition of the allotment.

In 2008 we initiated a landscape-scale replicated grazing experiment to examine ponderosa pine understory community response to livestock grazing two years after fire. In an extensive wildfire burn north of Grand Canyon, AZ, we tested the following hypotheses: 1. Livestock grazing two years after fire alters relative plant species abundance between years; and 2. Livestock grazing two years after fire increases incidence of the invasive exotic Bromus tectorum (cheatgrass) between years. We present preliminary results from an ongoing study. This research contributes new ecological information to the highly pertinent topic of grazing-fire interactions in Southwestern forests.

## METHODS

### Study Area

This study was conducted in the Warm Fire burn on the North Kaibab Ranger District of Kaibab National Forest (Figure 3.1a). The Kaibab Plateau, a forested sky-island north of Grand Canyon National Park, AZ, spans elevations from 1676 m to 3175 m. Average annual temperatures at Jacob Lake, AZ, elevation 2376 m, range from January lows of -8.4° C to July highs reaching 26.2° C (Western Regional Climate Center, www.

wrcc.dri.edu). Average annual precipitation is 58.7 cm, falling in a bimodal seasonal distribution, the majority as snow during winter months and much of the remainder during the summer monsoon. Ponderosa pine forests extend from 2195 to 2590 m, grading into pinyon-juniper at lower elevations and mixed-conifer forests at their upper extent.

As in many Arizona forests, logging, livestock grazing and fire suppression have shaped contemporary forest dynamics (For a thorough review, see Covington and Moore 1994). The Kaibab Plateau has been continuously grazed by livestock since 1871. Intensive grazing, primarily by cattle but also by sheep and horses, began in 1877 and continued through the twentieth century. In the late twentieth century the US Forest Service initiated extensive range modifications including forest thinning (1,175 ha) and disking, drilling and seeding (6,224 ha) with non-native species in an effort to improve poor range conditions (Trudeau 1996). Stocking was dramatically reduced to near zero from 2000 to 2005. In 2005, the Grand Canyon Trust purchased the Kane and Two Mile permits, and livestock grazing was resumed at 400-600 head. In June 2006 the Warm Fire burned approximately 24,281 hectares of ponderosa pine forest within the Kane Ranch grazing allotment on the Kaibab Plateau, including approximately 95% of the area of the Kane Ranch North Pasture.

### Experimental Design

In May 2008 we initiated a Before-After/ Control-Impact (BACI) replicated grazing enclosure experiment in low and moderate-high severity areas of the burn (Figure 3.1b). We identified 20 initial points for the purposes of establishing treatment plots within the Warm Fire burn on the Kaibab Plateau. Ten points each were selected within a) low and b) moderate to high fire severity areas, as determined from a remotely sensed Differenced Normalized Burn Ratio (DNBR) map, to assess the effects of livestock grazing after fire across a mosaic of fire severities. Because moderate and high severity burns resulted in similar ecological effects, as reflected in Composite Burn Index (CBI) Burn Severity parameters, we grouped moderate and high fire severity for this study (henceforth, "high severity"; Key and Benson 2006). All twenty initial points were located 100-800 m from lightly traveled roads in ponderosa pine vegetation type. Slopes 30% and greater were excluded. We used the CBI to verify burn severity in the field at each of the twenty points. Archaeological clearance was obtained for ten of the original points, five each within low and high fire severity, through consultation with the U.S. Forest Service. The final ten points selected for the project were those for which we had verified burn severity and were able to obtain archaeological clearance.

From each point that met burn severity and archaeological clearance criteria, we located a grazing treatment plot at a random azimuth and random distance within a 100-m radius. To be included in the study a plot had to meet a minimum initial understory cover value of twenty-five percent. Additionally, plots that encompassed steep drainages were eliminated, and in these cases a new plot location was selected at a second random azimuth and distance from the original point.

For each of the ten treatment plots established, we selected a paired control plot. Although burn severity was already accounted for among plot selection criteria, a visual comparison of burn severity, understory vegetation, slope, aspect, and canopy cover between potential control plots and established treatment plots allowed for maximum initial similarity of control and treatment plots. Control plots were located not closer than 14 m to treatments nor were they located downhill from treatments, in order to minimize the possibility of treatment effects confounding data from control plots. We georeferenced plot corners using GPS and marked them for ease of future location.

Following the Warm Fire, livestock grazing was deferred in the North Pasture of the Kane Ranch grazing allotment for the remainder of the 2006 growing season and throughout the 2007 growing season. Grazing was deferred in the North Pasture an additional two years, in 2008 and 2009, in part to facilitate this experiment. In July 2008, five cow-calf pairs were introduced into the ten, 64 x 34 m treatment plots, over a 12-day period. Cattle remained in each enclosure for 20-24 hours to mimic a heavy grazing impact, as qualitatively assessed by professional cattlemen. Grazing was initially observed by researchers and ranch managers to determine how many cattle were required to graze for how many hours to achieve the desired plot impacts. The treatment was judged to be approximately equivalent to the maximum grazing intensity that would occur in heavily utilized areas of a large pasture, based on local experience, and as such represented the extreme upper extent of the impact cattle could be expected to exert on the understory over a typical season of grazing. To achieve a more consistent grazing intensity and to avoid causing severe and lasting damage to plots, cattle were removed from plots with low initial cover after a shorter duration than from plots with high initial cover. Adjacent control plots were not grazed.

### Percent Cover Surveys

We used a 20x50m modified Modified-Whittaker nested plot design (Stohlgren et al. 1995) to locate sixteen $1m^2$ (0.5 x 2 m) subplots in each treatment and control unit (Figure 3.2). We chose the Modified-Whittaker footprint to allow us to observe changes in specific locations over time, and for the potential to analyze data at multiple scales. We added six subplots to the original Modified-Whittaker plot to better characterize the vegetation community and capture the high spatial variability observed. The subplots were surveyed before the grazing treatment in June 2008, two weeks after grazing in August 2008, and again in June 2009 one growing season after grazing.

We estimated percent cover by plant species in each subplot. Field technicians were trained in visual estimation and evaluated on relative accuracy and similarity of estimates prior to collecting data. Laminated plastic squares equivalent to one-percent cover were used throughout training and sampling to aid in accurate estimation. Accuracy of estimation was regularly calibrated among field technicians. The same four-person field crew sampled both periods in 2008, with turnover of one field technician in 2009, reducing the likelihood of observer bias. Plant species identifications were confirmed against specimens in the Deaver Herbarium collection and vouchers were retained there. Species newly differentiated in 2009 surveys were analyzed as originally identified in 2008. Species for which identification at the species level was uncertain due to phenological stage were identified to genus.

Data analysis

BACI design allows experimental sample populations to be compared across time intervals, as well as allowing for comparison between treatment and control groups before the treatment is applied, to determine the initial similarity of experimental pairs, and after the treatment is applied, to discern new differences in trend between treatment and control groups resulting from the experimental treatment. To confirm the initial similarity of treatment and control plots we compared mean values for percent cover between treatment and control plots in June 2008, before the grazing event. To elucidate how the understory community in grazed plots diverged from the understory community in ungrazed plots following the grazing treatment, we compared the differences of mean percent cover values in treatment and control plots among three time

intervals: June 2008 (baseline data before grazing), August 2008 (after grazing), and June 2009 (one growing season after grazing). To establish a significant grazing effect we compared the differences of mean percent cover values from treatment and control plots between June 2008 and August 2008. To measure inter-annual understory response to grazing we compared the differences of mean percent cover values from treatment and control plots between June 2008 and June 2009. For each time interval we tested the hypothesis that differences in percent cover between grazed and ungrazed plots were greater following the grazing treatment and one growing season after grazing than initial differences in percent cover between plots prior to the grazing treatment. We subtracted grazed from ungrazed values for percent cover, which we expected to decrease in grazed plots. We tested plant species in six groups: total herbaceous cover, cover of three functional groups (grass cover, forb cover, perennial bunchgrass cover) and non-native cover, in general, as well as cover of the invasive exotic cheatgrass (*Bromus tectorum*).

Because some distributions of sample values did not meet assumptions of normality and our sample size was small (n = 10) we used a univariate exact permutation test to compare differences of mean percent cover values, using R open source statistical software (Hothorn and Hornik 2006, R Core Development Team 2007). An exact permutation test compares the observed statistic from the experimental groups to the expected statistic under a permutation distribution created by randomly resampling from all possible permutations of the data from treatment and control groups. Here, for the comparison of baseline data, the observed statistic is the mean difference between grazed and ungrazed plots in June 2008; the observed statistic for all other comparisons is the difference of the mean difference between grazed and ungrazed plots at two time periods. The permutation distribution is created by resampling from all possible permutations of the data from two paired groups and calculating a new mean difference between the paired groups for each resample; altogether these new, resample mean values form the permutation distribution, with each value in the permutation distribution representing a mean calculated from the difference between paired resamples randomly selected from the two paired groups. Under the null hypothesis of no difference between two mean differences of grazed and ungrazed plots at two time periods we expect the mean value of the permutation distribution to be zero. Under the alternative hypotheses that a) there is a greater difference between grazed and ungrazed plots after grazing than before grazing and b) there is a greater difference between grazed and ungrazed plots in the growing season following the grazing treatment than before grazing, we expect the mean difference between times to be a positive value. The p-value for a permutation test is calculated as the proportion of the number of resamples that result in a value at least as large as the observed statistic (Figure 3.3).

## RESULTS

Permutation tests indicated no significant differences in mean percent cover between treatment and control plots in June 2008 before grazing for any of the groups analyzed (Table 3.1).

Differences in mean cover between grazed and ungrazed plots were significantly greater after the grazing treatment in August 2008 than before grazing in June 2008 for total herbaceous cover, grass cover, forb cover, and non-native cover. Differences of mean percent cover values for perennial bunchgrass and cheatgrass were not significantly different in August 2008 following grazing than in June 2008 prior to grazing (Table 3.1; Figure 3.4a). Differences in grass cover between grazed and ungrazed plots were significantly greater one growing season after grazing in June 2009 than before grazing in June 2008 (Table 3.1; Figure 3.4b). Mean grass percent cover in grazed plots decreased in 2009 by 2% from 2008 values, from 5.68% to 5.55%, while mean grass cover in ungrazed plots increased in 2009 by 19%, from 4.92% to 5.90% (Table 3.2). There was no significant change in differences of mean percent cover between grazed and ungrazed plots from June 2008, prior to the grazing treatment, to June 2009, one growing season after grazing, for total herbaceous cover, perennial bunchgrass cover, forb cover, non-native cover or cover of cheatgrass (Table 3.1; Figure 3.4b). However, trends for total herbaceous cover, forb cover, and perennial bunchgrass cover followed the same pattern as that observed for grass cover, with very slight (.09-2%) decreases in values from grazed plots but 1-24% increases in percent cover in ungrazed plots (Table 3.2). Conversely, mean percent cover of non-native species in grazed plots increased by 13%, from 1.09% to 1.24%, in contrast with a 6% decrease in mean non-native percent cover in ungrazed plots, from 1.26% to 1.19%.

Although cheatgrass abundance in grazed and ungrazed plots and difference in cheatgrass abundance in grazed and ungrazed plots between time intervals did not change significantly, cheatgrass occurrence increased in both grazed and ungrazed plots from June 2008 to June 2009 (Figure 3.5). The increase in cheatgrass occurrence was markedly greater in grazed plots than in ungrazed plots: cheatgrass was detected in over 3 times as many grazed subplots (1-m2) in 2009 as in 2008 and in 1.5 times as many ungrazed subplots. In June 2009 cheatgrass was present at low abundance ($\leq 4\%$) in 8% of all subplots (1-m$^2$), as compared to 4% of subplots in June 2008 prior to grazing. Cheatgrass was newly detected in one grazed plot (100-m$^2$) in 2009, with no new detections in ungrazed plots.

| Plant cover (%) | June 2008 | | | August 2008 | | | June 2009 | | |
|---|---|---|---|---|---|---|---|---|---|
| | Grazed | Ungrazed | Difference | Grazed | Ungrazed | Difference | Grazed | Ungrazed | Difference |
| Total Herbaceous | 15.5±8.3 | 14.9±8.4 | 0.6±3.4 | 13.2±9.5 | 19.8±12.7 * | 6.6±7.6 | 15.1±7.7 | 16.1±8.7 | 0.9±6 |
| Grass | 5.7±4.2 | 4.9±3.7 | 0.8±2.0 | 4.5±3.9 | 6.0±4.8 * | 1.5±2.3 | 5.6±3.3 | 5.9±3.5 | 0.3±1.9 * |
| Perennial Bunchgrass | 3.0±2.2 | 2.9±1.8 | 0.1±2.5 | 2.7±2.9 | 3.6±3.2 | 0.8±2.6 | 3.0±1.8 | 3.6±3.0 | 0.6±1.9 |
| Forb | 9.8±6.0 | 10±7.1 | 0.2±1.4 | 8.7±6.9 | 13.8±10.1 * | 5.1±6.8 | 9.6±5.4 | 10.2±7.0 | 0.6±5.2 |
| Non-native | 1.1±1.2 | 1.3±1.9 | 0.2±0.1 | 0.8±1.3 | 1.3±2.0 * | 0.5±1.1 | 1.2±1.3 | 1.2±1.2 | 0.1±0.9 |
| Bromus tectorum | 0.01±.02 | 0.04±0.06 | 0±0.05 | 0.01±0.04 | 0.03±0.07 | 0±0.07 | 0.05±0.09 | 0.09±0.18 | 0±0.16 |

Table 3.1. Mean and standard deviation for total percent herbaceous plant cover, grass cover, perennial bunchgrass cover, forb cover, and non-native cover, and differences between grazed and ungrazed plots for these groups in June 2008 before grazing, in August 2008 after grazing, and in June 2009 one growing season after grazing. Asterisks denote statistically significant results of exact permutation tests (p≤0.05).

| Plant cover (%) | Grazed | | | | Ungrazed | | | |
|---|---|---|---|---|---|---|---|---|
| | June 2008 | June 2009 | Difference | % Change | June 2008 | June 2009 | Difference | % Change |
| Total Herbaceous | 15.52±8.33 | 15.13±7.71 | 0.38 | -2 | 14.93±8.36 | 16.06±8.67 | 1.13 | 8 |
| Grass | 5.68±4.15 | 5.55±3.31 | 0.13 | -2 | 4.92±3.72 | 5.90±3.49 | 0.97 | 20 |
| Perennial Bunchgrass | 3.01±2.17 | 2.98±1.80 | 0.03 | -1 | 2.89±1.85 | 3.61±2.97 | 0.72 | 25 |
| Forb | 9.84±5.99 | 9.58±5.42 | 0.26 | -3 | 10.00±7.05 | 10.16±6.96 | 0.16 | 2 |
| Non-native | 1.09±1.16 | 1.24±1.26 | 0.15 | 14 | 1.26±1.90 | 1.19±1.16 | 0.07 | -6 |
| Bromus tectorum | 0.01±.02 | 0.05±0.09 | 0.04 | 300 | 0.04±0.06 | 0.09±0.18 | 0.06 | 161 |

Table 3.2  Mean percent cover and standard deviation for total herbaceous plant cover, cover by functional group, cover of non-native species, differences of mean percent cover from June 2008 before grazing to June 2009 one growing season after grazing within grazed and ungrazed plots for these groups, and percent change in mean percent cover for each group from June 2008 to June 2009. Differences reported below represent mean change in percent cover within grazed and ungrazed plots from one year to the next and were not used in statistical analyses, therefore standard deviations are not reported. Percent cover values are shown to two decimal places to accurately represent the change in percent cover from one year to the next.

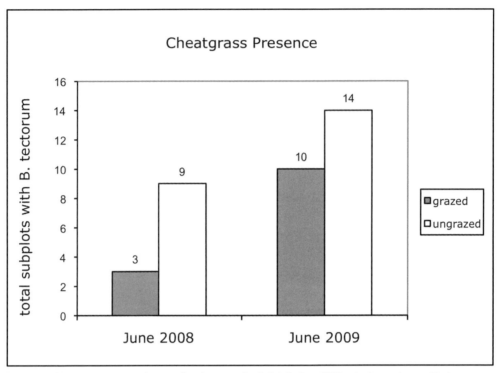

Figure 3.5 Cheatgrass presence in grazed and ungrazed subplots in June 2008, prior to grazing, and in June 2009, one year after the grazing treatment. Cheatgrass was detected in over 3 times as many grazed subplots (1-m²) in 2009 as in 2008 (3 vs. 9 plots), and in 1.5 times as many ungrazed subplots (10 vs. 14 plots).

## DISCUSSION

### Grazing treatment

Differences in percent herbaceous cover between grazed and ungrazed plots before and after grazing demonstrate – not surprisingly -- that grazing treatments had a measurable effect on the grazed understory community (Table 3.1). Differences in percent herbaceous cover between grazed and ungrazed plots in August 2008 were greater than differences between grazed and ungrazed plots in June 2008, before livestock were introduced to treatment plots. This outcome points to the grazing treatment as the source of the significantly greater differences between grazed and ungrazed plots in August 2008 given that grazed and ungrazed plots experienced otherwise similar environmental influences in the interim subsequent to the June 2008 baseline sampling.

Trends for herbaceous cover in grazed plots in 2009 reveal an initial divergence from ungrazed vegetation (Table 3.1, Table 3.2). While differences in total herbaceous, forb, and perennial bunchgrass cover did not change significantly from June 2008, prior to grazing, to June 2009, one year after the grazing treatment, the difference in grass cover between grazed and ungrazed plots was significantly greater in June 2009. When considered in light of the 2009 increases in mean percent cover for these groups in ungrazed plots, the small declines we observed in percent cover suggest that grazing in June 2008 likely compromised 2009 growth for both grasses and forbs in grazed plots (Table 3.1, Table 3.2).

Several studies of the effects of livestock grazing on understory cover document two major patterns: 1. Relatively lower herbaceous cover in grazed areas as compared to ungrazed areas, and 2. decrease in cover of grasses, and especially perennial bunchgrasses, with concurrent increases in forbs (Arnold 1950; Rummell 1951; Smith 1967; Zimmerman and Neuenschwander 1984; Bakker and Moore 2007). These studies track long-term vegetation response to livestock grazing or protection from grazing. A single growing season following utilization comparable to one season of grazing is likely too short an interval to produce the results observed following many years of persistent grazing. Yet, the decrease in grass cover in grazed plots observed in June 2009 and the significantly greater difference in grass cover in grazed and ungrazed plots between June 2008 and June 2009 suggests the potential for greater shifts in future plant community composition and structure. Likewise, the relatively smaller increase in total herbaceous cover in grazed plots compared to ungrazed plots in 2009 may represent an incremental decrease in total herbaceous cover as observed in other studies of grazing in ponderosa pine forests.

We observed a small but important increase in non-native species abundance in grazed plots. Although differences in mean cover of non-native species in grazed and ungrazed plots were not significant across time intervals and overall abundance of non-native species remained low in all plots, the increase in non-native cover in grazed plots in contrast with a concurrent net decrease in ungrazed plots implicates grazing as the source of this difference. This result complements the decrease in grass and forb cover in grazed plots, and indicates that livestock grazing in post-fire herbaceous communities may advantage non-native species over natives. The greater increase in cheatgrass presence in grazed plots versus ungrazed plots supports this conclusion.

And while the relatively close proximity of ungrazed control plots to grazed treatment plots has the potential to confound the observations, the opposing trends for non-natives in grazed versus ungrazed plots, and the magnitude of the difference in cheatgrass occurrence in grazed and in ungrazed plots leads us to conclude that treatment effects clearly outweigh any influence of treatment on control plots. This result is consistent with other studies from the Kaibab Plateau that document low non-native species richness and abundance after fire (Laughlin et al. 2004; Huisinga et al. 2005); yet it simultaneously provides preliminary experimental evidence for the potential of livestock grazing to increase abundance of exotic and invasive species at the expense of native species after wildfire. Additional data and further analyses provided by future phases of this study will help to elucidate the relationship between livestock grazing and increases in non-native and invasive species in post-fire herbaceous understory communities.

Ongoing and predicted future changes in climate regimes, restoration of a high-frequency fire regime, and reduction in ponderosa pine canopy cover prescribed by forest management objectives may exacerbate the effects of wildfire and livestock grazing on cheatgrass invasion. Some evidence suggests climatic factors may limit cheatgrass success at high elevations, and interactions between disturbance and climatic conditions may ameliorate microsite conditions for cheatgrass (Leger et al. 2009). Thus, as climate in the Southwest becomes hotter and drier, and disturbances such as wildfire remain frequent or increase, cheatgrass may be better able to exploit high-elevation sites. Several studies have shown survivorship to be the key factor in cheatgrass population establishment in marginal habitats (Pierson and Mack 1990; Rice and Mack 1991a, 1991b; Leger et al. 2009), and therefore any factor that increases cheatgrass survival at high elevations is likely to increase cheatgrass density and the potential for invasion, especially in combination with increased opportunities for establishment as a result of changing climate, fire, and grazing.

Interannual variability in precipitation may lead to reduced grazing impacts on plant production in relatively wetter years. Accordingly, the modest effects of grazing on plant production observed in our study may be due in part to higher spring precipitation prior to sampling in 2009 than in 2008. Fuhlendorf et al. (2001) showed episodic climatic events to influence the short-term rate and trajectory of vegetation compositional and structural change, while grazing intensity determined the long-term direction of these fundamental community characteristics. In their study of a 44-year dataset, interannual precipitation strongly affected total plant basal area and was also correlated with plant density, while total annual precipitation also influenced vegetation change. These authors observed gradual change in vegetation dynamics on a prolonged temporal scale, suggesting that short-term fluctuations in response to climatic variability may not reveal long-term grazing effects. Models of fire-grazing systems in the Serengeti also point to high precipitation as diminishing the effects of grazing (Archibald 2003). Relatively higher precipitation on the Kaibab Plateau preceding the 2009 growing season, 5.74 cm more from October to May than over the same months in 2008, may accordingly have increased herbaceous growth prior to June 2009 sampling, compensating for the effects of grazing.

We expect ongoing research to provide further insights into the understory response to livestock grazing two years after fire. We plan to conduct additional analyses on percent cover data to examine community-level compositional differences between grazed and ungrazed plots. We will analyze plant biomass by weight to examine differences in total abundance and abundance by functional group one year after grazing. We will also evaluate soil compaction effects through a comparison of soil bulk density between grazed and ungrazed plots. Near-term monitoring of plots will continue at least through summer 2010.

## CONCLUSIONS

Balancing livestock grazing with post-fire herbaceous community recovery is a management challenge faced with increasing regularity in forests throughout the West, however, there is little empirical scientific evidence to guide range managers. Our preliminary results show that livestock grazing two years after fire had a small but measurable effect on relative abundance of grasses, forbs and non-native species. Trends for these three variables match previously documented shifts in grazed understory communities, namely a decline in total herbaceous cover and cover of grasses, accompanied by an increase in non-native species. Importantly, increases in incidence of cheatgrass were greater in grazed than in ungrazed plots at the subplot and plot levels. These early changes, especially the observed greater increase in cheatgrass frequency in grazed plots, may forecast biologically significant changes in the understory community, and suggest the need for careful quantitative monitoring by range managers where livestock grazing is resumed two growing seasons after fire. As the area of western forests burned continues to increase, more long-term, landscape-scale field experiments are needed in multiple locations in the arid West to increase understanding of the combined impact of fire and grazing, and to provide a scientific basis for sustainable grazing practices. Our results represent preliminary experimental evidence of potentially strong impacts of livestock grazing on post-fire understory community dynamics.

## ACKNOWLEDGMENTS

I am grateful to the Grand Canyon Trust and to the Doris Duke Conservation Fellowship Program for the funding that made this research possible. John Heyneman, Justin Jones, and J.R. Jones provided expertise and assistance with fencing and livestock management. Frankie Coburn, Ryan Meszaros, Kirsten Novo, Wynne Geikenjoyner, and volunteers from the Grand Canyon Trust and Earlham College contributed valuable assistance in the field. The North Kaibab Ranger District of the U.S. Forest Service provided access to research sites, archeological clearance and guidance on numerous fronts; particular thanks to Dustin Berger, Rangeland Management Specialist, and Tim Short, District Ranger.

## LITERATURE CITED

Archibald, S. 2003. Modeling interactions between fire, rainfall and grazing. Proceedings of the VII International Rangelands Congress 2003, Durban, South Africa.

Arnold, J. F. 1950. Changes in ponderosa pine bunchgrass ranges in northern Arizona resulting from pine regeneration and grazing. Journal of Forestry 48:118-126.

Arnold, J. F. 1955. Plant life-form classification and its use in evaluating range conditions and trend. Journal of Range Management 8:176-181.

Bakker, J. D. and M. M. Moore. 2007. Controls on vegetation structure in southwestern ponderosa pine forests, 1941 and 2004. Ecology 88:2305-2319.

Beaulieu, J. T. 1975. Effects of fire on understory plant populations in a northern Arizona ponderosa pine forest. Masters thesis, Northern Arizona University, Flagstaff, AZ, 38 p.

Belsky J. A. and D. M. Blumenthal. 1997. Effects of livestock grazing on stand dynamics and soils in upland forests of the interior west. Conservation Biology 11:315-327.

Beschta R. L., J. J. Rhodes, J. B. Kauffman, R. E. Gresswell, G. W. Minshall, J. R. Karr, D. A. Perry, F. R. Hauer, C. A. Frissell. 2004. Postfire management of forested public lands of the western United States. Conservation Biology 18:957-967.

Clary, W. P. 1975. Range management and its ecological basis in the ponderosa pine type of Arizona: the status of our knowledge. Res. Pap. RM-158. Fort Collins, CO: U.S. Department of Agriculture, Forest Service, Rocky Mountain Forest and Range Experiment Station. 35 p.

Clary, W. P. 1969. Increasing sampling precision for some herbage variables through knowledge of the timber overstory. Journal of Range Management 22:200-201.

Costello, D. F., and G. T. Turner. 1941. Vegetation changes following exclusion of livestock from grazed ranges. Journal of Forestry 39:310-315.

Covington, W. W. and M. M. Moore. 1994. Southwestern ponderosa forest structure: changes since Euro-American settlement. Journal of Forestry 92:39-47.

Crawford, J. A., C.-H. A. Wahren, S. Kyle, W. H. Moir. 2001. Responses of exotic plant species to fires in Pinus ponderosa forests in northern Arizona. Journal of Vegetation Science 12:261-268.

D'Antonio, C. M. and P. M. Vitousek. 1992. Biological invasions by exotic grasses, the grass/fire cycle, and global change. Annual Review of Ecology and Systematics 23:63 – 87.

Foxx, T. 1996. Vegetation succession after the La Mesa fire at Bandelier National Monument in Allen, C. D. (editor) Fire effects in southwestern forests. Proceedings of the second La Mesa fire symposium, Los Alamos, New Mexico, March 29-31, 1994. US Department of Agriculture, Forest Service General Technical Report RM-286.

Franklin, J. F., and C. T. Dyrness. 1973. Natural vegetation of Oregon and Washington. General technical report PNW-8. U.S. Forest Service, Pacific Northwest Research Station, Portland, Oregon.

Fuhlendorf, S. D., D. D. Briske, and F. E. Smeins. 2001. Herbaceous vegetation change in variable rangeland environments: the relative contribution of grazing and climatic variability. Applied Vegetation Science 4:177-188.

Hobbs, R. J. and L. F. Huenneke. 1992. Disturbance, Diversity, and Invasion: Implications for Conservation. Conservation Biology 6:324 – 337.

Hothorn, T. and K. Hornik (2006). exactRankTests: Exact Distributions for Rank and Permutation Tests. R package version 0.8-16.

Huisinga, K. D., D. C. Laughlin, P. Z. Fule, J. D. Springer, and C. M. McGlone. 2005. Effects of an intense prescribed fire on understory vegetation in a mixed conifer forest. Journal of the Torrey Botanical Society 132:590-601.

Johnson, C. G., Jr., R. R. Clausnitzer, P. J. Mehringer, and C. D. Oliver. 1994. Biotic and abiotic processes of eastside ecosystems. General Technical ReportPNW-GTR-322. U. S. Forest Service, Pacific Northwest Research Station, Portland, Oregon.

Keeley J. E., D. Lubin, and C. J. Fotheringham. 2003. Fire and grazing impacts on plant diversity and alien plant invasions in the southern Sierra Nevada. Ecological Applications 13: 1355-1375.

Keeley J. 2006. Fire management impacts on invasive plants in the western United States. Conservation Biology 20:375-384.

Key, C.H. and N.C. Benson. 2006. Landscape assessment: ground measure of severity, the Composite Burn Index, and remote sensing of severity, the Normalized Burn Index. In 'FIREMON: Fire Effects Monitoring and Inventory System'. (Eds. D.C. Lutes, R.E. Keane, J.F. Caratti, C.H. Key, N.C. Benson, S. Sutherland, L.J. Gangi.) USDA Forest Service, Rocky Mountain Research Station, General Technical Report RMRS-GTR-164-CD: LA1-51. Ogden, UT.

Krenetsky, J. C. 1971. Effects of controlled grazing and complete protection on New Mexico rangelands after twenty-five years. Las Cruces: University of New Mexico. 176 p. Dissertation.

Laughlin, D. C., J. D. Bakker, M. T. Stoddard, M. L. Daniels, J. D. Springer, C. N. Gildar, A. M. Green, W. W. Covington. 2004. Toward reference conditions: wildfire effects on flora in an old-growth ponderosa pine forest. Forest Ecology and Management 199:137-152.

Leger, E. A., E. K. Espeland, K. R. Merrill, and S. E. Meyer. 2009. Genetic variation and local adaptation at a cheatgrass (Bromus tectorum) invasion edge in western Nevada. Molecular Ecology 18:4366-4379.

McGlone, C. M., J. D. Springer, and W. W. Covington. 2009. Cheatgrass encroachment on a ponderosa pine forest ecological restoration project in Northern Arizona. Ecological Restoration 27:37-46.

McKenzie, D., Z. Gedalof, D. L. Peterson, and P. Mote. 2004. Climatic change, wildfire, and conservation. Conservation Biology 18:890-902.

Milchunas, D. G. 2006. Responses of plant communities to grazing in the southwestern United States. Gen. Tech. Rep. RMRS-GTR-169. Fort Collins, CO: U.S. Department of Agriculture, Forest Service, Rocky Mountain Research Station. 126 p.

Pierson, E. A. and R. N. Mack. 1990. The population biology of Bromus tectorum in forests – effect of disturbance, grazing, and litter on seedling establishment and reproduction. Oecologia 84:526-533.

Potter, L. D. and J. C. Krenetsky. 1967. Plant succession with released grazing on New Mexico rangelands. Journal of Range Management 20:145-151.

R Development Core Team (2007). R: A language and environment for statistical computing. R Foundation for Statistical Computing, Vienna, Austria. ISBN 3-900051-07-0, URL http://www.R-project.org.

Rice, K. J. and R. N. Mack. 1991a. Ecological genetics of Bromus tectorum 1. A hierarchical analysis of phenotypic variation. Oecologia 88:77-83.

Rice, K. J. and R. N. Mack. 1991b. Ecological genetics of Bromus tectorum 2. Intraspecific variation in phenotypic plasticity. Oecologia 88:84-90.

Rummell, R. S. 1951. Some effects of livestock grazing on ponderosa pine forest and range in central Washington. Ecology 32:594-607.

Schmutz, E. M., C. C. Michaels, and B. I. Judd. 1967. Boysag Point: a relict area on the North Rim of Grand Canyon in Arizona. Journal of Range Management 20:363-368.

Smith, D. R. 1967. Effects of cattle grazing on a ponderosa pine-bunchgrass range in Colorado. USDA Department of Agriculture Technical Bulletin 1371, Washington, D.C., USA.

Stohlgren, T. J., M. B. Falkner, and L. D. Schell. 1995. A Modified-Whittaker nested vegetation sampling method. Plant Ecology 117:113-121.

Trudeau, J. M.1996. An environmental history of the Kane and Two Mile Ranches in Arizona: a report prepared for the Grand Canyon Trust, Flagstaff, AZ. 167 p.

Westerling, A. L., H. G. Hidalgo, D. R. Cayan, T. W. Swetnam. 2006. Warming and Earlier Spring Increase Western U.S. Forest Wildfire Activity. Science 313:940 – 943.

Wisdom, M. J., M. Vavra, J. M. Boyd, M. A. Hemstrom, A. A. Ager, and B. K. Johnson. 2006. Understanding ungulate herbivory-episodic disturbance effects on vegetation dynamics: knowledge gaps and management needs. Wildlife Society Bulletin 34:283 – 292.

Zimmerman, G. T. and L. F. Neuenschwander. 1984. Livestock grazing influences on community structure, fire intensity, and fire frequency within the Douglas-fir/ninebark habitat type. Journal of Range Management 37:104-110.

Clary, W. P. 1975. Range management and its ecological basis in the ponderosa pine type of Arizona: the status of our knowledge. Res. Pap. RM-158. Fort Collins, CO: U.S. Department of Agriculture, Forest Service, Rocky Mountain Forest and Range Experiment Station. 35 p.

# CHALLENGES AND OPPORTUNITIES FOR ECOSYSTEM SERVICES SCIENCE AND POLICY IN ARID AND SEMIARID ENVIRONMENTS

*Ken Bagstad, Darius Semmens, and Charles van Riper III*

## ABSTRACT

Ecosystem services – the economic benefits that nature provides to people – are gaining recognition in the research and policy communities as a means of better supporting sustainable resource management. Yet for arid and semiarid environments, including the Colorado Plateau, research and application of ecosystem services concepts has lagged behind the more populous temperate, humid, and coastal regions of the US. Here we explore three important issues for the Colorado Plateau research and policy communities related to the ecology, economics, and geography of ecosystem services. These include: 1) the critical importance of the temporal and spatial distribution of water in supporting the ecosystems that provide these services, 2) how the location of human beneficiaries within watersheds and airsheds affects the value attributable to the ecosystem service, and 3) how low population densities contribute to long distances between beneficiaries and the ecosystems providing key services, which can reduce public perceptions of the value of these ecosystems. We elaborate on these three issues, citing examples from the Colorado Plateau and other parts of the Intermountain West, along with science and policy implications. While ecosystem services research and application toward policy are at a nascent stage on the Colorado Plateau, as this field continues to advance increased attention to these issues can advance the research agenda and identify barriers and opportunities for applying ecosystem services to decision making.

## INTRODUCTION

The science of ecosystem services - quantifying and valuing the coupled ecological and economic production of the benefits nature provides to humans – is increasingly used to frame tradeoffs in conservation and economic development (Farber et al. 2006; Daily et al. 2009; Tallis et al. 2009). In recent years, interest in ecosystem services has grown among the academic, public, private, and nonprofit sectors and has potential for use in resource management on the Colorado Plateau. While several approaches exist for ecosystem services-based resource management (Salzman 2005), payments for ecosystem services (PES) programs remain the most well publicized (Engel et al. 2008). In the United States, PES has a 25-year history as part of the Farm Bill, as well as through early carbon and watershed credit trading programs. Recent Federal initiatives, including creation of the USDA Office of Environmental Markets, may provide leadership in incentivizing the protection and restoration of ecosystems and the services that they generate.

In this chapter, we argue that past research and policy applications of ecosystem services in the United States have received greater focus in temperate, humid, and coastal regions, with less attention paid to the Intermountain West and North American Desert regions, and particularly the Colorado Plateau. This situation is slowly changing (Melis et al. 2010), but there are three key issues related to the ecology, geography, and economics of ecosystem services that

# CHALLENGES AND OPPORTUNITIES FOR ECOSYSTEM SERVICES SCIENCE AND POLICY IN ARID AND SEMIARID ENVIRONMENTS

*Ken Bagstad, Darius Semmens, and Charles van Riper III*

## ABSTRACT

Ecosystem services – the economic benefits that nature provides to people – are gaining recognition in the research and policy communities as a means of better supporting sustainable resource management. Yet for arid and semiarid environments, including the Colorado Plateau, research and application of ecosystem services concepts has lagged behind the more populous temperate, humid, and coastal regions of the US. Here we explore three important issues for the Colorado Plateau research and policy communities related to the ecology, economics, and geography of ecosystem services. These include the critical importance of the temporal spatial distribution of water in suppor the ecosystems that provide these serv 2) how the location of human benefici within watersheds and airsheds affects value attributable to the ecosystem ser and 3) how low population dens contribute to long distances betv beneficiaries and the ecosystems provi key services, which can reduce pt perceptions of the value of these ecosyst We elaborate on these three issues, c examples from the Colorado Plateau other parts of the Intermountain West, a with science and policy implications. W ecosystem services research and applica toward policy are at a nascent stage or Colorado Plateau, as this field continue advance increased attention to these is can advance the research agenda and ide barriers and opportunities for appl ecosystem services to decision making.

## INTRODUCTION

The science of ecosystem services - quantifying and valuing the coupled ecological and economic production of the benefits nature provides to humans – is increasingly used to frame tradeoffs in conservation and economic development (Farber et al. 2006; Daily et al. 2009; Tallis et al. 2009). In recent years, interest in ecosystem services has grown among the academic, public, private, and nonprofit sectors and has potential for use in resource management on the Colorado Plateau. While several approaches exist for ecosystem services-based resource management

pose special challenges for their application in arid environments such as the Colorado Plateau.

First, we will show that water is a key driver of ecosystem services, particularly in arid and semiarid environments. For all uses, the absolute quantity and quality of water matters greatly. Yet in addition to water quality and quantity, the specific temporal and spatial distribution of water (i.e., groundwater vs. surface water, seasonal permanence, degree of flow regulation) matters in terms of ecosystem services provision. Second, the location of different groups of human beneficiaries within watersheds (and airsheds) matters tremendously in terms of provision and use of key hydrologic services. Third, we will demonstrate the consequences to beneficiaries of ecosystem services that are sparsely distributed in the Intermountain West, particularly by contrast with densely populated eastern and coastal regions. We begin by describing the historical roots of ecosystem services, then deal with research and applications in the Intermountain West, and conclude by discussing science and policy implications in an effort to create a foundation for ecosystem services in arid and semiarid environments of the Colorado Plateau.

## Background on ecosystem services

Although pioneering ecologists such as George Perkins Marsh and Aldo Leopold recognized the critical life-support functions played by nature as early as the late 19th-mid 20th century, the 1970s-1980s saw the emergence of modern ecosystem services conceptualizations (Mooney and Ehrlich 1997). Valuation of ecosystem services grew from the 1970s onward, as economic methods to value ecosystem services were developed and applied by environmental and later ecological economists, who produced "primary valuation" studies for locally important ecosystem services. With larger

populations and more universities in coastal and humid regions, local focus on these geographic areas led to fewer ecosystem services valuation studies for western region arid lands. Thus, early efforts to synthesize this work via meta-analysis, value transfer, and the development of ecosystem services tools also took place largely outside the arid lands of the U.S. Southwest. For example, Farber (1996) completed an early synthesis of ecosystem services studies for coastal Louisiana, Villa et al. (2002, 2009) developed valuation databases and assessment tools, Costanza et al. (2006) conducted large-scale value transfer exercises at the University of Vermont, and Chan et al. (2006) and Daily et al. (2009) led development of ecosystem services mapping and valuation tools at Stanford University in California. With more primary studies to draw upon, researchers could produce more comprehensive syntheses for wetlands, forests, and coastal ecosystems (Woodward and Wui 2001; Brander et al. 2007; Zandersen and Tol 2009) than for arid and semiarid environments.

Early efforts to synthesize the valuation literature placed minimal value on semiarid and arid systems. For example, Costanza et al. (1997) gave a value of $0/ac-yr to deserts. In a more recent study, Dodds et al. (2008) assigned deserts the lowest value of six North American ecoregions. They found the value of deserts to be 1-2 orders of magnitude lower ($166/ac-yr, versus $1,879-$25,229/ac-yr for other ecoregions) than all but one other ecoregion – western forested mountains (valued at $986/ac-yr), which receive less rainfall than the eastern temperate and west coast marine forests also included in their studies. Along with having fewer primary valuation studies to work with, few past ecosystem services studies have explored the importance of services like dust regulation and its implications for human health (Richardson 2008), further underestimating their value.

The historical pattern of the arid southwestern Colorado Plateau and the Intermountain West as less researched regions in the field of ecosystem services is rapidly ending (e.g., Jones et al.2010; Norman et al. 2010; Semmens et al. 2010). The interaction between the region's historical and projected population growth, the growing strain on water, energy, and other resources, and uncertain impacts of climate change is driving research efforts and the need to apply their results toward policy. The West's abundant public land, which has historically been used primarily for extraction of ecosystem goods such as forage and timber, is increasingly recognized as a key source of other valuable ecosystem services to be managed and protected. Federal researchers at organizations like the U.S. Environmental Protection Agency (USEPA), the U.S. Geological Survey (USGS), U.S. Department of Agriculture (USDA), and others are increasing their research efforts on ecosystem services, western universities are building stronger research efforts, and university-agency partnerships are moving from investigations of ecology and hydrology toward integrated assessments of ecosystem services. Examples of these efforts can be found at USEPA's Southwest place-based research program (http://www. epa.gov/ecology/quick-finder/southwest. htm), Arizona State's EcoServices research group (http://www.ecoservices.asu.edu/), and the interagency AGAVES research effort (http://rmgsc.cr.usgs.gov/agaves/).

Researchers have explored the link between ecosystem services and net primary production (NPP, Costanza et al. 2007; Richmond et al. 2007). Similar linkages have been proposed between biodiversity and ecosystem services (Hooper et al. 2005, Balvanera et al. 2006), but in both cases these linkages and their causality are not fully understood. Ecosystems in more mesic regions and with greater biomass could generally be expected to cycle matter and nutrients more quickly, have greater throughput of energy, and produce more ecosystem goods. From an economic perspective, however, demand must exist for ecosystem services to be valuable, and high demand (i.e., number of users) may exist in some arid regions while remaining low in humid regions that are sparsely populated. Depending on the ecosystem service, scarce services in resource-limited arid environments may have a higher marginal value than in resource-rich humid environments, although their total value could be lower if the quantity of services produced is low or where there are few beneficiaries. The importance of arid lands such as the Colorado Plateau in providing human well-being was noted by the Millennium Ecosystem Assessment (2005a), as several subglobal assessments focused on arid and semiarid regions. Finally, while highly productive ecosystems might be expected to produce more "regulating" and "provisioning" services, assuming adequate demand, there is no explicit reason why the quantity and value of "cultural" services (Millennium Ecosystem Assessment 2005b) would depend on the quantity and rate of matter, nutrient, and energy processing in an ecosystem. The spectacular natural features found in some desert environments and their historic role as "cradles of civilization" (Diamond 1997) suggests a high degree of cultural values for certain arid and semiarid lands. On the Colorado Plateau, such values are particularly important for numerous Native American cultures.

## Water as a driver of ecosystem services on the Colorado Plateau

Water is the primary limiting resource in arid and semiarid ecosystems, as it controls the rates and the timing of biological processes in dryland species and ecosystems (Webb et al. 2007). Prior to the development of long-distance aqueducts, food transport, and pumps capable of accessing deep groundwater, local water availability was the critical limiting resource on human populations in deserts. Riparian ecosystems on the Colorado Plateau depend on shallow groundwater or precipitation to feed perennial (year-round), intermittent (flowing for a portion of the year), or ephemeral (flowing only in response to precipitation or snowmelt events) streams and wetlands. Even in the absence of permanent surface water, shallow groundwater can sustain riparian and wetland ecosystems that provide biological oases in the surrounding desert. Upland ecosystems are maintained by infrequent but critically important rain pulses and by snowmelt at higher elevations. Colorado Plateau uplands provide a range of important ecosystem services including carbon sequestration and storage, dust and sediment regulation, and forage provision (Miller et al. 2011).

Springs are an important type of wetland on the Colorado Plateau, which has over 5,000 named springs that have played important cultural and biological roles in this region (Stevens and Nabhan 2002). Ecological studies of desert spring ecosystems have found 100-500 times the number of species relative to the surrounding arid lands (W.R. Ferren and F.W. Davis, California Department of Fish and Game, unpublished report 1991; Stevens and Nabhan 2002; Sada and Pohlmann 2003). Humans also utilize springs through diversion, irrigating pastures, channeling water to livestock, household use, and recreation. In the southern half of the Colorado Plateau, these anthropogenic activities have degraded an estimated 75 percent of the springs (Stevens and Nabhan 2002). In addition, the extraction of water tributary to springs by pumping groundwater has caused spring discharge to diminish by more than 50 percent in the majority of USGS-monitored springs on the Colorado Plateau (Natural Resources Defense Council 2001).

While the four Colorado Plateau states have lost a smaller percentage of their wetlands than the national average (41 percent vs. 53 percent; Dahl 1990), the rarity of these wetlands and the degradation of remaining wetlands suggests that services have been lost from these valuable ecosystems. Additionally, in many large cities within the arid Southwest, flow diversions and groundwater pumping have left dry riverbeds, which provide minimal ecosystem services (Webb et al. 2007). Since surface water permanence and flow regulation govern the biotic communities of western upland and riparian systems, they are also key drivers of the potential supply of ecosystem services (Figure 4.1; Table 4.1).

Nearly all major rivers in the American west have been impounded (Reisner 1986), impacting flow quantity and timing, water temperature, sediment and flood pulses, riparian vegetation communities, and fish migration (Stromberg et al. 2007a; Webb et al. 2007). Dams create reservoirs that can provide an array of benefits – hydroelectric power generation, flood control, and reservoir-based recreation (e.g., boating, fishing). Hydroelectric dams, including Glen Canyon Dam, have extended the length of the river-rafting season to the benefit of this user group. However, dams managed for agricultural irrigation, such as those on the Dolores and San Juan rivers, reduce the length of the rafting season by holding water in the reservoir unless irrigation delivery commitments have been met. Riparian vegetation on flow-regulated streams provides a similar basket of ecosystem services as on perennial streams. Flow regulation in the Southwest has had numerous effects on riparian vegetation structure and diversity – by creating stable flow conditions and reducing flood scour, favoring establishment of different woody species based on timing of flow releases, inundating former riparian zones within reservoirs, and diverting flows from the

Perennial flow.                          Intermittent flow.                                    Ephemeral flow (shallow groundwater).

*Photo credits:*
*Sharon Lite (Perennial,*
*Intermittent, Ephemeral*
*flow), Anne Phillips*
*(Flow-regulated river)*

Highly ephemeral flow (deep groundwater).        Flow-regulated river.

Figure 4.1.  Desert aquatic habitat types.

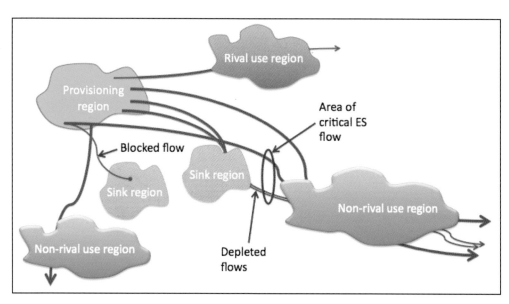

Figure 4.2.  Spatial dynamics of ecosystem services.  Sink regions are areas that block or absorb a matter, energy, or informational carrier of an ecosystem service (i.e., areas that absorb flood water, sediment, or nutrients or visual blight that degrades a high-quality view of nature).  Sinks or rival use of an ecosystem service carrier deplete the carrier quantity while non-rival use does not (Johnson et al. in Press).

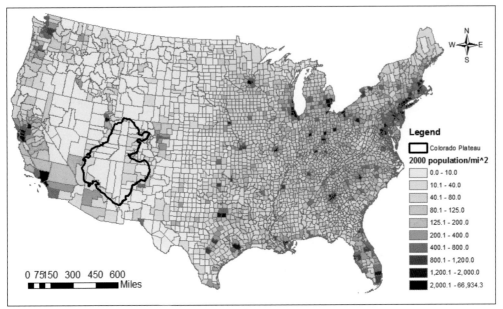

Figure 4.3.  Year 2000 continental U.S. population density by county.

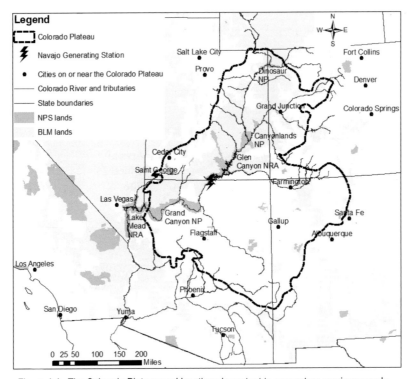

Figure 4.4.  The Colorado Plateau and locations important to ecosystem services supply, demand, and spatial flows.

Colorado River Delta (Webb et al. 2007). Flow regulation can trap heavy metals and other contaminants in reservoir sediments and has also provided salinity regulation on the Lower Colorado River, as releases from reservoirs can help reduce salinity for downstream users and water delivered to Mexico. Since flow regulation can lead to the endangerment of native fishes and shifts in vegetation and avian communities, certain habitat-derived ecosystem service values will differ in these systems (Osmundson et al. 2002; Stromberg et al. 2007b; van Riper et al. 2008).

The headwaters and smaller tributaries of many rivers originating on the Colorado Plateau have perennial flow, including Oak Creek, the East Verde, Tonto Creek, Nankoweap Creek, Clear Creek, and others. Consistent precipitation and snowmelt, along with shallow bedrock depth are important in maintaining perennial flow in these streams. In addition to providing water, perennial streams provide certain recreational benefits (i.e., rafting, fishing). Given their rarity, they may also have non-use value; that is, a value held by people who may never visit the ecosystem or derive direct benefit from it, but who value its continued existence and the right to pass it on to future generations, or the option to use the resource differently in the future (Bishop et al. 1987). Most importantly, unregulated perennial streams are more likely to support greater vegetation cover, species diversity, and native species dominance (Stromberg et al. 2005), which can combine to lead to greater carbon sequestration and storage, cooler microclimates, trapping of sediment and absorption of nutrients, greater aesthetic values, and other habitat-derived ecosystem services (e.g., wildlife watching).

Streams with intermittent flow like the Little Colorado River, Kanab Creek, and Paria Creek provide similar services as perennial reaches, with the major difference being the loss of riverine marshland near the active channel, as these plants require permanent flow and shallow groundwater to survive, and a shift in the dominance of tree species from more hydric to mesic species (e.g., from cottonwood-willow to nonnative species including tamarisk and Russian olive and native species including mesquite; Stromberg et al. 2005). Functionally, intermittent and ephemeral streams provide many of the same habitat and recreational benefits that are found along perennial streams, with the exception of those services depending on the presence of permanent surface water, riverine marsh, and cottonwood-willow vegetation types. In addition, intermittent and ephemeral channels are an important source of groundwater recharge because when water does flow during storm events it can recharge floodplain aquifers. Mountain-front recharge, which includes recharge from the mountain block system and stream channels, is considered to be the most significant form of groundwater recharge in arid and semiarid regions, but recharge in ephemeral stream channels also makes up a significant portion of the total (Goodrich et al. 2004; Coes and Pool 2005).

In regions where groundwater pumping, surface-water diversions, or naturally deep bedrock in low desert environments create ephemeral flow conditions, a lower diversity, less vegetated riparian ecosystem is often present. These conditions are also found in desert washes, which never had permanent flow but are still oases of productivity relative to the surrounding desert. Phreatophytes such as tamarisk and mesquite may still be able to access groundwater, providing greater and more seasonally permanent vegetation cover than the surrounding desert. These species still provide key ecosystem services, including carbon sequestration and storage, sediment regulation, groundwater recharge, and habitat-derived ecosystem service values. However, in the absence of riverine marsh and shallow-groundwater dependent phreatophytes like cottonwoods and willows, the ability of ephemeral streams to provide services like aesthetic values and microclimate regulation is typically less than for rivers with greater surface flow permanence (Hultine et al. 2010).

As groundwater is pumped from deeper and deeper depths, often in excess of natural recharge rates, the riparian system eventually collapses leading to the loss of nearly all phreatophyte plant cover. Such dewatering has been the norm in low deserts near population centers like Phoenix (Salt and Gila Rivers) and Tucson (Santa Cruz River) and also on the Colorado Plateau in regions subjected to diversion projects, such as Kanab Creek in Utah.

These water-related ecosystem service issues raise a paradox: while deep groundwater may still be available for human use in highly flow regulated or dewatered systems, it is the presence of shallow groundwater or seasonal to permanent surface water that is critical for generating ecosystem services. Unregulated, permanent surface water is one of the scarcest resources on the Colorado Plateau and elsewhere in the Intermountain West, potentially increasing the marginal value of the services generated by these systems. It has been common in the West to appropriate all water for irrigation, domestic use, or mining, and to impound rivers for flow regulation, flood control, hydroelectric generation, or water supply regulation such as was done with the Glen Canyon Dam. In many cases ecosystem services have been lost as a result of these decisions. By not fully accounting for the benefits of ecosystem services, society has come to solutions to the macroallocation problem – "How much ecosystem structure [e.g., water] should be apportioned toward the production of human-made goods and services and how much should be left intact to provide ecosystem services?" (Farley 2008) – thus leaving people with fewer of the benefits provided by free-flowing rivers.

Long-term economic flows of recreation, aesthetics, habitat-derived ecosystem services, and some regulating services (Table 4.1) have been traded off to satisfy demand for water, primarily for irrigation and urban growth. Western water law and settlement policies encouraged this through policies of "first in line, first in right" and "use it or lose it" (Glennon 2009). Surface flows, however, are the first thing to go, after which there may still be a relatively large amount of groundwater to pump. By recognizing that different types of aquatic systems provide different baskets of ecosystem services with different values, society could better recognize these trade-offs, ideally informing better water management.

### The importance of watershed and airshed position

The preceding discussion of ecosystem service provision by desert riparian and wetland ecosystems dealt only with the potential provision of ecosystem services. Ecosystem services are an inherently anthropocentric concept: without demand for a service, or clear human beneficiaries, there is no ecosystem service (Ruhl et al. 2007). Much of the spatial modeling of ecosystem services that has taken place in recent years has accounted only for the potential provision of ecosystem services (Eade and Moran 1996; Nelson et al. 2009; Raudsepp-Hearne et al. 2010), with minimal attention paid to the location of beneficiaries in relation to the ecosystems providing the service and the temporal and spatial flow characteristics for that service (Ruhl et al. 2007; Fisher et al. 2008; Tallis et al. 2008; Johnson et al. in Press).

In many situations where the benefits of watershed-based ecosystem services were large and obvious and the beneficiaries nearby, ecosystem services-based management took place well before the concept of ecosystem services was recognized. Cities like San Francisco, New York, Seattle, and Portland

| Ecosystem service | Unregulated streams and wetlands *(perennial-intermittent flow)* | Flow-regulated streams *(perennial-intermittent flow)* | Shallow groundwater *(intermittent-ephemeral flow)* | Deep groundwater *(highly ephemeral flow)* |
|---|---|---|---|---|
| Water supply | X | X | X | X |
| Recreation | X | X | x | x |
| Carbon sequestration & storage | **X** | **X** | X | x |
| Microclimate regulation | X | X | x | |
| Sediment & nutrient regulation | **X** | **X** | X | x |
| Aesthetic value | **X** | **X** | x | X |
| Other habitat-derived ecosystem services (including migration support) | **X** | X | x | x |
| Non-use value | **X** | X | x | |
| Hydroelectric generation | | X | | |
| Flood control | | X | | |
| Subsidence regulation | X | X | X | X |

Table 4.1. Ecosystem service provided by desert aquatic habitats. The hypothesized relative quantity of ecosystem service provision, all else being equal, is indicated, with blank boxes indicating no provision, a lower-case "x" indicating low levels of provision, an upper-case "X" indicating moderate levels of provision, and a bold upper-case "X" indicating the highest levels of provision.

protected their watersheds in order to avoid the costs of water filtration and purification (Chichilnisky and Heal 1998; Patterson and Coelho 2009), while in Hawaii watershed protection was undertaken to maintain agricultural water supplies especially for sugar cane. Many of these decisions were made a century ago or more. Contemporary examples such, as Quito, Ecuador where a watershed protection fund was created (Echavarria 2002), and similar payment for ecosystem services (PES) programs in Latin America, demonstrate that the PES concept can be quite powerful when there are large downstream cities with much to gain from

protecting land surrounding their water supplies.

Similar decisions have taken place in arid regions. Theodore Roosevelt established the Tonto National Forest for watershed protection purposes in 1905, when downstream Phoenix's population was less than 10,000 residents. Today the Tonto plays a critical role in watershed protection for the Salt and Verde Rivers, which supply 54 percent of Phoenix's current water supply (City of Phoenix 2005). Other large western cities similarly derive water-supply benefits from open space: Flagstaff from the Coconino National Forest, Salt Lake

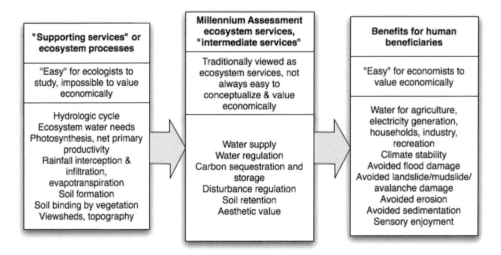

Figure 4.5.  A beneficiaries-based conceptualization of ecosystem services.

City from the Wasatch Range managed by Salt Lake County and the U.S. Forest Service, Tucson from the Bureau of Land Management's (BLM) Las Cienegas National Conservation Area and other protected lands on the Sonoita Plain, and Denver and other Colorado Front Range cities from national forests along the Rocky Mountains. These "spontaneous" uses of ecosystem services in land management took place when the benefits from ecosystem services were large and the beneficiaries obvious and transparent. However, these cases do not necessarily demonstrate new ways forward for ecosystem services-based management when the beneficiaries and benefits are less obvious – where beneficiaries are distant and/or not located in the downstream portion of a watershed or airshed. In the sparsely populated Intermountain West, where downstream beneficiaries are few or are located a far distance away, this may be the norm rather than the exception to the rule.

On the Colorado Plateau, researchers and managers are increasingly recognizing the importance of accounting for provision and beneficiaries (supply and demand) of ecosystem services and their spatial and temporal flow patterns in preparing rigorous assessments of the managed landscape. For hydrologic services (i.e., water regulation and supply, flood, sediment, and nutrient regulation), the location of human beneficiaries within a watershed matters tremendously.  The flow of surface and groundwater carries the benefit of water (or avoided detriment of flood water, excessive nutrients, sediment, or pathogens) toward human recipients.  The ecosystem service is the provision of a beneficial carrier (i.e., water supply in surface or groundwater) or the prevention of a detrimental carrier (i.e., absorption of flood water, sediment, nutrients, or pathogens; Johnson et al. in Press) from the landscape to people within that ecosystem.  Thus, if beneficiaries are physically located downstream from an ecosystem providing a benefit, there can be high ecosystem service value, particularly in arid regions where water is scarce. If there are few to no downstream human beneficiaries, there is likely minimal economic value associated with these ecosystem processes (Figure 4.2).  Similarly, the location of people and recreation sites such as the Grand Canyon within airsheds dictates the value of dust and noise regulation: without human beneficiaries, values are likely to be low, while the presence of a large number of beneficiaries is likely to yield higher values (Schulze et al. 1983).

The arrangement of water users and supplies in the Southwest has evolved dramatically over the last century and is likely to continue to do so in the future. Initially, the movement of water through interbasin transfers and agricultural irrigation canals, accompanied by regulation of flows from dam construction, brought water to burgeoning southwestern cities and agricultural lands while reducing flows to the Colorado River Delta in Mexico. This spatial reallocation of water and NPP across the Southwest has provided ecosystem services to the region's cities and agricultural lands while at the same time reducing them in the delta and along low desert rivers such as the Santa Cruz, Salt, and Gila Rivers. Climate change, water availability, and demographic trends, all influenced by the national and regional economy, will continue to influence population trends in the Southwest in coming years, also influencing allocation of water and ecosystem services.

## Long-distance beneficiary flows

Like the case where precipitation from the Rocky Mountains ultimately provides water to distant Los Angeles, Las Vegas, Phoenix, and Tucson, low population densities in the Intermountain West mean that spatial flows of ecosystem service benefits often cross long distances, particularly relative to densely settled coastal regions (Figure 4.3). Census Bureau population density estimates of the Intermountain West states (Arizona, Colorado, Idaho, Montana, New Mexico, Nevada, Utah, and Wyoming) in 2010 were 25.8 persons/mi$^2$, versus 87.3 persons/mi2 for the U.S. as a whole and 135.6 persons/mi2 for states outside the Intermountain West and excluding Alaska. Population density on the Colorado Plateau is even

lower, with 11.2 persons/mi$^2$ estimated to be living in counties on the Colorado Plateau.

Important examples of long-distance beneficiary flows on the Colorado Plateau include: 1) the dependence of downstream cities on upstream snowpack and land-use practices as they affect water quality and quantity (e.g., residents of Los Angeles, Phoenix, Tucson, and Las Vegas depend on Colorado snowpack that feeds the Colorado River reservoirs); 2) migration-derived ecosystem services (Semmens et al. 2011) that cross the continent; 3) recreation and non-use values for charismatic species and landscapes (e.g., iconic western species and national parks) that are valued by Americans and international visitors; 4) impacts of dust transport on Rocky Mountain snowpack and Colorado River runoff (Painter et al. 2010). For rare but poorly-known species, ecosystems, or places it can be difficult to reconcile low estimates of willingness to pay (particularly for non-use value) with the greater value attributed to equally rare but otherwise better known resources. For instance, Brookshire (University of New Mexico, Albuquerque, New Mexico, personal communication) found steep distance decay in willingness to pay for non-use values for southeast Arizona's San Pedro River, likely in part because it is not a nationally-known resource like some of the well-recognized national parks on the Colorado Plateau. Distance decay functions may be much less steep for charismatic species or sites, and must be taken into account when aggregating estimates of willingness to pay (Pate and Loomis 1997; Loomis 2000; Bateman et al. 2006). For example, visibility and quiet in national parks like the Grand Canyon are highly valued not just by visitors but by Americans at large, indicating substantial non-use value (Schulze et al. 1983). These values played a role in the decisions to limit helicopter overflights of the canyon and to upgrade pollution control at the nearby Navajo Generating Station, which has improved wintertime visibility in the park (Green et al. 2005).

Low population densities and long-distance beneficiary flows present several challenges to ecosystem services valuation on the Colorado Plateau. All else being equal, smaller beneficiary populations should lead to less demand for and value of an ecosystem service. However, long-distance beneficiary flows can lead to underestimates of ecosystem service values in two ways. First, a lack of public awareness, and hence value may be placed by the public on distantly-derived ecosystem services. Second, economists may aggregate values over an inappropriately small number of beneficiaries relative to those who actually derive benefits from the services. This illustrates the potential value of new tools to map the full extent of provision, beneficiaries, and spatial flows of ecosystem services (Johnson et al. in Press): for both economists doing survey work, to properly survey and aggregate results across the entire beneficiary population, and for survey respondents, who are often unaware of their dependence on distant ecosystems for economic benefits. For decision makers, ignoring certain beneficiary populations can lead to underrepresentation of key stakeholders in decision making, leading to less equitable choices, such as the early-20th century decision to limit flows of the Colorado River into Mexico, which effectively undervalued the benefits provided by the Colorado River Delta ecosystem.

Two examples from the Colorado Plateau illustrate the importance of considering the watershed and airshed position of beneficiaries across long geographic distances (Figure 4.4). First, rising energy costs and increasing demand for domestic energy sources is leading to renewed interest in oil shale and tar sands development in the Upper Colorado Basin, and uranium mining on the north rim of Grand Canyon National Park (Alpine 2010). Many of these proposed extraction areas are on BLM land adjacent to or upstream of national parks including Canyonlands and Dinosaur, and Colorado River reservoirs including Lakes Powell and Mead. The high water demand associated with oil shale and tar sands development, coupled with associated land disturbance and water pollution, could potentially impact recreational, water supply, and non-use values for a large number of distant beneficiaries (BLM 2008, http://www.blm.gov/wo/st/en/prog/energy/oilshale_2.html).

A second example links land use to Rocky Mountain snowpack and regional water supplies via regional dust transport. Recent research has linked grazing history and the presence of biological soil crusts and perennial grassland plants to the amount of dust generated during windstorms (Painter et al. 2010; Miller et al. 2011). Sites with a long history of grazing or off-road vehicle use can act as sources of windblown dust, which can then cause faster snowmelt of the Rocky Mountain snowpack due to decreased albedo. Models of dust transport, snowmelt, runoff, and evapotranspiration suggest that dust loading may be responsible for an approximately 5% reduction in runoff in the Upper Colorado Basin (Painter et al. 2010). For dispersed downstream water users both on and downstream of the Colorado Plateau, maintaining the dust regulation service provided by Colorado Plateau rangelands would produce substantial benefits in water supply, human health (Richardson 2008), and other ecosystem services. Managing these services well would require improved public awareness, political will, scientific capacity, and institutions to manage resources on the Colorado Plateau.

## Implications for science and policy

### Scientific implications

Aside from recreational values and ecosystem goods, ecosystem services on the Colorado Plateau have rarely been quantified for either upland or aquatic systems. Emerging tools to map and quantify ecosystem services offer a way forward in comparing resource management tradeoffs. Incorporating locally relevant ecological and hydrologic process models within an ecosystem services assessment and valuation framework is a shared goal of many of these modeling tools, and would substantially improve the scientific validity and managerial relevance of these efforts. Their initial development would account for locally relevant ecological processes and economic preferences over the Colorado Plateau. A forthcoming USGS review of ecosystem services mapping and valuation tools is intended to inform public land managers as to the ability of these tools to quantify and value ecosystem services, thus adding value to established agency decision-making processes (Bagstad et al. in Press).

Further research on ecological and economic production functions related to ecosystem services provision would improve our understanding of how services are produced and valued across all ecosystems (Nelson et al. 2009), including those on the Colorado Plateau. Two other recent trends in ecosystem services research – a move toward recognizing services as concrete, spatially explicit benefits (Boyd and Banzhaf 2007; Wallace 2007) rather than abstract classes of ecosystem services (Millennium Assessment 2005b), and the mapping of provision, use, and spatial service flows (Johnson et al. in Press), offer a path forward in addressing the challenges of beneficiary locations and long-distance benefit flows to ecosystem services applications on the Colorado Plateau. A move toward a beneficiary-focused framework for mapping and valuing ecosystem services can help to avoid double counting of ecosystem services, while more clearly identifying different values held by various beneficiary groups (Figure 4.5). The "static value maps" (Tallis et al. 2008) that have predominated recent ecosystem

services literature (e.g., Eade and Moran 1996; Nelson et al. 2009; Raudsepp-Hearne et al. 2010) often fail to account for areas where services are provided, used, and the spatial flow characteristics that connect these regions of provision and use. As such these maps typically better represent potential provision of an ecosystem service, rather than its actual use. While work is underway to fully map spatial dynamics of ecosystem services, in the interim, static maps often remain more feasible to generate in the absence of support to further develop case studies of ecosystem service flows.

Recent advances in conceptualizing ecosystem service flows (Semmens et al. 2011) can better inform stakeholders and beneficiaries ranging across wide distances as to their spatial dependencies on ecosystems, while ensuring that benefits are aggregated correctly (Figure 4.2). Relevant long-distance flows on the Colorado Plateau include water supply, dust regulation, migration-derived services, aesthetics, recreation, and non-use values. A full understanding of spatial flow dynamics also allows planners to make better decisions about spatial planning of conservation, restoration, development, and extractive resource use – by avoiding or targeting activities that protect ecosystem services and enable their spatial flow to beneficiaries. With a full view of beneficiaries and regions that reduce or enhance service provision, maps of the spatial dynamics of ecosystem services over the Colorado Plateau can provide guidance on structuring economic incentives to promote more sustainable management and use of ecosystem services, such as polluter pays versus beneficiary pays approaches to PES programs.

*Policy implications*

The difficulty of linking water quantity, quality, location, and timing to ecosystem service provision, watershed position, and long-distance beneficiary flows partially explains why ecosystem services to date have failed to impact watershed management decisions over the Colorado Plateau on a meaningful level. Ecosystem services like outdoor recreation are relatively easy to include in traditional economic assessments. Yet the reason for this chapter and the promise of ecosystem services assessments lies in reframing decision making, particularly on public lands in the Western United States, as more than just a tradeoff between recreation and resource extraction. This issue will continue to grow in importance as population growth, climate change, and demand for domestic energy development place increasing strain on natural resources, and as Federal agencies continue to explore the ecosystem services paradigm as a way to better balance public lands management for both private and public goods.

For managers, ecosystem services provide more information about stakeholders that can be incorporated into resource management decisions. Public lands supply numerous ecosystem services to beneficiaries ranging from local to global scales, and are themselves impacted by upstream management practices. A more sophisticated understanding of who these impacts influence can provide better management guidance. While managers may be accustomed to thinking about movement of wildlife and visitors within protected areas, methods to visualize and map other ecosystem services can offer additional means of engaging new stakeholders and adjacent landowners that influence ecosystems and ecosystem services.

Improved accounting for the spatial dynamics and beneficiaries of ecosystem services is an important step in bringing ecosystem services into policy. Yet since most ecosystem services are not market goods, they require institutions to manage them, through alternative means such as assignment of property rights, extraction quotas, market structures, and social norms

for common-pool resource management (Ostrom et al. 1999; Salzman 2005). Western institutions were historically created to facilitate settlement and resource extraction – from 19th century homesteading and mining laws, to modern water law, to the use of state trust lands for funding public education. Recent decades have seen major Federal landowners like the BLM and the US Forest Service move from "multiple use-sustained yield" toward an "ecosystem management" paradigm to better balance extractive resource use with other benefits derived from public lands. The Glen Canyon Adaptive Management Program is a recent example from the Colorado Plateau that seeks to maintain and enhance ecosystem processes and services of the Grand Canyon by changing operations of Glen Canyon Dam (Melis 2011). Yet typical western patterns of land ownership – divided between Federal, state trust, and private land – many times intermixed in a checkerboard pattern – rarely facilitate easy environmental management and will require continued coordination of multiple actors to maintain ecosystem service flows.

Institutions can help foster trust, bring Native American tribes and small landowners into new incentive systems, spatially target incentives, and monitor and verify gains in ecosystem services provisioning. In the United States, perhaps the longest running ecosystem service-based institution pays farmers to provide environmental benefits, dating to the 1986 Farm Bill. The recently created USDA Office of Environmental Markets seeks to develop the standards and infrastructure for market-based conservation. Western farms and ranches are generally larger than the national average but have lower productivity on a per-acre basis (USDA 2007), which may mean lower ecosystem services production, depending on their location and number of ecosystem service beneficiaries. Quantification of services is thus important, and potential value matters relative to "opportunity costs," the potential economic return from an alternative extractive resource use. For instance, if the provision of carbon sequestration or other commoditized ecosystem services are lower in arid regions, would incentives for carbon sequestration help ranchers improve their management practices (de Steiguer 2008)? The answer to this question depends on quantifying service values and opportunity costs, developing institutions to facilitate ecosystem services-based management, and disseminating these concepts beyond academia.

Finally, while there is currently great interest in valuing ecosystem services, it is not always appropriate or possible to value services in monetary terms (USDA 2008). While dollar values provide a common currency, shoehorning all public preferences for the environment and ecosystem services into dollars or a cost-benefit framework should not be the only goal of ecosystem services research and policy on the Colorado Plateau. Non-monetary estimates of value or preferences may be highly appropriate for some values (e.g., Native American cultural values), less controversial than using dollars in other cases (e.g., endangered species), or useful and well established alternatives to economic approaches in some fields (e.g., recreation management).

To bring ecosystem services into decision making on the Colorado Plateau, further evolution of their underlying science and policy is needed. However, the needed conceptual frameworks, spatial models, and policy tools are rapidly developing and offer promise for enhancing resource management both in general, and for the unique cultural and biological resources of the Colorado Plateau.

## ACKNOWLEDGEMENTS

We thank Ted Melis for a review of an early draft of this chapter and comments on management impacts of ecosystem services particularly for Glen Canyon Dam, Grand Canyon National Park, and the Lower Colorado River. Ed de Steiguer, Jason Kreitler, and Carl Shapiro also provided constructive editorial comments on earlier drafts of this chapter. Partial funding for this work was provided by the U.S. Geological Survey Mendenhall Research Fellowship Program.

## LITERATURE CITED

Alpine, A. E., editor. 2010. Hydrological, geological, and biological site characterization of breccia pipe uranium deposits in northern Arizona. U.S. Geological Survey Scientific Investigations Report 2010–5025, 353 pp.

Bagstad, K. J., D. J. Semmens, R. Winthrop, D. Jordahl, and J. Larson. In Press. Ecosystem services valuation to support decision making on public lands: A case study for the San Pedro River, Arizona. U.S. Geological Survey Scientific Investigations Report.

Balvanera, P., A. B. Pfisterer, N. Buchmann, J. He, T. Nakashizuka, D. Raffaelli, and B. Schmid. 2006. Quantifying the evidence for biodiversity effects on ecosystem functioning and services. Ecology Letters 9: 1146-1156.

Bateman, I. J., B. H. Day, S. Georgiou, and I. Lake. 2006. The aggregation of environmental benefit values: Welfare measures, distance decay, and total WTP. Ecological Economics 60: 450-460.

Bishop, R. C., K. J. Boyle, and M. P. Welsh. 1987. Toward total economic valuation of Great Lakes fishery resources. Transactions of the American Fisheries Society 116 (3): 339-345.

Boyd, J. and S. Banzhaf. 2007. What are ecosystem services? The need for standardized environmental accounting units. Ecological Economics 63: 616-626.

Brander, L. M., P. van Beukering, and H. S. J. Cesar. 2007. The recreational value of coral reefs: A meta-analysis. Ecological Economics 63: 209-218.

Bureau of Land Management (BLM). 2008. Proposed oil shale and tar sands resource management plan amendments to address land use allocations in Colorado, Utah, and Wyoming and Final Programmatic Environmental Impact Statement. U.S. Department of Interior Bureau of Land Management, 8971 pp.

Chan, K. M. A., M. R. Shaw, D. R. Cameron, E. C. Underwood, and G. C. Daily. 2006. Conservation planning for ecosystem services. PLOS Biology 4 (11): 2138-2152.

Chichilnisky, G. and G. Heal. 1998. Economic returns from the biosphere. Nature 391: 629-630.

City of Phoenix. 2005. Water Resources Plan, 2005 update. City of Phoenix, Phoenix, AZ.

Coes, A. L. and D. R. Pool. 2005. Ephemeral-Stream Channel and Basin-Floor Infiltration and Recharge in the Sierra Vista Subwatershed of the Upper San Pedro Basin, Southeastern Arizona. U.S. Geological Survey Open File Report 2005-1023. 59 pp.

Costanza, R, R. d'Arge, R. de Groot, S. Farber, M. Grasso, B. Hannon, K. Limburg, S. Naeem, R. V. O'Neill, J. Paruelo, R. G. Raskin, P. Sutton, and M. van den Belt. 1997. The value of the world's ecosystem services and natural capital. Nature 387: 253-260.

Costanza, R., M. Wilson, A. Troy, A. Voinov, S. Liu, and J. D'Agostino. 2006. The value of New Jersey's ecosystem services and natural capital. New Jersey Department of Environmental Protection, Trenton, NJ.

Costanza, R., B. Fisher, K. Mulder, S. Liu, T. Christopher. 2007. Biodiversity and ecosystem services: A multi-scale empirical study of the relationship between species richness and net primary productivity. Ecological Economics 61 (2-3): 478-491.

Dahl, T. E. 1990. Wetland losses in the United States, 1780s to 1980s. U.S. Department of Interior, Fish and Wildlife Service, Washington, D.C. 13 pp. Accessed February 8, 2012 at: http://www.npwrc.usgs.gov/resource/wetlands/wetloss/.

Daily, G. C., S. Polasky, J. Goldstein, P. M. Kareiva, H. A. Mooney, L. Pejchar, T. H. Ricketts, J. Salzman, and R. Shallenberger. 2009. Ecosystem services in decision making: Time to deliver. Frontiers in Ecology and the Environment 7 (1): 21-28.

de Steiguer, J. E. 2008. Semi-arid rangelands and carbon offset markets: A look at the economic prospects. Rangelands 30: 27-32.

Diamond, J. 1997. Guns, germs and steel. W. W. Norton, NY 480 pp.

Dodds, W. K., K. C. Wilson, R. L. Rehmeier, G. L. Knight, S. Wiggam, J. A. Falke, H. J. Dalgleish, and K. N. Bertrand. 2008. Comparing ecosystem goods and services provided by restored and native lands. BioScience 58 (9): 837-845.

Eade, J. D. O. and D. Moran. 1996. Spatial economic valuation: Benefits transfer using geographical information systems. Journal of Environmental Management 48: 97-110.

Echavarria, M. 2002. Financing watershed conservation: The FONAG water fund in Quito, Ecuador. In Selling forest environmental services: Market-based mechanisms for conservation and development, edited by S. Pagiola, J. Bishop, and N. Landell-Mills, pp. 91-101. Earthscan, London.

Engel, S., S. Pagiola, and S. Wunder. 2008. Designing payments for environmental services in theory and practice: An overview of the issues. Ecological Economics 65: 663-674.

Farber, S. 1996. Welfare loss of wetlands disintegration: A Louisiana study. Contemporary Economic Policy 14: 92-106.

Farber, S., R. Costanza, D. Childers, J. Erickson, K. Gross, M. Grove, C.S. Hopkinson, J. Kahn, S. Pincetl, A. Troy, P. Warren, and M. Wilson. 2006. Linking ecology and economics for ecosystem management. Bioscience 56: 121-133.

Farley, J. 2008. The role of prices in conserving critical natural capital. Conservation Biology 22 (6): 1399-1408.

Fisher, B., K. Turner, M. Zylstra, R. Brouwer, R. de Groot, S. Farber, P. Ferraro, R. Green, D. Hadley, J. Harlow, P. Jefferiss, C. Kirkby, P. Morling, S. Mowatt, R. Naidoo, J. Paavola, B. Strassburg, D. Yu, and A. Balmford. 2008. Ecosystem services and economic theory: Integration for policy-relevant research. Ecological Applications 18 (8): 2050-2067.

Glennon, R. 2009. The conflict between law and science in the San Pedro River. In Ecology and conservation of the San Pedro River, edited by J. C. Stromberg and B. Tellman, pp. 407-414. University of Arizona Press, Tucson, AZ.

Goodrich, D. C., D. G. Williams, C.L. Unkrich, J. F. Hogan, R. L. Scott, K. R. Hultine, D. Pool, A. L. Coes, and S. Miller. 2004. Comparison of methods to estimate ephemeral channel recharge, Walnut Gulch, San Pedro River Basin, Arizona. In: Groundwater Recharge in a Desert Environment: The Southwestern United States. Edited by J.F. Hogan, F.M. Phillips, and B.R. Scanlon, Water Science and Applications Series, Vol. 9, American Geophysical Union, Washington, D.C., pp. 77-99.

Green, M., R. Farber, N. Lien, K. Gebhart, J. Molenar, H. Iyer, and D. Eatough. 2005. The effect of scrubber installation at the Navajo Generating Station on particulate sulfur and visibility levels in the Grand Canyon. Journal of the Air and Waste Management Association 55: 1675-1682.

Hooper, D. U., F. S. Chapin, III, J. J. Ewel, A. Hector, P. Inchausti, S. Lavorel, J. H. Lawton, D. M. Lodge, M. Loreau, S. Naeem, B. Schmid, H. Setala, A. J. Symstad, J. Vandermeer, and D. A. Wardle. 2005. Effects of biodiversity on ecosystem functioning: A consensus of the current knowledge. Ecological Monographs 75 (1): 3-35.

Hultine, K. R., J. Belnap, C. van Riper III, J. R. Ehleringer, P. E. Dennison, M. E. Lee, P. L. Nagler, K. A. Snyder, S. M. Uselman, and J. B. West. 2010. Tamarisk biocontrol in the western United States: ecological and societal implications. Frontiers in Ecology and the Environment 8 (9): 467-474.

Johnson, G. W., K. J. Bagstad, R. Snapp, and F. Villa. In Press. Service Path Attribution Networks (SPANs): A network flow approach to ecosystem service assessment. International Journal of Agricultural and Environmental Information Systems.

Jones, K. B., E. T Slonecker, N. S. Nash, A. C. Neale, T. G. Wade, and S. Hamann. 2010. Riparian habitat changes across the continental United States (1972-2003) and potential implications for sustaining ecosystem services. Landscape Ecology 25 (8): 1261-1275.

Loomis, J. B. 2000. Vertically summing public good demand curves: An empirical comparison of economic versus political jurisdictions. Land Economics 76 (2): 312-321.

Melis, T. S., J. F. Hamill, L. G. Coggins, Jr., G. E. Bennett, P. E. Grams, T. A. Kennedy, T. A. Kubly, and B. E. Ralston. 2010. Proceedings of the Colorado River Basin Science and Resources Management Symposium, November 18-20, 2008, Scottsdale, Arizona: U.S. Geological Survey Scientific Investigations Report 2010-5135, 372 pp.

Melis, T. S., editor. 2011. Effects of three high-flow experiments on the Colorado River ecosystem downstream of Glen Canyon Dam, Arizona. U.S. Geological Survey Circular 1366, 147 pp.

Millennium Ecosystem Assessment. 2005a. Ecosystems and human well-being: Multiscale assessments, Volume 4. World Resources Institute, Washington, D.C.

Millennium Ecosystem Assessment. 2005b. Millennium Ecosystem Assessment: Living beyond our means - Natural assets and human well-being. World Resources Institute, Washington, D.C.

Miller, M. E., R. T. Belote, M. A. Bowker, and S.L. Garman. 2011. Alternative states of a semiarid grassland ecosystem: Implications for ecosystem services. Ecosphere 2 (5): 1-18.

Mooney, H. A. and P. R. Ehrlich. 1997. Ecosystem services: A fragmentary history. In Nature's services: Societal dependence on natural ecosystems, edited by G. C. Daily, pp. 11-19. Island Press, Washington, DC.

Natural Resources Defense Council. 2001. Groundwater mining of Black Mesa. NRDC, New York.

Nelson, E., G. Mendoza, J. Regetz, S. Polasky, H. Tallis, D. R. Cameron, K. M. A. Chan, G. C. Daily, J. Goldstein, P. M. Kareiva, E. Lonsdorf, R. Naidoo, T. H. Ricketts, and M. R. Shaw. 2009. Modeling multiple ecosystem services, biodiversity conservation, commodity production, and tradeoffs at landscape scales. Frontiers in Ecology and the Environment 7 (1): 4-11.

Norman, L., N. Tallent-Halsell, W. Labiosa, M. Weber, A. McCoy, K. Hirschboeck, J. Callegary, C. van Riper III, and F. Gray. 2010. Developing an Ecosystem Services Online Decision Support Tool to Assess the Impacts of Climate Change and Urban Growth in the Santa Cruz Watershed; Where We Live, Work, and Play. Sustainability 2 (7): 2044-2069.

Osmundson, D. B., R. J. Ryel, V. L. Lamarra, and J. Pitlick. 2002. Flow-sediment-biota relations: Implications for river regulation effects on native fish abundance. Ecological Applications 12(6): 1719-1739.

Ostrom, E., J. Burger, C. B. Field, R. B. Norgaard, and D. Policansky. 1999. Revisiting the commons: Local lessons, global challenges. Science 284 (5412): 278-282.

Painter, T. H., J. S. Deems, J. Belnap, A. F. Hamlet, C. C. Landry, and B. Udall. 2010. Response of Colorado River runoff to dust radiative forcing in snow. Proceedings of the National Academy of Sciences 107 (40): 17125-17130.

Pate, J. and J. Loomis. 1997. The effect of distance on willingness to pay values: A case study of wetlands and salmon in California. Ecological Economics 20: 199-207.

Patterson, T. M. and D. L. Coelho. 2009. Ecosystem services: Foundations, opportunities, and challenges for the forest products sector. Forest Ecology and Management 257: 1637-1646.

Raudsepp-Hearne, C., G. D. Peterson, and E. M. Bennett. 2010. Ecosystem service bundles for analyzing tradeoffs in diverse landscapes. Proceedings of the National Academy of Science 107 (11): 5242-5247.

Reisner, M. 1986. Cadillac Desert: The American West and its disappearing water. Penguin Books, New York.

Richardson, R.B. 2008. Conceptualizing the value of ecosystem services in deserts. In Creating Sustainability Within Our Midst: Challenges for the 21st Century, edited by R. L. Chapman, pp. 225-248. Pace University Press, New York.

Richmond, A., R. K. Kaufmann, and R. B. Myneni. 2007. Valuing ecosystem services: A shadow price for net primary production. Ecological Economics 64: 454-462.

Ruhl, J. B., S. E. Kraft, and C. L. Lant. 2007. The law and policy of ecosystem services. Island Press, Washington, D.C.

Sada, D. W. and K. F. Pohlman. 2003. Draft U.S. National Park Service Mojave Inventory and Monitoring Network spring survey protocols: Level I, 19 November 2003. Desert Research Institute, Inc., Reno.

Salzman, J. 2005. Creating markets for ecosystem services: Notes from the field. New York University Law Review, 80 N.Y.U.L. Rev. 870.

Schulze, W. D., D. S. Brookshire, E. G. Walther, K. K. MacFarland, M. A. Thayer, R. L. Whitworth, S. Ben-David, W. Malm, and J. Molenar. 1983. The economic benefits of preserving visibility in the National Parks of the Southwest. Natural Resources Journal 23 (1): 149-173.

Semmens, D.J., W. Kepner, and D. Goodrich. 2010. Assessment of goods and valuation of ecosystem services (AGAVES) San Pedro River Basin, United States and Mexico: U.S. Geological Survey Fact Sheet 2010-3082, 4 pp.

Semmens, D. J., J. E. Diffendorfer, L. Lopez-Hoffman, and C. S. Shapiro. 2011. Accounting for the ecosystem services of migratory species: Quantifying migration support and spatial subsidies. Ecological Economics 70 (12): 2236-2242.

Stevens, L. E., and G. P. Nabhan. 2002. Hydrological diversity: water's role in shaping natural and cultural diversity on the Colorado Plateau. Pages 33-40 in Center for Sustainable Environments Terralingua and Grand Canyon Wildlands Council, editors. Safeguarding the uniqueness of the Colorado Plateau: an ecoregional assessment of biocultural diversity. Center for Sustainable Environments, Northern Arizona University, Flagstaff.

Stromberg, J. C., K. J. Bagstad, J. M. Leenhouts, S. J. Lite, and E. Makings. 2005. Effects of stream flow intermittency on riparian vegetation of a semiarid region river (San Pedro River, Arizona). River Research and Applications 21: 925-938.

Stromberg, J. C., V. B. Beauchamp, M. D. Dixon, S. J. Lite, and C. Paradzick. 2007a. Importance of low-flow and high-flow characteristics to restoration of riparian vegetation along rivers in the arid south-western United States. Freshwater Biology 52: 651-679.

Stromberg, J. C., S. J. Lite, R. Marler, C. Paradzick, P. B. Shafroth, D. Shorrock, J. M. White, and M. S. White. 2007b. Altered stream-flow regimes and invasive plant species: the Tamarix case. Global Ecology and Biogeography 16: 381-393.

Tallis, H., P. Kareiva, M. Marvier, and A. Chang. 2008. An ecosystem services framework to support both practical conservation and economic development. Proceedings of the National Academy of Sciences 105 (28): 9457-9464.

Tallis, H., R. Goldman, M. Uhl, and B. Brosi. 2009. Integrating conservation and development in the field: Implementing ecosystem services projects. Frontiers in Ecology and the Environment 7 (1): 12-20.

U.S. Department of Agriculture (USDA). 2007. USDA Census of Agriculture. Accessed February 8, 2012 at: http://www.agcensus.usda.gov/.

U.S. Department of Agriculture (USDA). 2008. Counting all that matters: Recognizing the value of ecosystem services. PNW Science Update 16. Pacific Northwest Research Station: USDA Forest Service.

van Riper III, C., K. L. Paxton, C. O'Brien, P. B. Shafroth, and L. J. McGrath. 2008. Rethinking avian response to Tamarix on the Lower Colordo River: A threshold hypothesis. Restoration Ecology 16 (1): 155-167.

Villa, F., M. A. Wilson, R. de Groot, S. Farber, R. Costanza, and R. M. J. Boumans. 2002. Designing an integrated knowledge base to support ecosystem services valuation. Ecological Economics 41: 445-456.

Villa, F., M. Ceroni, K. J. Bagstad, G. Johnson, and S. Krivov. 2009. ARIES (Artificial Intelligence for Ecosystem Services): A new tool for ecosystem services assessment, planning, and valuation. Proceedings of the 11th Annual BIOECON Conference on Economic Instruments to Enhance the Conservation and Sustainable Use of Biodiversity, Venice, Italy, September 2009.

Wallace, K. J. 2007. Classification of ecosystem services: Problems and solutions. Biological Conservation 139: 235-246.

Webb, R. H., S. A. Leake, and R. M. Turner. 2007. The ribbon of green: Change in riparian vegetation of the southwestern United States. University of Arizona Press, Tucson, AZ.

Woodward, R.T. and Y. Wui. 2001. The economic value of wetland services: A meta-analysis. Ecological Economics 37: 257-270.

Zandersen, M. and R. S. J. Tol. 2009. A meta-analysis of forest recreation values in Europe. Journal of Forest Economics 15: 109-130.

# Assessing Natural and Man-Made Threats to Ecological Systems

# REVISITING TRENDS IN VEGETATION RECOVERY FOLLOWING PROTECTION FROM GRAZING, CHACO CULTURE NATIONAL HISTORIC PARK, NEW MEXICO

*M. Lisa Floyd, David D. Hanna, Thomas L. Fleischner, and Brad Shattuck*

## ABSTRACT

Livestock grazing has serious ecological consequences in the arid Southwest, leading to management dilemmas that become more problematic with global climate change projections. Management challenges can be illuminated by long-term monitoring studies, especially under varying climatic conditions. We revisited sites in Chaco Culture National Historic Park, where we had previously analyzed historic livestock grazing impacts under drought conditions (1999-2000). Two ages of grazing exclosures were created by fencing projects in the 1930s and 1990s, such that long-term protection (>60 years of exclosure), recent protection (≤12 years), and current grazing treatments were immediately adjacent at six sites. We compared plant species richness, cover of biological soil crusts, shrub density, vegetative cover, and plant community composition at these six sites. Our recent resurvey (2006) took place during a period of higher summer precipitation and temperature. The greatest differences between surveys were in forb characteristics. In all grazing treatments, forb density and cover was higher during the wetter season than during drought, and 77 forb species were recorded that had not been present in our earlier surveys (Floyd et al. 2003). Plant species richness continued to be significantly greater under long-term protection at all six sites. No differences among treatments in invasive plant diversity or cover were detected. The cover of black biological soil crusts was significantly higher under long-term protection at three of the four sites monitored; in the fourth, cover was highest under short-term protection. On Menefee Shales at Kin Klizhin, crust cover was six times higher with long-term protection than where currently grazed. These results affirm our earlier assertion that recovery of soil crusts can proceed rapidly with protection from grazing. Post-grazing trends were variable at the six sites with Fajada Gap (a grassland) having a significantly greater shrub density and cover in the currently grazed treatment than under protection. In contrast, at Mockingbird Mesa-top (a low shrubland) there was a significantly greater shrub density and cover in protected treatments. Thus plant community structure re-established itself differentially with protection from grazing depending on the inherent biotic potential of each site. This variable trend is an important management consideration.

## INTRODUCTION

Livestock were introduced into the American Southwest in the early sixteenth century from Mexico (Stewart 1936; Stoddart and Smith 1943). Grazing by domesticated livestock, primarily cattle, has become the most ubiquitous land use in the western United States. Approximately 70% of the 11 westernmost states in the USA (those including and west of the Rocky Mountains) is grazed by livestock at least part of the year (CAST 1974; Longhurst et al. 1982; Crumpacker 1984), including approximately 90% of federal land in these states (Armour et al. 1991). Further, livestock grazing occurs in >75% of the ecoregions delineated by the World Wildlife Fund (Ricketts et al. 1999) in the American West. As such, it represents a primary ecological influence in more than half of these ecoregions (Fleischner, 2010).

Because of the ubiquity of livestock grazing, its ecological impacts can be difficult to discern. The primary method of studying grazing impacts has been comparison of areas exclosed to livestock (exclosures) and adjacent rangelands. These exclosures tend to be relatively small ($\leq$80 ha), and a preponderance of them are in riparian habitats (e.g., Krueper et al. 2003). The lack of landscape-scale exclosures was noted by Bock et al. (1993), who called for an ecologically representative system of large livestock exclosures. Information on post-grazing recovery of dry uplands has been scarce because of the relative dearth of exclosures within these sites, and because recovery is inherently slower than in riparian habitats (Fleischner 1994).

Chaco Culture National Historic Park (CCNHP), New Mexico, situated in one of the longest continuously grazed regions of North America, is one of the largest and longest-term livestock exclosures in western North America (see Floyd et al. 2003 for details of grazing and management history). The U.S. National Park Service (NPS) began fencing the boundaries of Chaco Canyon National Monument (8600 ha) in 1936, completing the task in 1948 (NPS 1995, 1998). In 1980 the monument was expanded and redesignated as Chaco Culture National Historic Park. Fencing of the four new parcels (amounting to 5000 ha) was completed from 1995 to 1999. Thus, 8600 ha have been protected from grazing for $\geq$60 years, and an additional 5000 ha have been protected for $\leq$12 years. The entire 13,600 ha exclosure is surrounded by lands that continue to be grazed by Navajo ranchers although it is not possible to accurately determine frequency, duration, and intensity

of grazing, or class of livestock, on these lands.

During 1999-2000 we studied historic grazing impacts at CCNHP (Floyd et al. 2003). We compared the effects of three different livestock grazing treatments (long-term protection, short-term protection, and current grazing) on the cover of plants, biological soil crusts, and plant species richness at six sites with different potential natural vegetation. Plant species richness and cover of black (nitrogen-fixing) biological soil crusts were higher under long-term protection than under current grazing at all six sites. Trends in shrub and grass response varied with the site's potential. Shrub cover increased with long-term protection at four upland sites, and grass cover increased with protection at four sites.

Monitoring vegetation through cycles of climatic variation can provide more complete and accurate perspective on long-term trends. The National Park Service has recently implemented long-term monitoring in all parks as part of the Inventory and Monitoring Project. Because our data at CCNHP were collected during two years of drought (1999-2000), we were interested in comparing vegetation at the same sites during a year with greater precipitation. During May to September 2006, precipitation was roughly three times as great as during this period in 2000. Such biological comparison during different climatic regimes is especially important in the face of predicted global climate change--especially because vegetation change in the Colorado Plateau region is currently following the predictions of global climate change models (Breshears et al. 2005, IPCC 2007).

We are particularly concerned with the condition of biological soil crusts, which provide critical ecosystem functions (Belnap and Lange 2003), including fixing carbon in sparsely vegetated areas. Such carbon contributions help keep interspaces between vascular plants fertile and support other microbial populations (Beymer and Klopatek 1991). The availability of nitrogen is an important factor limiting primary production in arid habitats throughout the world. In the Great Basin Desert of the western USA, nitrogen is second only to moisture in importance (James and Jurinak 1978). In desert shrub and grassland communities that support few nitrogen-fixing plants, biological soil crusts can be the dominant source of nitrogen (Rychert et al. 1978; Harper and Marble 1988; Evans and Ehleringer 1993; Evans and Belnap 1999). Nitrogen inputs are highly dependent on temperature, moisture, and species composition of the crusts (Belnap and Lange 2003); therefore, both prevailing climate and the legacy of disturbances influence fixation rates (Belnap 1995, 1996). Additionally, crusts stabilize soils (Belnap and Gillette 1997, 1998; Warren 2003), retain moisture, and provide seed germination sites. Soil crusts are effective in capturing eolian dust deposits, contributing to a 2- to 13-fold increase in nutrients in southeastern Utah (Reynolds et al. 2001). The presence of soil crusts generally increases the amount and depth of rainfall infiltration (Loope and Gifford 1972; Brotherson and Rushforth 1983; Harper and Marble 1988; Johansen 1993). Thus, biological soil crusts play critical roles regarding the two most important limiting factors in arid landscapes: water and nitrogen.

We compared the effect of three grazing treatments—long-term protection (≥60 years), recent protection (< 12 years), and currently grazed--on plant diversity, capacity for nutrient cycling, and vegetation structure and composition in Chaco Culture National Historic Park. Within these three broad categories, we attempted to determine if livestock grazing leads to a difference in: 1) shrub and forb density, 2) grass and forb cover, 3) bare soil cover, 4) plant community composition, 5) plant species richness, or 6) cover of biological soil crusts. We then compared these vegetation patterns from the wetter field season (2006) to data from earlier drought years (1999-2000) at the same sites to begin to tease out effects of climatic pattern on post-grazing ecosystem recovery.

## METHODS

We sampled at the six sites reported in Floyd et al. (2003): Mockingbird Mesa-top, Clys Mesa-top, Northern Side Canyons, Fajada Gap, East Canyon, and Kin Klizhin (Figure 5.1). At each location, removal of fencing due to acquisition of NPS land had created T-junctions where 3 grazing treatments juxtaposed: long term protection (greater than or equal to 60 years), short term protection (less than or equal to 12 years) and currently grazed land under jurisdiction of the Navajo Nation.

Within each treatment we sampled 6-10 points that were randomly selected with GIS software. At each sample point, we laid out 2 parallel 30 m transects, 10 meters apart. Shrubs were counted and tallied by species in the 10m x 30m area created by the parallel lines. Shrub cover was measured using the transects for line intercept sampling (Mueller-Dombois and Ellenburg 2003). We placed a point frame at each 10 m interval. This frame create 25 points and intersections of two lines defined ground substrate (bare soil, forb, shrub, grass, litter, biotic crust, physical crust, invasive forb, invasive grass, and dung). This was repeated 12 times at each sample point. Finally, a releve analysis was done once at each point, recording all plant species and their cover/abundance values using the Braun Blanquet scale (Mueller-Dombois and Ellenburg 2003). Data were analyzed using an ANOVA to determine significant differences in each cover and density variable between the three different treatments.

## CHACO CULTURE NATIONAL HISTORIC PARK
## GRAZING EXCLOSURE COMPARISON STUDY: SAMPLE LOCATIONS

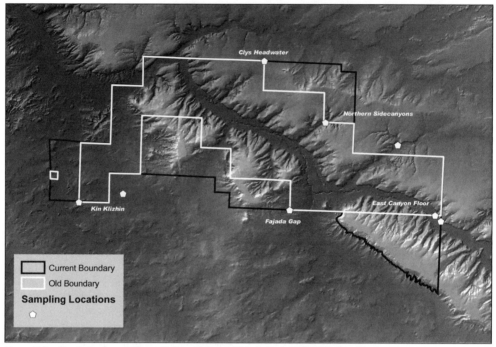

Figure 5.1    Six study sites at Chaco Culture National Historic Park (CCNHP), New Mexico.

|  | 2000 | 2006 |
|---|---|---|
| **Fajada** | 2.01 (5.1) | 5.56 (14.1) |
| **Pueblo Bonito** | 1.62 (4.1) | 4.02 (10.2) |
| **Visitor Center** | 1.64 (4.2) | 5.97 (15.2) |

Table 5.1   Total precipitation from May through August 2000 and 2006  in inches (cm) at three gauging stations, CCNHP.

| Site Location | Current | Short-Term | Long-Term | Significance |
|---|---|---|---|---|
| **Clys Mesa-top** | 31 (2) | 33 (3) | 31(3) | ns |
| **East Canyon** | 62 (5) | 44 (5) | 32(5) | F= 7.5, p=.001 |
| **Fajada Gap** | 72 (3) | 77 (4) | 61 (5) | F=5.4, p=.006 |
| **Kin Klizhin** | 67 (5) | 49 (6) | 23 (4) | F=19.5, p<.001 |
| **Mockingbird Mesa-top** | 39 (3) | 24 (3) | 35 (4) | F=7.0, p=.001 |
| **North Side Canyons** | 62 (6) | 42 (7) | 40 (6) | F=3.7, p=.03 |

Table 5.2   Bare soil cover, mean % (SE) of bare soil at 6 study sites in 3 treatments (currently grazed, short-term protection, and long-term protection)  CCNHP, September 2006.

## RESULTS

### Precipitation and Floristic Trends in 2006

While the annual precipitation in 2006 was not appreciably different from 1999 and 2000, Chaco received far greater precipitation during the summer months of May-August (Table 5.1).

This promoted germination of 77 forb species that had not been present in our plots in the earlier study (91 species were recorded in 2000, 168 in 2006).

Vegetative recovery after protection from grazing can be assessed by monitoring residual bare soil (unvegetated) surfaces. In five of the six sites there was significantly greater cover of bare soils in currently grazed treatments (Table 5.2). In the one site, Clys Mesa-top, where no significant difference occurred, the primary cover was provided by Bouteloua gracilis and Gutierriza sarothrae, species known to be tolerant of grazing (and diversity was low at the grazed treatment).

### Shrub Density and Cover

At two sites we found significant differences in shrub density and cover (Table 5.3). At Fajada Gap (a grassland with sandy, often-shifting soils, on level terrain), there was a significant difference in shrub density (F=7.4, p<.008) and cover (F=8.0, p=001); the highest shrub density occurred in the currently grazed treatment. In contrast at Mockingbird Mesa-top (a low shrubland) there was a significant difference in shrub density (F=7.56, p=.005) and cover (F=4.89, p=.009); the greatest shrub density occurred in the long term treatment and the highest cover in the short-term treatment. At the other four sites we could detect no difference in shrub density. However, there was a significant difference in shrub cover at Clys Mesa-top (F=4.4, p<=.014) and Kin Klizhin sites (F=4.3, p=.016); in both sites, the greatest shrub cover occurred in long-term protected sites.

### Forb and Grass Cover

At Fajada Gap, there was a significant difference (F=3.8, p=.026) in forb cover between all treatments , with cover highest in the long-term treatment (Table 5.4). A similar trend occurred in grass cover (F=4.6, p=.012) at this site. At Mockingbird Mesa-top, forb cover was also significantly higher (F=3.6, p=.03), in the long-term protection treatment (Table 5.4). At this site, grass cover was also significantly different across treatments (F=4.14, p=.018) with cover highest in the short-term protection treatment. No differences in forb or grass cover were detected at the other four sites.

### Biological Soil Crusts

The cover of black, nitrogen fixing, biotic crusts differed across treatments in three of the six study sites (Table 5.5). At East Canyon, a significant difference in black crust cover occurred (F=12.9, p<.001) with crusts most heavily developed in short and long term treatment with no differences between these protected sites. At Fajada Gap, significant differences in black crust cover occurred (F=30.9, p<.001) with the highest cover in the short-term protected sites. At Kin Klizhin, a site with the most prolific biotic crust growth in Chaco Canyon NHP, significant differences occurred (F=31.4, p<.001) with the greatest crust cover in the long term protection treatment, where crust cover was six times higher than on the adjacent currently grazed treatment.

### Plant Species Richness

In all sites, considered collectively, there were greater numbers of plant species found in long-term protected treatments; overall, we recorded 124 species in currently grazed treatments, 123 species in short-term protected areas, and 168 species in long-term protected sites. At each of the six study sites in Chaco Canyon, we tallied the total number of plant species in three grazing treatments

| Site Location | Current | Short-Term | Long-Term | Significance |
|---|---|---|---|---|
| Shrub Density (#/300 m$^2$) | | | | |
| **Clys Mesa-top** | 91 (23.2) | 95 (45.6) | 151 (37) | ns |
| **East Canyon** | 48 (4.9) | 41 (4.2) | 44 (2.9) | ns |
| **Fajada Gap** | 121 (41.9) | 1 (43.9) | 12 (19) | F=7.4, p=.008 |
| **Kin Klizhin** | 82 (26.4) | 72 (24.0) | 58 (16.6) | ns |
| **Mockingbird Mesa-top** | 37 (15.2) | 37 (15.2) | 48 (15.7) | F=7.5, p=.005 |
| **North Side Canyons** | 51 (5.3) | 79 (17.9) | 62 (9.7) | ns |
| Shrub cover | | | | |
| **Clys Mesa-top** | 4.2 (.9) | 3.0 (.4) | 7.2 (1.2) | F=4.4, p=.014 |
| **East Canyon** | 24.8 (3.4) | 17.8 (2.0) | 21.0 (2.5) | ns |
| **Fajada Gap** | 7.6 (.9) | 0 | 15.2 (4.2) | F=8.0, p=.001 |
| **Kin Klizhin** | 3.4 (.6) | 5.9 (.6) | 6.8 (.8) | F=4.3, p=.016 |
| **Mockingbird Mesa-top** | 4.5 (.8) | 7.8 (1.2) | 4.7 (.6) | F=4.8, p=.009 |
| **North Side Canyons** | 7.7 (2.5) | 16.8 (2.3) | 19.6 (2.8) | F=2.9, p=.056 |

Table 5.3   Shrub density and shrub cover, mean % (SE) in 6 sites in 3 treatments (currently grazed, short and long term protection from grazing), CCNHP, September, 2006.

| Site Location | Current | Short-Term | Long-Term | Significance |
|---|---|---|---|---|
| **Clys Mesa-top** | 1.1 (.4) | .44 (.26) | .11 (.1) | ns |
| **East Canyon** | 0 | 27 (.18) | 1.3 (.7) | ns |
| **Fajada Gap** | 3.3 (1.4) | .67 (.3) | 4.5 (1.3) | F=3.8, p=.026 |
| **Kin Klizhin** | 1.7 (.7) | 2.1 (.9) | 3.4 (1.5) | ns |
| **Mockingbird Mesa-top** | .22 (.15) | .22 (.2) | 1.3 (.5) | F=3.6, p=.03 |
| **North Side Canyons** | 0 | 8.0 (3.0) | 5.3 (5) | ns |

Table 5.4   Forb cover, mean % (SE) in 6 sites in 3 treatments (currently grazed, short and long term protection from grazing), CCNHP, September, 2006.

| Site Location | Current | Short-Term | Long-Term | Significance |
|---|---|---|---|---|
| **East Canyon** | 0 | 5.7 (1.5) | 11.3 (1.6) | F=12.9, p<.001 |
| **Fajada Gap** | .16 (.1) | 10.5 (1.9) | 0 | F=30.9, p<.001 |
| **Kin Klizhin** | 4 (.9) | 8 (1.3) | 24 (2.5) | F=31.4, p<.001 |
| **North Side Canyons** | 4 (1.9) | .4 (.4) | 1.7 (.8) | ns |

Table 5.5   Biotic crust cover.  Mean % ( SE), shown for 4 of the 6 sites where crusts occur at CCNHP in 2006.

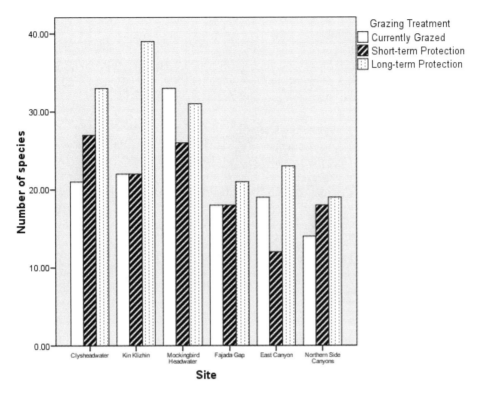

Figure 5.2    Plant species richness in three treatments at six sites, CCNHP.

(Figure 5.2). While the greatest number of species was found consistently in long-term protected sites, no significant differences were detected among the short-term and currently grazed treatments, suggesting that conditions suitable for establishment of many species (especially forbs in 2006) require decades of recovery after grazing disturbance ceases.

### Invasive plant species

We encountered only four invasive species in the study plots: the grasses *Bromus tectorum* and *Chloris varigatus*, and the forbs *Kochia scoparia*, *Halogeton glomeratus*, and *Salsola kali*. There were no detectable differences in the distribution of these species across the treatments. Nearly every treatment included the ubiquitous *Salsola kali* among its flora, and each

treatment at each site had present at least one invasive plant species.

### DISCUSSION

An earlier study examining trends in post-grazing recovery at Chaco Canyon National Historic Park (Floyd et al. 2003) found different trends in vegetation trajectories depending on biotic potential of the site-- a potential dictated by geology, soils, and other site conditions. Lacking a "one size fits all model," some sites returned to grasslands when grazers removed, while in others shrub density increased with release from grazing pressure. However, we were unable to investigate trends in forb populations because the original study took place during extreme drought (1999-2000).

Due to increased summer precipitation, forb germination in 2006 was prolific in all treatments relative to the previous study. We observed prolific growth of plant species which had been absent for many years, such as *Solanum jamesii*, a wild potato perhaps facilitated by prehistoric farmers in the region (Yarnell 1965). However, in only two of the four sites can we attribute increases in the cover of forbs to protection from grazing. Long-term protected grasslands at Fajada Gap and protected low shrublands at Mockingbird Mesa-top supported greater forb cover than at currently grazed sites. And, as in Floyd et al. 2003, the diversity of plant species was consistently greater in the long-term protected treatments at all six study areas, including an additional 44 plant species in the long-term protection treatments (124 species in currently grazed, 123 species in short-term protected, and 168 in long-term protected). Many of these additional species were annual forbs that were absent in the earlier study; in 1999/2000 Floyd et al. (2003) documented 91 species, while in 2006 we documented 168 species. Most of the new species were forbs whose appearance is tied to increased precipitation.

While we hypothesized that there would be a potential reduction in invasive species in long-term protected sites, this was not the case in 2006. There was no difference in invasive plant diversity or cover when we compared the currently grazed sites to those with long or short term protection.

One of the most startling effects documented by Floyd et al. (2003) attributed to removal of grazers was the re-establishment of black, potentially nitrogen-fixing, biotic crusts within 5-6 years of protection. This was especially pronounced on Menefee Shale substrates at the Kin Klizhin site. We continue to see this pattern in 2006 at East Canyon and Kin Klizhin locations where there was greater crust cover in protected (long and short term) than in currently grazed treatments but no significant difference among the crust cover in the two protection treatments. This suggests rapid recovery in the short-term, followed by stability of these crust populations thereafter. The importance of these microbial communities cannot be overstated (Belnap 1995, 1996, Belnap and Gillette 1997, 1998, Belnap and Lange 2003, Reynolds et al. 2001).

## SUMMARY

After 60 years, the legacy of grazing at Chaco Culture National Historic Park continues to influence patterns of plant distribution and abundance. However, there are signs of ecological recovery. While trends in shrub and perennial grass cover are similar to drought conditions in 1999-2000, we detected greater forb diversity and cover with the increased summer moisture in 2006. Another sign of recovery was that of black crust abundance and plant biodiversity, which were significantly greater in long-term protected treatments. While several species of invasive grasses and forbs occur in Chaco Canyon, their abundance and cover is similar in protected and currently grazed treatments. Long-term monitoring of vegetation communities at Chaco Canyon National Historic Park (effectively a 13,600 ha grazing exclosure) will contribute to our understanding of the influence of changing climate (drought and rising temperatures) in high desert shrubland and grasslands unhindered by the overlying influences of grazing disturbance.

## ACKNOWLEDGMENTS

This paper is based on work undertaken for the National Park Service. We are grateful to Dustin Hanna, Patrick Calloway, and the Prescott College students in "Field Biology Studies: Colorado Plateau" for their help with data collection, and other significant contributions to this work.

## LITERATURE CITED

Armour, C.L., D.A. Duff, and W. Elmore. 1991. The effects of livestock grazing on riparian and stream ecosystems. Fisheries 16: 7-11.

Belnap, J. 1995. Surface disturbances: their role in accelerating desertification. Environmental Monitoring and Assessment 37:39-57.

Belnap, J. 1996. Soil surface disturbances in cold deserts: effects on nitrogenase activity in cyanobacterial-lichen soil crusts. Biology and Fertility of Soils 23:362-367.

Belnap, J., and D. A. Gillette. 1997. Disturbance of biological soil crusts: impacts on potential wind erodibility of sandy desert soils in southeastern Utah. Land Degradation and Development 8:355-362.

Belnap, J., and D. A. Gillette. 1998. Vulnerability of desert biological soil crusts to wind erosion: the influences of crust development, soil texture and disturbance. Journal of Arid Environments 39:133-142.

Belnap, J., and O. L. Lange. 2003. Biological soil crusts: structure, function, and management. Ecological Studies Series, Springer-Verlag, Berlin.

Beymer, R J., and J. M. Klopatek. 1991. Potential contribution of carbon by microphytic crusts in pinyon-juniper woodlands. Arid Soil Research and Rehabilitation 5:187-198.

Bock, C.E., J. H. Bock, and H. M. Smith. 1993. Proposal for a system of federal livestock exclosures on public rangelands in the western United States. Conservation Biology 7: 731-733.

Breshears, D.D. and 12 co-authors. 2005. Regional vegetation die-off in response to global-change-type drought. Proceedings of the National Academy of Sciences 102: 15144-15148.

Brotherson, J. D., and S. R. Rushforth. 1983. Influence of cryptogamic crusts on moisture relationships of soils in Navajo National Monument, Arizona. Great Basin Naturalist 43:73-78.

CAST (Council for Agricultural Science and Technology ). 1974. Livestock grazing on federal lands in the 11 western states. Journal of Range Management 27: 174-181.

Crumpacker, D.W. 1984. Regional riparian research and a multi-university approach to the special problem of livestock grazing in the Rocky Mountains and Great Plains. Pages 413-422 in R.E. Warner and K.M. Hendrix, editors. California riparian systems: ecology, conservation, and productive management. University of California Press, Berkeley.

Evans, R. D., and J. Belnap. 1999. Long-term consequences of disturbance on nitrogen dynamics in an arid ecosystem. Ecology 80:150-160.

Evans, R. D., and J. R. Ehleringer. 1993. A break in the nitrogen cycle in aridlands? Evidence from N15 of soils. Oecologia 94:314-317.

Fleischner, T. L. 1994. Ecological costs of livestock grazing in western North America. Conservation Biology 8: 629-644.

Fleischner, T.L. 2010. Livestock grazing and wildlife conservation in the American West: historical, policy, and conservation biology perspectives. Pages 235-265 in J. DuToit, R. Kock, and J. Deutsch, eds. Wild Rangelands: Conserving Wildlife While Maintaining Livestock in Semi-Arid Ecosystems. Zoological Society of London/ Blackwell Publishing Ltd., Oxford, UK.

Floyd, M.L., T.L. Fleischner, D. Hanna, and P. Whitefield. 2003. Effects of historic livestock grazing on vegetation at Chaco Culture National Historic Park, New Mexico. Conservation Biology 17: 1703-1711.

Harper, K. T., and J. R. Marble. 1988. A role for nonvascular plants in management of arid  and semiarid rangeland. pp. 135-169 in P.T. Tueller editor. Vegetation science applications for rangeland analysis and management. Kluwer Academic Publishers, Dordrecht, the Netherlands.

IPCC (Intergovernmental Panel on Climate Change). 2007. Summary for policymakers.  In: S. Solomon and 7 co-editors, Climate change 2007: the physical science basis. Contribution of Working Group I to the Fourth Assessment Report of the Intergovernmental Panel on Climate

Change. Cambridge University Press, Cambridge, UK.

James, D.W., and J.J. Jurinak. 1978. Nitrogen fertilization of dominant plants in the northeastern Great Basin Desert. Pages 219-231 in N.E. West and J. Skujins, editors. Nitrogen in desert ecosystems. Dowden, Hutchinson, and Ross, Inc., Stroudsburg, Pennsylvania.

Johansen, J. R. 1993. Cryptogamic crusts of semiarid and arid lands of North America. Journal of Phycology 29:140-147.

Krueper, D., J. Bart, and T.D. Rich. 2003. Response of vegetation and breeding birds to the removal of cattle on the San Pedro River, Arizona (U.S.A.). Conservation Biology 17: 607-615.

Longhurst, W.M., R.E. Hafenfeld, and G.E. Connolly. 1982. Deer-livestock relationships in the western states. Pages 409-420 in L. Nelson, J.M. Peek, and P.D. Dalke, eds. Proceedings of the wildlife-livestock relationships symposium. Forest, Wildlife, and Range Experiment Station, University of Idaho, Moscow, Idaho.

Loope, W. L., and G. F. Gifford. 1972. Influence of a soil microfloral crust on select properties of soils under pinyon-juniper in southeastern Utah. Journal of Soil Water and Conservation 27:164-167.

Mueller-Dombois, D., and H. Ellenburg. 2003. Aims and methods in vegetation ecology. Blackwell Press. New York. 547 pp.

NPS (National Park Service). 1995. Resource management plan—Chaco Culture National Historic Park, Nageezi, New Mexico.

NPS (National Park Service). 1998. Chaco Culture National Historic Park—grazing history. Natural Resource file. Chaco Culture National Historic Park, Nageezi, New Mexico.

Reynolds, R., J. Belnap, M. Reheis, P. Lamothe, and F. Luiszer. 2001. Eolian dust in Colorado Plateau soils: nutrient inputs and recent change in source. Proceedings of the National Academy of Sciences 98:7123-7127.

Ricketts, T.H. et al. 1999. Terrestrial ecoregions of North America: a conservation assessment. Island Press, Washington, D.C.

Rychert, R.C., J. Skujins, D. Sorensen, and D. Porcella. 1978. Nitrogen fixation by lichens and free-living microorganisms in deserts. Pages 20-30 in N.E. West and J. Skujins, editors. Nitrogen in desert ecosystems. Dowden, Hutchinson, and Ross, Inc., Stroudsburg, Pennsylvania.

Stewart, G.. 1936. History of range use. Pages 119-133 in U.S. Forest Service. The Western range. 74th Congress, 2nd session, Senate Document 199.

Stoddart, L.A., and A.D. Smith. 1943. Range management. McGraw-Hill, New York.

Warren, S.D. 2003. Biological soil crusts and hydrology in North American deserts. Pages 327-337 in J. Belnap and O. Lange, eds. Biological soil crusts: structure, function, and management. Ecological Studies Series, Springer-Verlag, Berlin.

Yarnell. R.A. 1965. Implications of distinctive flora on Pueblo Ruins. American Antiquity 67(3): 662-674.

# THE EFFECTS OF SUDDEN ASPEN DECLINE ON AVIAN SPECIES COMPOSITION AND BIODIVERSITY IN SOUTHWESTERN COLORADO

*Sara P. Bombaci and Julie E. Korb*

## ABSTRACT

Aspen (*Populus tremuloides*) stands in southwestern Colorado have recently experienced sudden aspen decline (SAD), which is a unique form of large-scale aspen mortality associated with the rapid loss of entire aspen stands. Aspen forests are biologically diverse, and a comparatively high diversity and abundance of birds are associated with aspen habitat. Yet, studies have not yet evaluated avian community changes associated with SAD-affected aspen forests. Therefore, from early June to early July 2009, we conducted avian surveys and evaluated stand structure and forest understory in aspen stands located in the Dolores Ranger District of the San Juan National Forest. We classified different SAD levels that included: 1) low SAD (0-29%), 2) moderate SAD (30-70%), and 3) high SAD (71-100%). We used ordination analysis to compare avian species composition and abundance among different SAD levels, measured species richness and diversity, and performed an indicator species analysis to determine species that were particular indicators for different SAD levels. Patterns in the avian community produced good discrimination in ordination analysis between low and high SAD stands. Avian species richness was greater in stands with high SAD than in stands with low SAD, and the diversity indices were greater in stands with both moderate and high SAD than in low SAD stands. There was a greater number (4 of 5) of indicator species identified for the high SAD category. Our data suggests that the initial changes in avian community structure associated with SAD are distinct between aspen forest with low and high SAD, and that stands experiencing high SAD support avian biodiversity.

## INTRODUCTION

Aspen (*Populus tremuloides*) stands in southwestern Colorado have experienced the greatest occurrence of sudden aspen decline (SAD) in the state (almost 10 percent of aspen cover, Worrall et al. 2008). SAD is defined as a synchronous pattern of rapid aspen branch dieback and crown thinning associated with landscape-scale tree mortality (Worrall et al. 2008). SAD loss is exacerbated by an associated lack of regeneration beneath failing stands; whereas normal regeneration would usually serve to replace SAD-related losses (Worrall et al. 2008). Aspen regeneration typically occurs vegetatively after a disturbance, via a large interconnected root system that sends up root suckers (Schier et al. 1985). Although sexual reproduction can occur, specific combinations of conditions are necessary, and vegetative regeneration may be the only means of reproduction in regions with challenging environmental conditions (McDonough 1985). With a lack of suckering from healthy root systems to regenerate stands, forest cover will likely convert to other conifer dominated vegetation types, which will replace tree cover in areas where aspen stands have been eliminated by SAD (Shepperd et al. 2001).

Several causal factors have been identified with sudden aspen decline occurrence in this region. Given aspen's minimal drought tolerance, heat stress from recent drought conditions (2000-2005) is considered to be the initial destabilizing factor that left aspen stands susceptible to certain biotic agents which are not typical agents of aspen mortality (Worrall et al. 2008). Evidence for the climate-related influence lies in the fact that aspen mortality was found to be greater at lower elevations, on southern to western aspects, and in mature stands with lower densities, where existing heat levels are comparatively higher (Worrall et al. 2008). The secondary biotic agents identified as causal factors of the decline include aspen bark beetles (*Trypophloeus populi* and *Procryphalus mucronatus*), poplar borer beetles (*Saperda calcarata*), bronze poplar borer beetles (*Agrilus liragus*), and cytospora canker (*Valsa sordida*) (Worrall et al. 2008).

Aspen has the greatest distribution of any native tree in North America (Little 1971; Jones 1985). Within Colorado, there are more than one million acres of aspen forest (Jones 1985), much of which can be found on high mesas and on middle elevation mountain zones (Miller and Choate 1964); although it has been reported to exist in Colorado up to 3,650 m in elevation (Sudworth 1934). Aspen is commonly found amongst mixed conifer forest in the southern Rocky Mountains (Jones 1974) with a preference for cool, moist environments, especially northern aspects (Jones 1985). Due to these preferences, aspen is often limited to higher elevations in Colorado and other states at the southern end of its range (Jones 1985).

Aspen habitat is considered one of the most biologically diverse ecosystems in the west (Kay 1997). Others have found greater plant diversity in aspen than in surrounding habitats (Mueggler 1985; Chong et al. 2001) and this is due, in part, to the multilayer structure of aspen forests. This structure allows variable light to reach the undergrowth,

leading to increased understory diversity (Mueggler 1985), and can create conditions favorable to faunal diversity. Thus, aspen is a unique component of conifer-dominated ecosystems in the west because it provides a haven of plant and animal diversity that is not comparable in surrounding vegetation types. Consequently, research has shown significant biodiversity loss associated with conversion from aspen to other forest cover (Bartos and Campbell 1998a, b).

Studies on bird populations indicate a general larger species richness and abundance associated with aspen forests than compared with most other habitats in the intermountain west of North American (Salt 1957; Flack 1976; Winternitz 1980; Griffis-Kyle and Beier 2003), and studies have found that these diversity values are significantly higher in aspen than in neighboring conifer stands (Turchi et al. 1995; Richardson and Heath 2004). Data from the Breeding Bird Survey indicates that aspen habitat holds a species diversity ranking of eight out of 95 amongst the different habitats evaluated (Robbins et al. 1986). Griffis-Kyle and Beier (2005) also found that aspen habitat is particularly important to Neotropical migrants, especially stands with high densities of large aspen. Aspen habitat is also important to many birds in both isolated patches (Griffis-Kyle and Beier 2003) and in mixed conifer forest (Finch and Reynolds 1987). DeByle (1985) identified 134 species of birds that utilize aspen habitat, and aspen may provide critical habitat for red-naped sapsuckers (Sphyrapicus nuchalis), warbling vireos (Vireo gilvus) and MacGillivray's warblers (Oporornis tolmiei) (Reynolds and Finch 1988). Also, cavity nesting birds demonstrate a strong relationship with aspen stands (Pinkowski 1981; Daily et al. 1993; Dobkin et al. 1995; Martin et al. 2004), especially dead or dying aspen (Dobkin et al. 1995; Martin et al. 2004) and may be dependent upon this habitat (Pinkowski 1981). Verner (1988) has identified several advantages that aspen habitat provides to avian communities, including 1) a dense herbaceous understory that confers food and shelter benefits, 2) increased insect diversity and abundance, 3) the vulnerability of aspen to heart rot which cavity nesters favor, and 4) the tendency for aspen habitat to remain moist and provide water sources for both birds and insects.

Studies of avian species are particularly important when examining the impacts of SAD on ecosystems because these studies are cost-effective and because birds move easily amongst different habitat patches. Previous literature regarding SAD has either focused individually on the causes and distributions of sudden aspen decline or in general on the biodiversity associated with aspen ecosystems. Therefore, an assessment of changes in avian species composition and biodiversity associated with sudden aspen decline within the Colorado Plateau region will provide a valuable measure of preliminary avifaunal changes within this ecosystem and a basis for long term monitoring of SAD-related avian community changes.

The primary objective of our study was to determine the effects of sudden aspen decline on avian communities in southwestern Colorado. To accomplish this objective, our research had two main goals: 1) to provide a quantitative evaluation of the level of SAD that is occurring within the survey areas and an assessment of the associated vegetation structure and 2) to compare the vegetative data with avian survey data to provide a measure of initial avian changes associated with SAD within the region. We hypothesized that our results would indicate a decrease in avian biodiversity associated with SAD-affected stands when compared with healthy aspen stands, given the affinity of many avian species for aspen habitat.

# METHODS

## Study Sites and Selection Criteria

We conducted our study within the Mancos-Dolores Ranger District of the San Juan National Forest (SJNF), given that a larger percentage of aspen cover has been lost here due to SAD (~10%) than in any other location in Colorado (Worrall et al. 2008). This portion of the SJNF is located in the southwestern-most extent of the San Juan mountain range of southwest Colorado (Figure 6.1). It is bordered to the east by the La Plata Mountains and is topographically composed of foothills that transition into extensive mesas, which are then carved by river valleys. The combination of high mesas and middle-elevation zones provided by the foothills create conditions that allow for substantial aspen forest habitat (~53,000 ha). Aspen forest dominates an elevation band from approximately 2200 m to 3300 m, with site-specific deviation occurring due to topographic variability. The forest is predominantly managed by the SJNF, Dolores Public Lands Office, with exception of a few large private in-holdings. Some of the activities managed include grazing allotments, recreation, and timber harvesting. The previous 30 years of timber harvests have produced a patch mosaic of aspen forest habitat, with mature stands bordered by immature regenerating stands or open meadows.

We employed a stratified random sampling design. We initially selected stands composed of >95% aspen overstory to maintain homogeneity among stands. We also controlled for slope, aspect, and elevation (2600 m to 3000 m range) during stand selection. Stands were defined as any area of contiguous aspen forest with relative similarities in tree species composition and height. A total of 60 sample points were randomly located within these stands to establish a plot center for both the avian and the vegetation circle plots. Randomization

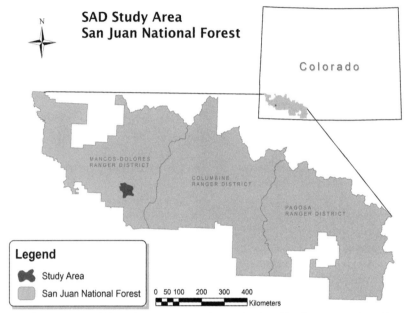

Figure 6.1  Study location within the Mancos-Dolores Ranger District of the San Juan National Forest, Montezuma County, Colorado.

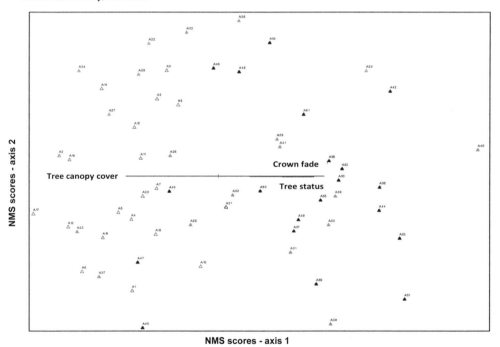

Figure 6.2 NMS ordination of avian species composition and abundance data for each of the three different SAD levels: low (△), moderate (▲), and high (▲). Each symbol represents one plot (n = 60). Points closer together indicate greater similarity in avian species composition and abundance. The diagram was rotated in the direction of the SAD gradient; thus vector lengths are proportional to significant correlations between vegetation variables and percentage of SAD. The final solution had two dimensions, final stress = 25.0231.

of plot centers was achieved using ArcGIS software (ArcGIS: Release 9.3. Redlands, California: Environmental Systems Research Institute, 1999-2009) and Hawth's Analysis Tools (Beyer 2004) to randomly locate an initial plot within each stand from which a transect was paced out and plot centers were located every 200 m along the transect. Transects were terminated at distances ≥ 50 m from stand perimeters. The sample point locations were classified based on the percentage of SAD in each plot (stems with SAD/total stems in plot) into one of 3 categories: 1) low SAD (0-29%), 2) moderate SAD (30-70%), and 3) high SAD (71-100%). There were 20 sample points for each category.

Vegetation Survey Protocol

At the center of each sample point, two circular plots were established to record vegetation data, based on plot sizes used by the San Juan National Forest (outer plot diameter = 16 m, inner plot diameter = 7.2 m). This protocol was used to maintain consistency between monitoring strategies and to allow for data compatibility, since the SJNF is performing monitoring as part of restoration treatments (coppice harvesting and prescribed fire treatments) that will be occurring in the vicinity (Aspen Forest Health and Restoration Project, Dolores Public Lands Office, Montezuma County, Colorado).

Within the 16 m diameter circular plot we gathered data on all trees ≥ breast height (1.37 m) and with a diameter at breast height (dbh) ≥ 5cm, including: 1) species, 2) tree condition (evaluated based on the US Forest Service tree status scale {1 = live, 2 = declining, 3 = recent snag, 4 = loose bark snag, 5 = clean snag, 6 = broken above breast height, 7 = broken below breast height}), 3) dbh in centimeters (measured for all tree species present, stands >99% aspen), 4) tree canopy height in meters (measured with a hypsometer), 5) crown fade (assessed by ocular estimation of the percentage of crown loss in an individual tree, which was categorized into a 1-9 scale (0 = 0-9% fade,…9 =90-99% fade), 6) SAD agent presence (0 = not present, 999= present, codes used by USFS for SAD monitoring), and 7) type of SAD agent when present (agents recorded: aspen bark beetles, poplar borer beetles, bronze poplar borer beetles and Cytospora canker). We used a 50-m transect intersecting the circle plot centers (transects placed parallel to the environmental gradient, with 0-m at the top of the slope, 25-m at plot center, and 50-m at the slope base) to record canopy cover (0 = no canopy, 1= canopy) with a mirror densitometer at 16 points every 3 meters along the transect. The overstory data collected were also used to calculate percentage of SAD (stems with SAD/total stems in plot), stand density (total stems/ha), and stand basal area (based on dbh) for each plot.

In the smaller 7.2-m diameter circular plot, aspen regeneration was tallied (and converted to regeneration/ha), and in a northeast quadrant of this circle plot, shrub understory was identified and tallied (and converted to stems/ha) for each species. Since 95 percent of the shrub understory was snowberry (*Symphoricarpos rotundifolia*), understory shrubs were divided into two categories (snowberry and other shrubs) for analysis. The northeast subplot was also used to estimate ground cover using the Braun-Blanquet scale (Braun-Blanquet 1932). Additionally, at the center point of each circular plot, aspect, slope (measured with a hypsometer), and elevation (measured with a Garmin 360 SCX GPS unit, Garmin International, Inc., Olathe, KS, USA) were recorded. The GPS unit was also used to ascertain the coordinates of the plot center for future monitoring. Plot centers were marked with flagging tape for re-measurements.

## Avian survey protocol

We conducted avian censuses during the breeding season from 05 June to 09 July, 2009. Surveys were conducted from approximately 0600 until 1100. Two avian surveys were completed for each sample point location, including the initial surveys beginning in early June, and a second set of surveys beginning two weeks later. This spacing ensured that we captured early season and late season breeders. Data from these separate counts were then averaged for analyses. To prevent biases that could occur due to time-of-day bird activity levels, locations that were initially surveyed in the early half of the morning (0600 to 0830) were randomly chosen to be surveyed the second time during the latter half of the morning (0830 to 1100) and vice versa.

Surveys were conducted at each of the 60 sample point locations using a modified version of the standard point count method (Ralph et al. 1993). From a fixed center position, five minute interval counts were used to record all species detected visually or aurally, except when it was evident that an individual was using the habitat for hunting only (e.g. flyovers). Only unique individual detections were recorded, and detection distance was also estimated and plotted on data maps for each observation to aid in awareness of previous detections. Point counts were initiated immediately upon arrival, and any species that flushed upon approach were included in the point count data if their location of first detection fell within the plot boundary. Observations were all made by the same observer (SPB) in 100 m fixed-radius plots. Given the patchy distribution of continuous aspen forest habitat, we were not able to maintain the recommended minimum distance in wooded areas of 250 m (Ralph, et al. 1993) between sample points for every survey. However, a 200 m minimum distance should be sufficient, given that 98 percent of our detections occurred at distances less than 75m, and 92 percent occurred at distances less than 50 m. Also, due to the discontinuous aspen forest habitat of the region, we were limited to a 50 m minimum distance to stand edges in some plots. To control for species using external habitat, detections that were known to originate outside stand perimeters were omitted in the cases when our 100 m radius plot extended beyond stand edges. During periods of rain, fog, or high winds (> 13 km/hr), surveys were not conducted. However, cloud cover and temperature were not considered as survey restrictions, given that temperatures during our study were relatively consistent and that cloud cover conditions were continually variable in this mountainous terrain.

Point counts are the most efficient and favorable method for forest avian surveys, yet they can misrepresent quiet, loud, nocturnal, or flocking species (Ralph, et al. 1993). However, we assumed equal detection of these species for the following reasons: 1) flocking species were not detected during any of our surveys, 2) ninety-two percent of detections occurred at distances of 50 m and closer, which would increase detectability of quiet species and allow for better discrimination of louder species, and 3) we are not generalizing our results to include nocturnal species.

## Statistical Analysis

Vegetation data from individual plots were averaged for analysis among the three SAD categories (n = 20). Mean vegetative structure changes among the different categories of SAD were analyzed using multiple Kruskal-Wallis one-way analysis of variance (ANOVA) tests. A Bonferroni post-hoc pairwise comparison of the different SAD categories was also used to assess whether vegetation variables were significantly different across all SAD categories. We also used nonmetric multidimensional scaling (NMS) ordination analysis to demonstrate the relationship between percentage of

SAD and other vegetative variables, and calculated a Pearson's correlation for each vegetative variable using percentage of SAD as our main axis (PC-ORD software version 5.10, McCune and Mefford 2006). Vectors were graphed for significant vegetation variables, with vector lengths proportional to correlations with percentage of SAD, and with vector direction representing the relationship between the variable and the percentage of SAD.

We examined differences in avian distributions between different SAD levels using a Permutation-based nonparametric multivariate analysis of variance (PerMANOVA) (Anderson 2001, McArdle and Anderson 2001). PerMANOVA uses common ecological distance measures (Bray–Curtis for this study) to examine multivariate datasets and calculates *P-values* using permutations, rather than tabled *P-values* that assume normality. We used PerMANOVA to quantify differences in avian composition and abundance amongst different SAD levels. We used a one way design with the level of SAD as our main effect (PC-ORD software version 5.10, McCune and Mefford 2006). Only species that were present in > 5% of the plots were analyzed, as recommended by McCune and Grace (2002), since ordination analyses are very sensitive to infrequently detected species.

To evaluate different biodiversity measures across SAD categories, we calculated two diversity indices, the Shannon-Weaver Diversity Index (calculated as $e^H$) and the Simpson's Index of Diversity, and determined species richness values for each plot. These values were averaged amongst the three SAD categories for analysis (n = 20). To analyze mean differences in species richness and the two diversity indices between the different categories of SAD, we used Kruskal-Wallis one-way ANOVA tests, followed by Bonferroni post-hoc comparison tests.

To examine species-specific changes associated with SAD, we performed a MONTE CARLO test of significance of observed maximum indicator values for each species detected (McCune and Grace 2002), which uses indicator values based on abundance and frequency to identify species that are particularly consistent indicators for different SAD levels. Indicator values were compared to random trials (1000 Monte Carlo randomizations) to determine *P-values* for each species (McCune and Grace 2002). Indicator species were defined as species where $P \leq 0.05$ and indicator value > 25 (INDVAL = relative abundance x relative frequency, INDVAL range = 0 to 100) (Dufrene and Legendre 1997).

## RESULTS

### Vegetation

We found significant differences in several vegetation and site variables (10 of 18) amongst the different SAD levels (Table 6.1). Canopy cover, crown fade, tree status and stand density were highly significantly different between all SAD categories and had similar *P-values* (Table 6.1), suggesting autocorrelation between these variables. Canopy cover decreased, and the related variable of crown fade increased, as SAD level increased. Tree status (more advanced snag condition) increased with SAD level increases (Table 6.1). Stand density and basal area were lower in areas of high SAD and significantly different between low to moderate SAD and low to high SAD categories (Table 6.1). DBH differed significantly between low and high SAD categories and was greater in high SAD stands (Table 6.1). Aspen regeneration, woody groundcover and graminoid cover were all greater in high SAD plots and differed significantly between low and high SAD (Table 6.1). Other ground cover types (shrubs, forbs, rock, bare ground, and litter) did not demonstrate significant differences

| Vegetation variable | Low SAD | Moderate SAD | High SAD | H | P |
|---|---|---|---|---|---|
| % Canopy cover | 81.54±2.00[a] | 59.54±5.00[b] | 33.10±4.00[c] | 35.34 | <0.000 |
| Crown fade | 1.53±0.18[a] | 4.26±0.16[b] | 7.42±0.35[c] | 50.31 | <0.000 |
| Tree status | 1.46±0.05[a] | 2.43±0.08[b] | 3.51±0.08[c] | 50.95 | <0.000 |
| Basal area (m²/ha) | 54.01±5.05[a] | 39.35±2.32[b] | 41.32±3.15[b] | 5.84 | 0.052 |
| Stand density (stems/ha) | 1887.26±136.69[a] | 1248.27±130.59[b] | 1069.94±84.91[b] | 18.85 | <0.000 |
| Tree height (meters) | 11.96±0.52[a] | 11.78±0.50[a] | 13.45±0.80[a] | 5.14 | 0.077 |
| DBH (cm) | 17.19±0.97[a] | 18.55±0.93[ab] | 21.74±0.97[b] | 10.17 | 0.006 |
| Regeneration (stems/ha) | 1087.27±408.17[a] | 2149.82±717.55[ab] | 4484.97±1238.20[b] | 9.56 | 0.008 |
| Shrubs/ha (snowberry) | 13836.50±2195.69[a] | 16889.39±3168.67[a] | 17726.48±3373.85[a] | 0.86 | 0.650 |
| Shrubs/ha (other) | 295.44±295.44[a] | 570.14±393.60[a] | 725.65±375.36[a] | 1.98 | 0.373 |
| % Graminoid cover | 11.72±3.20[a] | 24.675±6.99[ab] | 32.17±5.53[b] | 8.36 | 0.015 |
| % Forb cover | 31.375±4.85[a] | 28.125±5.36[a] | 22.72±3.10[a] | 2.17 | 0.339 |
| % Woody debris cover | 2.025±0.76[a] | 6.925±2.52[b] | 10.17±1.50[b] | 16.67 | <0.000 |
| % Rock cover | 0.001±0.00[a] | 0.25±0.01[a] | 0.13±0.01[a] | 2.05 | 0.358 |
| % Bare ground cover | 0.75±0.01[a] | 2.05±0.01[a] | 4.18±0.01[a] | 1.41 | 0.495 |
| % Litter cover | 4.68±0.21[a] | 4.05±0.09[a] | 3.08±0.07[a] | 0.02 | 0.991 |
| Aspect | 258.60±8.98[a] | 207.25±19.52[a] | 225.00±19.79[a] | 1.90 | 0.388 |
| Slope | 4.95±0.61[ab] | 5.62±0.89[a] | 3.02±0.37[b] | 10.50 | 0.005 |

Table 6.1   Mean (± SE) differences in vegetation variables between low, moderate, and high SAD categories. Categories with the same letters are not significantly different at p ≤ 0.05 by Bonferroni pairwise comparison tests. *P*-values determined using a non-parametric Kruskal Wallis test, *n* =20. Tree status evaluated based on the US Forest Service tree status scale (1 = live, 2 = declining, 3 = recent snag, 4 = loose bark snag, 5 = clean snag, 6 = broken above breast height, 7 = broken below breast height). Crown fade assessed by ocular estimation of the percentage of crown loss in an individual tree, which was categorized into a 1-9 scale (0 = 0-9% fade...9 =90-99% fade).

| SAD Level | Species | Common Name | Indicator Value | P |
|-----------|---------|-------------|-----------------|---|
| High | *Colaptes auratus* | Northern Flicker | 45.6 | 0.0012 |
| High | *Contopus sordidulus* | Western wood peewee | 40.2 | 0.0090 |
| Low | *Vireo gilvus* | Warbling Vireo | 41.9 | 0.0004 |
| High | *Tachycineta thalassina* | Violet-green swallow | 49.3 | 0.0004 |
| High | *Troglodytes aedon* | House Wren | 42.6 | 0.0190 |

Table 6.2  Indicator species associated with different SAD levels in the San Juan National Forest.  Indicator species analyzed with a Monte Carlo test of significance of observed indicator values, which identifies species that are consistent indicators for different SAD levels (Indicator value = species abundance x species frequency).

between different SAD levels (Table 6.1). Slope differed significantly between moderate and high SAD categories (Table 6.1). This was an expected outcome because one plot in the moderate SAD category had a relatively high gradient (21%), which may have skewed the results for slope. However, all other slopes had a grade less than 10 percent, which should have minimal impacts on avian distributions. Other vegetation and site characteristics (tree height and aspect) were not significantly different between different categories of SAD (Table 6.1). Relationships between percentage of SAD and vegetation variables are also presented in the ordination graph (Figure 6.2). The ordination diagram was rotated so that the horizontal axis points in the direction of the SAD gradient. Percentage of SAD showed strong positive correlations with crown fade and tree status and strong negative correlations with tree canopy cover.

Avifauna

We detected a total of 24 species between all categories that had observations >5 percent (Table 6.2, Appendix 6.1). Of these, five species abundances' differed significantly between SAD categories and were determined to be indicator species for particular categories of SAD: the warbling vireo, northern flicker (*Colaptes auratus*), violet-green swallow (*Tachycineta thalassina*), house wren (*Troglodytes aedon*), and western-wood peewee (*Contopus sordidulus*) (Table 6.2). The warbling vireo had a higher indicator value in low SAD. The northern flicker, violet-green swallow, house wren, and western-wood peewee all had higher indicator values in high SAD. The purple martin (*Progne subis*) was the only other species in our dataset that showed near-significant results using indicator species analysis (Appendix 1). This species was expected to show results similar to the violet-green swallow, but it had low statistical power due to minimal observations during surveys.

Our ordination results indicated significant differences in avian species composition and abundance between the different SAD categories (Figure 6.2, $n = 20$, F = 5.1490, $P = 0.0002$, PerMANOVA). Pairwise comparisons (Bonferroni adjusted) revealed that significant differences in avian distributions occurred between low to moderate SAD ($n$ = 20, t = 2.2047, $P$ = 0.0002) and low to high SAD stands ($n$ = 20, t = 3.0712, $P$ = 0.0002), and not between moderate to high SAD stands ($n$ = 20, t = 1.3647, $P$ = 0.0708). In NMS ordination space, high SAD sample points plotted closer together, indicating greater similarity in avian species composition and abundance amongst high SAD stands. Low SAD sample points were also more spatially related due to greater likenesses in avian community characteristics, and were largely separately located on the diagram from the high SAD stands, indicating differences in avian communities between low and high SAD stands.

Species richness differed significantly between SAD categories ($n$ = 20, H = 12.1270, $P$ = 0.0020, Kruskal-Wallis test), with greater species richness occurring in stands with high SAD (Figure 6.3A). Bonferroni pairwise comparisons revealed that the differences were significant between low and high SAD levels ($P$ =0.0010). The two diversity indices (Shannon-Weaver Diversity Index {$e^H$} and Simpson's Index of Diversity {D}) were also significantly different between SAD levels ($e^H$: n = 20, H = 11.2130, $P$ = 0.0040, Kruskal-Wallis test,

D: $n$ = 20, H = 9.210, $P$ = 0.0100, Kruskal-Wallis test), and stands with higher SAD had higher diversity values (Figures 6.3B, 6.3C). Significant differences for these indices were detected between low to high SAD ($e^H$: P = 0.0020, D: $P$ = 0.0060) and low to moderate SAD levels ($e^H$: P = 0.0190, D: $P$ = 0.0050) by Bonferroni pairwise tests (Figure 6.3B, Figure 6.3C).

## DISCUSSION

Differences were detected in both overstory and understory vegetation structure among low, moderate and high SAD stands. Many of the changes observed are predictable responses to an increase in aspen mortality, found in the high SAD stands. For example, overstory increases in crown fade and tree status (snag level), and decreases in percent canopy cover are intuitive because these characteristics correspond with an increase in dead and dying aspen. The greater average DBH values and the observed lower stand density and basal area in high SAD stands are consistent with results presented by Worrall et al. (2008). Their results indicated that larger trees were more prone to SAD events perhaps due to decreased stress tolerance associated with maturity. Their study also found that stands with lower densities were

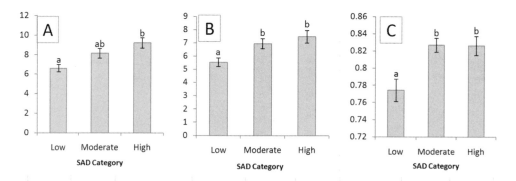

Figure 6.3 Mean species richness (A), mean Shannon-Weaver Diversity Index ($e^H$) (B), and mean Simpson's Index of Diversity (C) at different SAD levels, determined from five minute point counts conducted on the San Juan National Forest in southwest Colorado. Error is presented as ± SE. Bars with the same letters are not significant at $P \leq 0.05$ by Bonferroni pairwise comparison tests. $P$-values determined using a non-parametric Kruskal Wallis test, $n$ = 20.

more susceptible to SAD occurrences due to the increased propensity for heat stress to infiltrate more open stand structures. The open structure found in high SAD stands in our study was further intensified by an increase in felled snags (tree status in our study), which amplified the woody debris in the understory. Consequently, our results did indicate a higher percentage of woody debris groundcover in high SAD stands, when compared with low SAD stands.

Other understory differences detected between low and high SAD stands included an increase in aspen regeneration and percent graminoid cover in high SAD stands. These changes would be expected because decreases in overstory canopy cover will allow greater light infiltration in high SAD stands, resulting in more favorable conditions for graminoid growth and aspen regeneration. It is important to note that although we did see increases in aspen regeneration in stands with high SAD, the mean regeneration (4484.97 stems/ha) was lower than regeneration observed with other disturbances in the area. Shepperd (2004) found sucker densities within mixed aspen/conifer stands to be approximately 7,500 stems/ha in the first year after burn treatments in the San Juan National Forest in 1998.

Contrary to our hypothesis, avian biodiversity was greater in areas that have experienced high SAD, as indicated by associated higher mean species richness and diversity values. The relative similarity between the Shannon Index mean values and the mean species richness values also indicated that species were fairly evenly distributed across SAD levels.

The significantly greater biodiversity values found in high SAD stands may suggest that certain avian species are responding to the newly-available niches provided by changes in vegetation structure associated with SAD. The forest structure in SAD-affected stands appears to be more diverse, attracting a variety of species that have habitat requirements that are not equally available in the largely unaffected stands. Similar results were found in studies on shelterwood cuts, which promote forest structure diversity through harvesting methods that imitate natural disturbances, where avian diversity and abundance values were greater in shelterwood forests than in unmanaged forests (King and DeGraaf 2000; Goodale, et al. 2009). Although this method may not create the same conditions as SAD, it demonstrates that processes which promote diversity in forest structure are often associated with greater avian biodiversity.

One of the vegetative characteristics of high SAD stands that may be exerting a strong influence on avian distributions, especially on cavity nesting species, is the increase in available snags. Cavity nesters typically utilize snags for nest sites, and studies have indicated a strong association between cavity nesters and snags (Scott 1979; Raphael and White 1984; Zarnowitz and Manuwal 1985; Hollenbeck and Ripple 2006). Hollenbeck and Ripple (2006) found that six of 11 cavity nesting species were more abundant in areas of dead or dying aspen than in the healthy stands that occurred in their study area. Zarnowitz and Manuwal (1985) found that species richness, density, and diversity values for cavity nesting birds were greater in areas with higher concentrations of snags. Given the affinity of cavity nesting birds for snags, cavity nester population responses strongly influenced the species richness and diversity measures in our results. Indeed, three of the four species identified as indicator species for high SAD levels were primary or secondary cavity nesters—the northern flicker, the violet-green swallow, and the house wren. Since indicator analyses are based on abundance and frequency data for a particular species, our results suggest that these species' responses to snags and to other conditions associated with SAD may be playing a critical role in the observed differences in biodiversity between low and high SAD stands. However, for some cavity-nesting species, the number of available snags is not the only important characteristic to consider. The size of the snag may also play a critical role for woodpeckers according to Flack (1976), who found that woodpeckers primarily utilize snags with a DBH > 15 cm. Since the mean DBH in our stands for the low, moderate, and high SAD categories was > 15 cm, the snags identified in our study are likely to be suitable for cavity nesters. Future studies should evaluate nest hole frequency among the different SAD levels to confirm that the snags identified in our study are being utilized by cavity-nesting birds.

The house wren may not only be influenced by increases in snags. A study by Daily et al. (1993) found that house wren abundances increased with increases in the number of downed trees. Therefore, this species may also be responding to an increase in the percentage of woody debris found in the understory of high SAD stands.

Another important vegetative characteristic identified in high SAD stands was the associated decreased stand density and basal area. These measures imply a more open forest structure. This open habitat may particularly be influencing western-wood peewee populations, which is the other indicator species identified for the high SAD category. Western wood-peewees prefer open forests and woodlands, and previous studies on burn treatments and timber harvests indicate an increase in this species abundance associated with these types of disturbances (Garrison et al. 2005; Kirkpatrick et al. 2006). SAD typically affects areas with decreased stand density and basal area and can amplify this open structure by an increase in the number of felled trees. SAD may be a new type of disturbance that is generating a temporary new niche of open forests in the region that this species is currently exploiting. The

open conditions may also be influencing the swallow populations in our study. Previous work by Rendell and Robertson (1989) indicates that swallows frequently choose cavity nest sites in snags that are in open stands. Consequently, the violet-green swallows and purple martins (Pronge subis) detected in our study may also be responding to this vegetative condition.

Another forest structure difference detected between varying SAD levels that may be affecting avian populations was the decreased canopy cover in high SAD stands and the associated increase in crown fade. This loss of canopy may particularly impact warbling vireo populations. The warbling vireo was the only indicator for the low SAD category, as it was the only species that responded with a significant decrease in abundance as SAD increased. Warbling vireos are quite common in aspen habitat throughout western North America (Flack 1976) as well as in habitat provided by other deciduous trees. They commonly favor nest sites with canopy cover, which is believed to provide thermoregulation and protection from heat stress (Walsberg 1981). The leaf-fluttering associated with aspen is thought to provide evenness in the distribution of light below the canopy, creating greater temperature uniformity (Roden and Pearcy 1993). Therefore, the decreased canopy cover associated with areas of high SAD likely explains the apparent decrease in abundance of this species. Matson (2000) suggests that the warbling vireo may be used as an indicator species of aspen habitats, given their strong association with aspen.

Several avian species identified in this study did not seem to respond to SAD. These species may not have been greatly affected by the changes in vegetation structure found with the occurrence of SAD. More importantly, we had comparatively smaller sample sizes for many of these species which resulted in a relatively low statistical power to detect significant differences in abundance between the different SAD levels. Of particular interest is the purple martin, which was expected to show similar results to the violet-green swallow. However, the low statistical power for this species may not have allowed this pattern to emerge. The purple martin did show near-significant results and a trend similar to the violet-green swallow, indicating that it may be significantly responding to high SAD with associated greater abundances, which would become more evident with a larger sample size. This species is particularly important to mention, given that it is a species being monitored on the SJNF.

Our NMS analysis demonstrated good discrimination of differences in avian composition and abundance among different SAD levels. Patterns in the avian community produced better discrimination between low and high SAD stands, which were clearly separated by the ordination, while discrimination of avian communities in the moderate SAD stands was less apparent. This pattern was also reflected in our indicator analysis which distinguished indicator species for low and high SAD, but none were determined for moderate SAD. The species composition of low SAD stands was influenced by canopy-dependant species, namely the warbling vireo; whereas the composition of high SAD was largely determined by cavity-nesting species. Pairwise comparisons confirmed significant differences in avian community composition between low SAD and moderate SAD as well as low SAD and high SAD but not between moderate SAD and high SAD. This same pattern was detected in the results of our analyses on species diversity. This may imply that the forest structure characteristics which avian species are responding to in high SAD stands are more similar to the characteristics of moderate SAD stands, but that both have distinct forest structure characteristics from low SAD stands.

## Management Implications

Our study has identified that vegetative conditions associated with SAD events are favorable to certain species and may result in temporary changes in distributions for avian communities utilizing aspen forests in the San Juan National Forest. Amongst these changes are increases in biodiversity associated with areas that have experienced SAD. Since this increased diversity can be related to changes in forest structure associated with SAD, it is important for forest managers that are interested in sustaining this new habitat to maintain SAD-impacted stands within the district. This could be managed through progressive timber harvest treatments that aim to restore aspen regeneration, while leaving some standing snags on site for use by cavity nesting species. For example, shelterwood cutting may be a useful method for regenerating aspen stands while promoting both vegetative and avian diversity (King and DeGraaf 2000; Goodale et al. 2009). Northern flickers have been found to utilize regenerating aspen habitat that had lingering standing snags (Kirk et al. 1996), and this pattern may be reflected by other cavity nesting species.

Continued monitoring may be necessary for certain species identified in our study as indicator species for different SAD levels. Since warbling vireos seem particularly sensitive to increases in SAD, this population may be important to monitor in the SJNF, given the progression of SAD events.

Caution should be used when generalizing our results, especially to a landscape scale. Since all data were generated and analyzed at a site scale, landscape level changes were not addressed. Also, these results cannot be applied to nocturnal species of the region, since night surveys were not conducted. In addition, it is important to note that the list of indicator species identified in this study may not be an exhaustive list of species associated with changes in SAD. There is reason to believe that additional indicator

| Species | Common Name | P |
|---|---|---|
| *Selasphorus platycercus* | Broad-tailed hummingbird | 0.5529 |
| *Sphyrapicus nuchalis* | Red-naped sapsucker | 0.3185 |
| *Picoides pubescens* | Downy woodpecker | 0.3675 |
| *Picoides villosus* | Hairy woodpecker | 0.3045 |
| *Corvus brachyrhynchos* | American crow | 0.4777 |
| *Progne subis* | Purple martin | 0.0816 |
| *Poecile gambeli* | Mountain chickadee | 0.7786 |
| *Sialia mexicana* | Western bluebird | 0.7614 |
| *Sialia currucoides* | Mountain bluebird | 0.3243 |
| *Turdus migratorius* | American robin | 0.3697 |
| *Catharus guttatus* | Hermit thrush | 0.2893 |
| *Vermivora celata* | Orange-crowned warbler | 0.2621 |
| *Dendroica coronata* | Yellow-rumped warbler | 0.4147 |
| *Dendroica petechia* | Yellow warbler | 0.1696 |
| *Wilsonia pusilla* | Wilson's warbler | 0.3237 |
| *Piranga ludoviciana* | Western tanager | 1.0000 |
| *Melospiza lincolnii* | Lincon's sparrow | 0.3821 |
| *Junco hyemalis* | Dark-eyed junco | 0.8764 |
| *Carduelis pinus* | Pine siskin | 0.3561 |

Appendix 6.1
Species detected during point counts in the San Juan National Forest that were not identified as indicator species using a MONTE CARLO test of significance of observed indicator values (Indicator value = avian abundance x avian frequency).

species may be identified, provided a sufficient number of detections would lead to greater statistical power for these species. This is probably especially true for other cavity nesting species that were detected in our study, but had low statistical power due to minimal detections.

## CONCLUSION

Our study determined differences in avian species composition, abundance and biodiversity associated with sudden aspen decline in southwest Colorado. Our research may provide insight on responses of forest bird communities to future SAD disturbances. Climate models anticipate higher temperatures and increasing aridity in western North America; thus, it is possible that drought-induced aspen forest mortality will occur more frequently under future climate change scenarios. Therefore, our study may help wildlife managers anticipate future changes in avian community distribution patterns associated with SAD. However, it is important to consider that this is an initial measure of avian changes associated with this ephemeral event. The vegetative conditions that contributed to increased avian diversity in the high SAD stands in our study will not remain in stasis. These stands will likely convert to conifer-dominated forests through natural successional processes, given a lack of aspen regeneration (Crawford et al. 1998, Rogers 2002). These conifer-dominant forests will favor different avian communities with biodiversity levels that are distinct from current conditions. Future studies should investigate mixed aspen-conifer stands to provide land managers with an idea of how avian species might respond to the holistic changes that will likely occur as stands progress from aspen to mixed conifer, adding long-term value to this initial measure of avian changes associated with SAD.

## ACKNOWLEDGEMENTS

Funding for this project was provided by a Natural and Behavioral Sciences Grant, Fort Lewis College. We thank the San Juan National Forest, Dolores Public Lands Office for providing maps, data, and background information on the San Juan National Forest. We also thank Joseph Ortega and Erin Lehmer for their contributions of time and knowledge to this project.

## LITERATURE CITED

Anderson, M. J. 2001. A new method for non-parametric multivariate analysis of variance. Austral Ecology 26: 32–46.

Bartos, D. L., and R. B.Campbell, Jr. 1998a. Decline of quaking aspen in the Interior West: examples from Utah. Rangelands 20: 17–24.

Bartos, D. L., and R. B.Campbell, Jr. 1998b. Water depletion and other ecosystem values forfeited when conifer forests displace aspen communities. in Proceedings of AWRA specialty conference, rangeland management and water resources. (D. F. Potts, ed.) TPS-98-1: 427–434. American Water Resources Association, Herndon, VA.

Beyer, H. L. 2004. Hawth's Analysis Tools for ArcGIS. Available at http://www.spatialecology.com/htools.

Braun-Blanquet, J. 1932. Plant sociology. Translated by G.D. Fuller and H.S. Conrad. McGraw-Hill Book Co., Inc. New York and London. 439 pp.

Crawford, J. L., S. P. McNulty, J. B. Sowell, and M. D. Morgan. 1998. Changes in aspen communities over 30 years in Gunnison County, Colorado. American Midland Naturalist 140: 197-205.

Chong, G. W., S. E. Simonson, T. J. Stohlgren, and M. A. Kalkhan. 2001 Biodiversity: aspen stands have the lead, but will nonnative species take over? in: Sustaining aspen in western landscapes. (W. D. Shepperd, D. Binkley, D. L. Bartos, T. J. Stohlgren, and L. G. Eskew, comps.) 2000 June 13-15. Grand

Junction, CO. RMRS-P-18: 261-271. Rocky Mountain Forest and Range Experiment Station, USDA Forest Service, Fort Collins, Colorado.

Daily G. C., P. R. Ehrlich, and N. M. Haddad. 1993. Double keystone bird in a keystone species complex. Proceedings of the National Academy of Science 90: 592-594.

DeByle, N.V. 1985. Wildlife. in Aspen: ecology and management in the western United States. (N.V. DeByle and R.P. Winokur, Eds.). USDA Forest Service General Technical Report RM-119: 135-152. Rocky Mountain Forest and Range Experiment Station, USDA Forest Service, Fort Collins, Colorado.

Dobkin, D.S., A.C. Rich, J.A. Pretare, and W.H. Pyle. 1995. Nest-site relationships among cavity-nesting birds of riparian and snowpocket aspen woodlands in the northwestern Great Basin. Condor 97: 694-707.

Dufrene, M., and P. Legendre. 1997. Species assemblages and indicator species: the need for a flexible asymmetrical approach. Ecological Monographs 67: 345–366.

Finch, D. M., and R. T. Reynolds. 1987. Bird response to understory variation and conifer succession in aspen forests. in Proceedings of issues and technology in the management of impacted wildlife. (J. Emmerick, ed.): 177 pp. Thorne Ecological Institute, Colorado Springs, CO.

Flack, J. A. Douglas. 1976. Bird populations of aspen forests in western North America. Ornithological Monographs 19: iii-viii, 1-97.

Garrison, B. A., M. L. Triggs, and R. L. Wachs. 2005. Short-term effects of group-selection timber harvest on landbirds in montane hardwood-conifer habitat in the central Sierra Nevada. Journal of Field Ornithology 76: 72–82.

Goodale, E., P. Lalbhai, U. M. Goodale, and P. M. S. Ashton. 2009. The relationship between shelterwood cuts and crown thinnings and the abundance and distribution of birds in a southern New England forest. Forest Ecology and Management 258: 314-322.

Griffis-Kyle, K.L., and P. Beier. 2003. Small isolated aspen stands enrich bird communities in southwestern ponderosa pine forests. Biological Conservation 110: 375-385.

Griffis-Kyle, K.L., and P. Beier. 2005. Migratory strategy and seasonal patterns of bird diversity in relation to forest habitat. American Midland Naturalist. 153: 436-443.

Hollenbeck, J. P., and W. J. Ripple. 2006. Multi-scale relationships between aspen and birds in the northern yellowstone ecosystem. Unpublished doctoral dissertation, Oregon State University, Corvalis, OR.

Jones, J. R. 1974. Silviculture of southwestern mixed conifers and aspen: The status of our knowledge. USDA Forest Service Research Paper RM-122: 44 pp. Rocky Mountain Forest and Range Experiment Station, USDA Forest Service, Fort Collins, Colorado.

Jones, J. R. 1985. Distribution. in Aspen: ecology and management in the western United States. (N.V. DeByle and R.P. Winokur, Eds.). USDA Forest Service General Technical Report RM-119: 9-10. Rocky Mountain Forest and Range Experiment Station, USDA Forest Service, Fort Collins, Colorado.

Kay, C. E. 1997. Is aspen doomed? Journal of Forestry 95: 4–11.

King, D. I., and R. M. DeGraaf. 2000. Bird species diversity and nesting success in mature, clearcut and shelterwood forest in northern New Hampshire USA. Forest Ecology and Management 129: 227–235.

Kirk, D. A., A. W. Diamond, K. A. Hobson, and A. R. Smith. 1996. Breeding bird communities of the western and northern Canadian boreal forest: relationship to forest type. Canadian Journal of Zoology 74: 1749-1770.

Kirkpatrick, C., C. J. Conway, and P. B. Jones. 2006. Distribution and relative abundance of forest birds in relation to burn severity in southeastern Arizona. Journal of Wildlife Management 70: 1005-1012.

Little, E. L., Jr. 1971. Atlas of United States Trees:Vol. 1. Conifers and important hardwoods. U.S. Department of Agriculture Forest Service, Miscellaneous Publication 1146: 9 pp. Washington, D.C.

Martin, K., K. E. H. Aitken, and K. L. Wiebe. 2004. Nest sites and nest webs for cavity-nesting communities in interior British Columbia, Canada: Nest characteristics and niche partitioning. Condor 106: 5–19.

Matson, N. P. 2000. Biodiversity and its management on the National Elk Refuge, Wyoming. Yale Forestry & Environmental Studies Bulletin 104: 101-138.

McArdle, B. H., and M. J. Anderson. 2001. Fitting multivariate models to community data: a comment on distance-based redundancy analysis. Ecology 82: 290–297.

McCune, B., and Grace, J. B. 2002. Analysis of Ecological Communities. MjM Software Design, Gleneden Beach, Oregon, USA.

McCune, B., and M. J. Mefford. 2006. PCORD. In: Multivariate Analysis of Ecological Data. Version 5.10, MjM Software, Gleneden Beach, Oregon, USA.

McDonough, W. T. 1985. Sexual Reproduction, Seeds, and Seedlings. in Aspen: ecology and management in the western United States. (N. V. DeByle and R. P. Winokur, Eds.). USDA Forest Service General Technical Report RM-119: 25-28. Rocky Mountain Forest and Range Experiment Station, USDA Forest Service, Fort Collins, Colorado.

Miller, R. L., and G. A. Choate. 1964. The forest resource of Colorado. USDA Forest Service Resource Bulletin INT-3: 54 pp. Intermountain Forest and Range Experiment Station, Ogden, Utah.

Mueggler, W. F. 1985. Vegetation associations. in Aspen: ecology and management in the western United States. (N. V. DeByle and R. P. Winokur, Eds.). USDA Forest Service General Technical Report RM-119: 45-55. Rocky Mountain Forest and Range Experiment Station, USDA Forest Service, Fort Collins, Colorado.

Pinkowski, B. C. 1981. High density of avian cavity-nesters in aspen. Southwestern Naturalist 25: 560-562.

Ralph, C. J., G. R. Geupel, P. Pyle, T. E. Martin, and D. F. DeSante. 1993. Handbook of field methods for monitoring landbirds. USDA Forest Service General Technical Report PSWGTR-144-www. 41 pp.

Raphael, M. G., and M. White. 1984. Use of Snags by Cavity-Nesting Birds in the Sierra Nevada. Wildlife Monographs 86: 3-66.

Rendell, W. B., and R. J. Robertson. 1989. Nest-site characteristics, reproductive success and cavity availability for tree swallows breeding in natural cavities. Condor 91: 875–885.

Reynolds R. T. and D. M. Finch. 1988. Bird responses to understory variation and conifer succession in aspen forests. in (J. Emerick, S. Q. Foster, L. Hayden-Wing, J. Hodgson, J. W. Monarch, A. Smith, O. Thorne, and J. Todd, eds.). Proceedings III: issues and technology in the management of impacted wildlife: 87-95. Thorne Ecological Institute, Boulder, Colorado.

Richardson, T. W., and S. K. Heath. 2004. Effects of conifers on aspen-breeding bird communities in the Sierra Nevada. Transactions of the Western Section of the Wildlife Society 40: 68-81.

Robbins, C. S, D. Bystrak, and P. H. Geissler. 1986. The breeding bird survey: its first fifteen years, 1965-1979. U.S.D.I. Fish and Wildlife Service, Resource Publication 157. Washington, D.C.

Roden, J. S. and R. W. Pearcy. 1993. Effect of leaf flutter on the light environment of poplars. Oecologia 93: 201-207.

Rogers, P. 2002. Using forest health monitoring to assess aspen forest cover change in the southern Rockies ecoregion. Forest Ecology and Management 155: 233-236.

Salt, G. W. 1957. Analysis of avifaunas in the Teton Mountains and Jackson Hole, Wyoming. Condor 59: 373-393.

Schier, G. A., J. R. Jones, and R. P. Winokur. 1985. Vegetative regeneration. in Aspen: Ecology and management in the western United States. (N. V. DeByle and R. P. Winokur, Eds.). USDA Forest Service General Technical Report RM-119: 29–33. Rocky Mountain Forest and Range Experiment Station, USDA Forest Service, Fort Collins, Colorado.

Scott, V. E. 1979. Bird response to snag removal in ponderosa pine. Journal of Forestry 77: 26-28.

Shepperd, W. D., D. L. Bartos, and S. A. Mata. 2001. Above-and below-ground effects of aspen clonal regeneration and succession to conifers. Canadian Journal of Forest Restoration 31: 739-745.

Shepperd, W. D. 2004. Techniques to restore aspen forests in the western U.S. Transactions of the Western Section of the Wildlife Society 40: 52-60.

Sudworth, G. B. 1934. Poplars, principal tree willows and walnuts of the Rocky Mountain Region. U.S. Department of Agriculture Technical Bulletin 420: 111 p. Washington, D.C.

Turchi, G. M., P. L. Kennedy, D. Urban, and D. Hein. 1995. Bird species richness in relation to isolation of aspen habitats. Wilson Bulletin 107: 463-474.

Verner, J. 1988. Aspen. in A guide to wildlife habitats of California. (K. Mayer, W. F. Laudenslayer, Jr., eds.): 66-67. California Department of Forestry and Fire Protection, Sacramento, CA.

Walsberg, G. E. 1981. Nest-site selection and the radiative environment of the warbling vireo. The Condor 83: 86-88.

Winternitz, B. L. 1980. Birds in aspen. in Workshop proceedings on management of western forests and grasslands for nongame birds. (R. M. Degraff, Ed.). USDA Forest Service General Technical Report INT-86: 247-257.

Worrall, J. J., L. Egeland, T. Eager, R. A. Mask, E.W. Johnson, P. A. Kemp, and W. D. Shepperd. 2008. Rapid mortality of *Populus tremuloides* in southwestern Colorado, USA. Forest Ecology and Management 255: 686-696.

Zarnowitz, J. E., and D. A. Manuwal. 1985. The effects of forest management on cavity-nesting birds in northwestern Washington. The Journal of Wildlife Management 49: 255-263.

# AVIAN COMMUNITY RESPONSES TO VEGETATION STRUCTURE WITHIN CHAINED AND HAND-CUT PINYON-JUNIPER WOODLANDS ON THE COLORADO PLATEAU

*Charles van Riper III and Claire Crow*

## ABSTRACT

We investigated relationships between breeding birds and vegetation characteristics in fuels-reduction treatment areas within pinyon-juniper woodlands at locations over the Colorado Plateau. The goal of this study was to document differences in avian community responses to two types of pinyon-juniper fuels-reduction treatments (chained vs. hand-cut), relative to control sites. We selected 73 vegetation plots in southern Utah and northern Arizona, of which 33 had been previously thinned by handcutting or chaining, and 40 control plots in untreated pinyon-juniper woodlands. At the 73 locations we documented vegetation structure and counted birds within 3.1 ha circular plots during the 2005 and 2006 breeding seasons. We focused in particular on the effects of fuels-reduction treatments to 16 bird species that are considered pinyon-juniper obligates.

We found that density of pinyon pines was the most important variable in predicting bird species richness in all treatments and at control sites. Abundance of Brewer's Sparrow (*Spizella breweri*) was negatively related to chained, but positively related to hand cut areas. Vesper Sparrow (*Pooecetes graminius*) abundance was negatively related to both chaining and handcutting. Within 16 pinyon-juniper obligate bird species, abundance of five was positively related to pinyon-pine density, while two were positively related to juniper density. These responses, along with other bird-vegetation relationships influenced by treatment type, need to be considered by land managers when planning fuels reduction treatments in pinyon-juniper woodland habitat in the Colorado Plateau.

## INTRODUCTION

Pinyon (*Pinus spp.*) –juniper (*Juniperus spp.*) woodlands are estimated to cover 24 to 40 million ha in the Intermountain West and are distributed extensively across the Colorado Plateau (Samuels and Betancourt 1982). The pinyon-juniper woodland ecosystem complex is highly variable over the landscape (West et al. 1998; Bock and Block 2005), with either pinyon pines or junipers as the sole dominant, or as co-dominants. Associated shrubs, especially sagebrush (*Artemisia spp.*), and Gambel oak (*Quercus gambelii*), may also influence the dynamics of pinyon-juniper woodlands (Tausch and Hood 2007). Across western North America, the pinyon-juniper complex includes 70 plant associations and 230 ecological site types (Moir and Carton 1987).

Pinyon-juniper woodlands were historically maintained by fire in an open, savannah-like condition. As fire frequency has decreased largely due to a combination of fire suppression and livestock grazing, the resulting accumulation of fuels has shifted the fire regime to stand-replacing fires (West 1984; Miller and Rose 1999; Harris et al. 2003; Bock and Block 2005). Land managers have been directed to apply fuels reduction treatments in order to improve forest health, while simultaneously reducing fire risk to surrounding development (USDAFS and

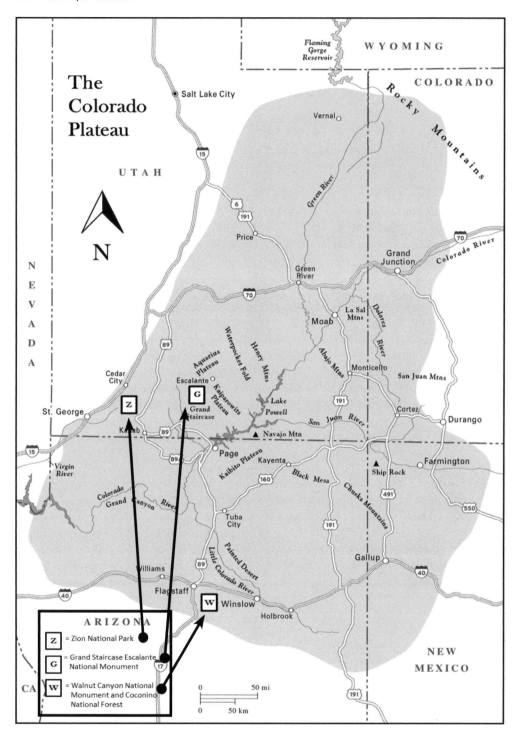

Figure 7.1 Three sites on the Colorado Plateau for which we modeled the relationships between bird community and vegetation characteristics in pinyon-juniper woodlands.

USDIBLM 2004, USDI and USDAFS 2006). In these directives managers are required to consider effects of management actions on multiple ecosystem attributes (Brunson and Shindler 2004; Dombeck 1996; Clark 1999), such as the status of neotropical migrant and other bird populations (USDAFS 1994, USDAFS 1996, USDAFS and USDIFWS 2001). Therefore, it is important to consider how change in vegetation structure caused by fuels reduction treatments might be expected to produce changes in avian communities within treated woodlands (Crow and van Riper 2010).

More than 70 species of birds breed in pinyon-juniper woodlands, but composition of avian communities varies considerably among woodland types (Balda 1987). Bird species identified as breeding, in at least part of their range, solely or preferentially in pinyon-juniper woodlands have been characterized as pinyon-juniper specialists (Pavlacky and Anderson 2001). Avian communities respond to changes brought about by either natural successional processes or management actions that alter forest structure in pinyon-juniper woodlands (Sedgwick 1987; Medin et al. 2000; Rosenstock and van Riper 2001; Knick et al. 2005). Because bird species with restricted habitat requirements are negatively impacted by loss of their specific habitat (Stauffer and Best 1980), it is likely that pinyon-juniper habitat alteration would most greatly influence pinyon-juniper specialists.

Contemporary methods of fuels reduction include various mechanical treatments, prescribed fire, or a combination thereof (Shindler 2003). There is considerable variation in the equipment used for mechanical treatments (e.g. chainsaws, wood shredding, chains), and disposition of felled trees. After felling, trees may be left, broadcast burned, removed, chipped and mulched on site, or piled and burned. Each of these methods might be expected to yield different modifications to vegetation, thus resulting in different potential changes to avian community structure.

The objectives of this study were to: (1) identify, across a range of natural pinyon-juniper woodland variation, relationships between vegetation characteristics and bird species richness and abundance that would likely be influenced by fuels reduction treatments, and (2) determine whether bird species richness or abundance differed between hand cut selectively thinned, or chaining treatments.

## METHODS

### Study Areas

Because environmental variables such as weather, soils, aspect, and vegetation composition vary considerably in pinyon-juniper woodlands, we collected data across a wide geographic range of locations. We selected pinyon-juniper woodlands at three federally-managed locations on the western Colorado Plateau in order to investigate avian-vegetation relationships (Figure 7.1). Our study areas were located in Walnut Canyon National Monument and on adjacent Coconino National Forest lands near Flagstaff, Arizona (hereafter called Walnut Canyon), and at Grand Staircase Escalante National Monument (GSENM) near Kanab, Utah (Table 7.1). We felt that the three locations were generally representative of pinyon-juniper woodlands over the Colorado Plateau.

### Sampling Design

To explore relationships between birds and vegetation, we installed 73 bird count stations, 33 as treatment areas that had previously been chained or hand cut at GSENM and Walnut Canyon, and 40 that served as non treated controls (Table 7.1). At all study areas we generated random coordinates for locating bird count stations, with all stations stratified by vegetation type. Bird count stations were separated by a minimum distance of 200 m and in

|  | GSENM | Walnut Canyon NM | Zion NP |
|---|---|---|---|
| Administration | BLM | NPS, USFS | NPS |
| Latitude | 37.04 | 35.17 | 37.3 |
| Longitude | -112.49 | -111.51 | -113.05 |
| State | Utah | Arizona | Utah |
| Mean elevation (m) | 2,000 | 2,000 | 2,000 |
| Annual precipitation (cm) | 25 | 51 | 38 |
| Months of precipitation | Nov-Mar | Jul-Oct, Dec-Mar | Mar-April, Jul-Sept |
| Dominant tree species | Utah juniper | Utah juniper, two-needle pinyon pine | Rocky mountain juniper |
| Secondary trees | Two-needle pinyon pine | - | Ponderosa pine, two-needle pinyon pine |
| Dominant shrub species | Sagebrush | Antelope bitterbrush | Gambel oak |
| Secondary shrubs | Rubber rabbitbrush | Rubber rabbitbrush | Utah serviceberry, manzanita, sagebrush |
| Total no. bird plots | 45 | 28 | 34 |
| No. plots chained | 8 | 8 | 0 |
| No. plots handcut | 9 | 8 | 0 |
| No. control plots untreated | 28 | 12 | 34 |

*(-) = not applicable  BLM=Bureau of Land Management  NPS=National Park Service*
*USFS=United States Department of Agriculture, US Forest Service*

Table 7.1  Descriptions of three pinyon-juniper woodland sites on the Colorado Plateau at which we investigated bird-vegetation and bird-fuels reduction treatment relationships during the 2005 and 2006 bird breeding seasons.

treated areas were a minimum of 100 m from treatment boundaries. We discarded coordinates on steep slopes and those located more than 30 min travel time from the closest station. Rebar was installed and GPS coordinates recorded at each location to mark the count station, and thereafter navigation to each set of coordinates was with handheld GPS receivers. We defined a 100 m radius circle centered on each station as a bird plot.

## Vegetation Sampling

Within each study area, coarse vegetation layers were initially identified from geographic information systems (ReGAP at GSENM, NPS-generated layers at Zion and Walnut Canyon). Detailed on-site vegetation data were then taken during the summers of 2005 and 2006. We based our vegetation sampling design on BBIRD (Martin et al. 1997), and here we discuss our departures from that protocol. From each bird count station, we measured the distance to the nearest juniper, the nearest pinyon, and the nearest tree if a third species was closer, in each of four directions (NE, SE, SW, NW). The nearest trees were found by walking in a spiral pattern from the point. Distances beyond 25 m were measured with handheld GPS receivers.

Four vegetation subplots were located from the bird count station in each of the cardinal directions, 50 m at the GSENM and Walnut Canyon study areas and 25 m from the bird count station at the Zion site. Within each subplot, we sampled trees within a 15 m radius, and shrubs within a 5 m radius. Gambel oak was considered a shrub, which is its common form at our study sites. We measured tree density by the number of individuals and shrub density by the number of stems. Diameter at breast height (DBH) was recorded for single-stemmed trees, diameter at root crown (DRC) for multi-stemmed trees. Diameter at stump height (ST), measured at 30.5 cm above the ground

(Bradshaw, Reveal 1943) was also recorded. All diameter measurements were made with a ruler held perpendicular to the trunk. The trunks of some living junipers lay prone on the ground, and for these we measured height of the tallest vertical "branch" and trunk diameter at 30.5 cm along the main stem from the original root crown. In each vegetation plot, we estimated percent ground cover and percent cover of plants <50 cm tall in a 1x1 m square relevé (Mueller-Dombois and Ellenberg 1974). The square was located by selecting from a random numbers table, a distance between 0 and 5 meters and an azimuth between 0 and 360 degrees. Canopy cover was estimated visually by a single observer in the field. We also used an overlay grid of 10m x 10m cells on aerial images (digital orthophotos [DOQ]) to calculate canopy cover of each bird plot (no. cells, to the nearest quarter cell, covered with tree canopy/total no. quarter cells in bird plot). We averaged vegetation measurements across the four vegetation subplots to describe vegetation characteristics at each plot.

## Bird Sampling

Bird counts, after Reynolds et al. (1980), were conducted within variable circular plots truncated at 100 m radius. Counts were undertaken at each station May-June on five occasions during the breeding seasons of 2005 and 2006. After arriving at each station, the observer waited for one minute to allow acclimation of birds. Over a 5-minute count period we recorded all birds that were detected visually or aurally within the 100 m radius bird plot. Birds that flew over plots were not included. We visited each plot at approximately 1-week intervals, and counted between sunrise and 10 AM. All hand cut plots were surveyed during the 2005 breeding season; two observers alternated visits to minimize observer bias. In 2006, birds were additionally counted at Walnut Canyon and at GSENM's chained

| Species Code | Common Name | Scientific Name | PJ Specific | Priority |
|---|---|---|---|---|
| ATFL | Ash-throated Flycatcher | *Myiarchus cinerascens* | Specialist | - |
| BCHU | Black-chinned Hummingbird | *Archilochus alexandri* | Specialist | - |
| BEWR | Bewick's Wren | *Thyomanes bewickii* | Specialist | - |
| BGGN | Blue-gray Gnatcatcher | *Polioptila caerulea* | Specialist | - |
| BRSP | Brewer's Sparrow | *Spizella brewerii* | - | UT |
| BTYW | Black-throated Gray Warbler | *Dendroica nigrescens* | Specialist | AZ, UT |
| GRFL | Gray Flycatcher | *Empidonax wrightii* | Exclusive | AZ |
| GRVI | Gray Vireo | *Vireo vicinior* | Exclusive | - |
| GTTO | Green-tailed Towhee | *Pipilo chlorurus* | Specialist | - |
| JUTI | Juniper Titmouse | *Baeolophus ridgwayi* | Specialist | AZ |
| HOFI | House Finch | *Carpodacus mexicanus* | Specialist | - |
| NOMO | Northern Mockingbird | *Mimus polyglottos* | Specialist | - |
| SPTO | Spotted Towhee | *Pipilo maculatus* | Specialist | - |
| VESP | Vesper Sparrow | *Pooecetes gramineus* | Specialist | - |
| VIWA | Virginia's Warbler | *Vermivora virginiae* | - | UT |
| WSJA | Western Scrub Jay | *Aphelocoma californica* | Exclusive | - |

*PJ specific: specialist=may nest in other vegetation communities, exclusive=nests only in pinyon-juniper woodlands. Priority: UT=listed by Utah Partners in Flight, AZ=listed by Arizona Partners in Flight, (-)=not a priority species*

Table 7.2 Bird species selected for modeling of relationships between relative abundance and vegetation characteristics, with level of pinyon-juniper specialization (PJ specific) (per Balda and Masters 1980) and state (Utah or Arizona) of priority population status (per Partners in Flight, Latta et al. 1999, Parrish et al. 2002).

| Category | Variable |
|---|---|
| Trees | % canopy cover |
| | % canopy pinyon pine and juniper |
| | % canopy non-PJ trees |
| | Non-PJ trees density |
| | Snag density |
| | Juniper height |
| | Pinyon pine height |
| | Juniper diameter |
| | Pinyon pine diameter |
| Shrub | Stem density of tall shrubs |
| | Stem density of short shrubs |
| | Density of small diameter (<2.5 cm) stems |
| | Density of large diameter (≥2.5 cm) stems |
| Ground Cover | % bare ground |

Table 7.3    Vegetation characteristics excluded from analysis due to low explanatory power or high correlation with other characteristics.

| Vegetation characteristic* | Degrees freedom | F ratio | Prob >F | Relationship direction |
|---|---|---|---|---|
| Juniper density | 1 | 0.7911 | 0.3778 | - |
| Pinyon density | 1 | 5.7972 | 0.0196 | P |
| PJ canopy height | 1 | 1.2903 | 0.2611 | - |
| Shrub density | 1 | 4.3757 | 0.0413 | N |
| Shrub species richness | 1 | 7.7832 | 0.0073 | P |
| % grass cover | 1 | 0.0151 | 0.9027 | - |
| Treatment | 2 | 1.1511 | 0.3241 | - |

*Density of dead shrub stems and percent cover of litter and forbs were excluded by mixed stepwise regression analysis.*

Table 7.4    Species Richness: results of effects tests for relationships to bird species richness for vegetation characteristics on 107 plots at 3 sites on the Colorado Plateau. Relationship direction P = positive, N = negative, (-) = no significant relationship.

plots. Birds were not counted during rain or high winds. Sampling order of plots was varied to minimize temporal bias.

## Data Analysis

Utilizing all 53 detected bird species, we estimated bird species richness by summing the number of species detected on each bird plot across all surveys (Appendix 7). We then excluded data on species that were observed on fewer than five total plots, and relative abundance was calculated at each plot as total no. detections/no. surveys. We stratified a subset of birds for further analysis by choosing 16 species that are categorized as pinyon-juniper specialists (Balda 1987; Pavlacky and Anderson 2001) or are considered priority species in Utah (Parrish et al. 2002) or Arizona (Latta et al. 1999; Table 7.2).

When assessing the effects of vegetation characteristics on species richness and abundance, we excluded vegetation characteristics that were absent from >60% of plots. We also excluded one variable from all pairs of explanatory variables that were correlated (r>0.75; Table 7.3). Data were transformed where appropriate and our level of significance was P<= 0.05. We used multiple linear regressions to explore relationships of species richness and abundance as influenced by vegetation characteristics and treatment (Ramsey and Schafer 2002).

In our mixed stepwise regressions analyses we considered vegetation characteristics in two groups: pinyon-juniper (pj) characteristics, which would likely be altered by fuels reduction treatments, and non-pj characteristics, which we utilized as covariates. Pinyon-juniper characteristics considered were juniper stem density, pinyon pine stem density, and mean height of pinyon pines and junipers combined. Non-pj characteristics included density of live shrub stems, density of dead shrub stems, shrub species richness, percent ground cover

of litter, percent ground cover of grass, and percent ground cover of forbs. We employed mixed stepwise regressions on the non-pj variables for each responsible variable, with the probability to enter the model fixed at 0.250 and the probability to be removed from the model fixed at 0.100. We then fit a model for each response variable with all pinyon-juniper characteristics, the selected non-pj variables, site, treatment and site treatment interaction.

We used extra sum of squares F tests to identify relationships between each vegetation characteristic and bird community response. The direction of the relationship was defined as either being positive, in which the response variable increased as the vegetation variable increased in value, or negative when the response variable decreased as the vegetation variable increased, or neutral.

## RESULTS

### Vegetation-Site relationships

Canopy cover on our plots ranged from 0-89% (Figure 7.2) and tree density from 0-387 junipers/ha and 0-541 pinyon pines/ha (Figure 7.3). Plots at GSENM included the highest juniper density, while the highest pinyon pine density and highest canopy cover were on Walnut Canyon plots. Shrub density ranged 0-1621 stems/ha (Figure 7.4) while tree height ranged 0-7 m (Figure 7.5) and was greatest at GSENM.

### Bird-Site relationships

Bird species richness at our study sites ranged from 4 to 20 species (x=11.8 species, SE=0.31), being positively related to density of pinyon pines and shrub species richness, and negatively related to shrub density (Table 7.4).

Of the 16 pinyon-juniper dependent bird species that we analyzed (Table 7.2), relative abundance of seven (44%) was positively related to one or more pinyon-juniper characteristics (Table 7.5). Conversely,

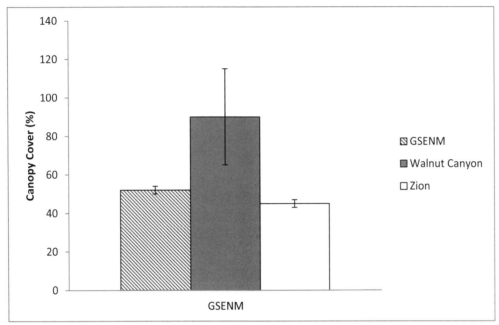

Figure 7.2  Ranges of canopy cover in three pinyon-juniper woodlands on the Colorado Plateau.

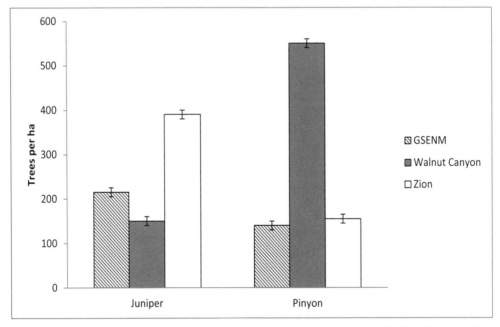

Figure 7.3  Ranges in density of juniper and pinyon pine trees in three pinyon-juniper woodlands on the Colorado Plateau.

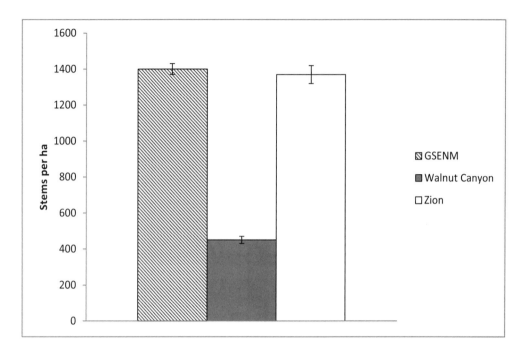

Figure 7.4  Ranges in density of shrub stems in three pinyon-juniper woodlands on the Colorado Plateau.

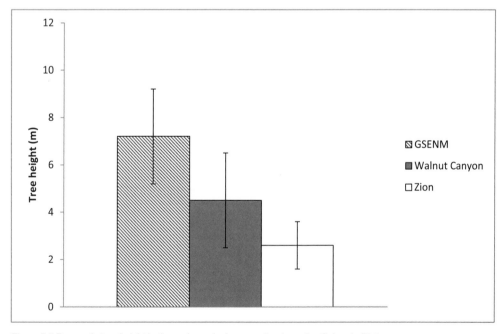

Figure 7.5 Ranges in tree height in three pinyon-juniper woodlands on the Colorado Plateau.

relative abundance of 4 bird species (25%) was negatively related to pinyon-juniper characteristics.   Of the seven species found to be related to one or more pinyon-juniper characteristic, five were positively (BCHU, BGGN, BTYW, HOFI, WSJA, Table 7.2), while only the Vesper Sparrow (VESP) was negatively related to pinyon pine density.   We found numbers of two species (BEWR, JUTI) positively affected by juniper density while numbers of five species (BCHU, BRSP, VESP, WSJA) were negatively associated with density of juniper trees (Table 7.5).   Overall, 12 of the 16 pinyon-juniper dependent bird species were influenced by tree density.

### Bird-treatment relationships

Hand cut plots included a higher range of canopy cover and density of pinyon pines and snag density than did chained plots (Tables 7.6 and 7.7).   The range of shrub stem density in hand cut plots also included higher values than chained plots.

Bird species presence and relative abundance was related to treatment type (Table 7.4).   Relative abundance of Brewer's Sparrow and Vesper Sparrow were related to treatment after accounting for differences among sites and for the effects of vegetation variables (Table 7.5).   Relative abundance of Brewer's Sparrow was negatively related to chaining and positively related to handcutting.   Vesper Sparrow was negatively related to chaining and handcutting.   For some species, effects of treatment were confounded with effects of differences among sites, particularly on the abundance of Bewick's Wren (*Thyomanes bewickii*) (p=0.0166 for interaction covariate) and Juniper Titmouse *(Baeolophus ridgwayi)* (p=0.0061 for interaction covariate).

## DISCUSSION

### Bird-vegetation relationships

At all sites that we examined on the Colorado Plateau, bird species richness was positively related to pinyon pine density and negatively related to shrub density, which is consistent with previous findings where bird species richness increased with maturity of woodlands (Rosenstock and van Riper 2001, Pavlacky and Anderson 2004).   Our findings also support those of Balda (1987), who found that bird species richness was related to density and foliage volume of pinyon pines.   Management actions such as fuels reduction treatments, that maintain some level of structural diversity, will result in greater species richness than will complete removal of vegetation by methods such as chaining (Wilson 1974).   However, vegetation complexity may be functionally reduced if some components of vegetation are rarely used, thus lowering species richness in spite of increased vegetation complexity (Laudenslayer and Balda 1976).

Woodlands in our study encompassed a range of variation in canopy cover, tree density and tree height, as well as shrub density (Figures 7.2-7.5).   We found more positive relationships with pinyon pine density than with any other vegetation characteristic (Table 7.5).   Insects were the preferred diet of most of the species positively related with pinyon pine density (60%, n=5) (Balda 1987, Ehrlich et al. 1988).   The greater foliage volume of pinyon pine trees compared to junipers might provide more foliage insects to foraging birds (Laudenslayer and Balda 1976).   However, two (50%) of the insect consumers that preferred areas of high pinyon pine density in the present study did not obtain their prey directly from trees.   Black-chinned Hummingbird (*Archilochus alexandri*) and Western Scrub-Jay (*Aphelocoma californicus)* forage in the air and on the ground, respectively.   Both of these species were negatively related to density of juniper, which suggests a strong preference for pinyon pines over junipers.   These two bird species might particularly benefit from selective fuels reduction where a higher proportion of pinyon pines are

| Species* | Juniper | Pinyon | Shrub | Dead shrub | % grass | Chained | Handcut |
|----------|---------|--------|-------|------------|---------|---------|---------|
| BCHU | N | P | - | - | - | - | - |
| BEWR | P | - | P | - | - | - | - |
| BGGN | - | P | P | - | N | - | - |
| BRSP | N | - | - | - | - | N | P |
| BTYW | - | P | - | - | N | - | - |
| JUTI | P | - | - | - | - | - | - |
| HOFI | - | P | N | - | - | - | - |
| VESP | N | N | - | N | P | N | N |
| WSJA | N | P | - | - | - | - | - |

*Juniper=juniper density, Pinyon=pinyon pine density, Shrub=density of live shrub stems, Dead shrub=density of dead shrub stems, % grass=percent ground cover grass.*
*\*See Table 7.2 for scientific and common names.*

Table 7.5 Direction of relationships with vegetation characteristics for bird species detected during 2005 and 2006 breeding seasons in pinyon-juniper woodlands across three sites on the Colorado Plateau. Bird response measured as relative abundance. (P) = positive relationship, (N) = negative relationship, (-) = no significant relationship.

| Vegetation variables | GSENM | | | Walnut Canyon | | |
|----------------------|-------|------|-----|---------------|------|-----|
|  | Range | Mean | SE | Range | Mean | SE |
| Canopy cover (%) | 0-12 | 3 | 1.8 | 0-20 | 6 | 2.3 |
| Pinyon pine density (trees/ha) | 0-14 | 4 | 2.3 | 0-230 | 105 | 36.5 |
| Juniper density (trees/ha) | 0-116 | 35 | 16.8 | 0-16 | 7 | 2.6 |
| PJ height mean | 6-10 | 8 | 0.5 | 4-13 | 6.5 | 1.6 |
| Snag density (snags/ha) | 0 | 0 | 0.0 | 0-7 | 1 | 0.9 |
| Shrub density | 0-12 | 3 | 1.8 | 37-297 | 151 | 34.2 |
| Shrub species richness (# species) | 1-3 | 1 | 0.3 | 1-4 | 2 | 0.4 |
| Grass (% cover) | 0-4 | 1 | 0.4 | 8-56 | 30 | 5.4 |
| Forbs (% cover) | 0-10 | 2 | 1.1 | 1-13 | 7 | 1.4 |
| Litter (% cover) | 13-71 | 31 | 6.9 | 4-30 | 9 | 3.0 |
| Bare ground (% cover) | 23-87 | 66 | 7.6 | 32-79 | 55 | 5.4 |

Table 7.6 Chained plots: ranges and means of vegetation variables on chained plots at Walnut Canyon and GSENM.

left.   The non-insectivorous House Finch (*Carpodacus mexicanus*) gleans seeds from the ground as well as seeds from foliage at all locations where this species is found (Bergtold 1913; van Riper 1976). We submit that the additional cover afforded by the higher foliage mass of pinyon pines might contribute to the preference of some species for areas of high pinyon pine density.

Black-chinned Hummingbird breeds in deciduous trees and shrubs (Wethington and Russell 2003), which seems contrary to the positive relationship we found with pinyon pine density. Previous research has shown that this species regularly travels farther than 200m from the nest to forage and concluded that nesting density for this species should not be estimated by searching for nests solely where individuals were detected (Brown 1992; Wethington and Russell 2003). The relationships modeled in the current study may reflect preferences in foraging habitat rather than for nest sites, as Black-chinned Hummingbird and Western Scrub-Jay were negatively related to juniper and positively related to pinyon pine density.

Bewick's Wren and Juniper Titmouse nest in live tree or snag cavities (Ehrlich et al. 1988; Pavlacky and Anderson 2004), but were positively related to juniper density. A pine beetle infestation on 96% of plots with Bewick's Wren detections (23 of 24 plots) and 83% of plots with Juniper Titmouse detections (20 of 24 plots) had killed many of the pinyon pines in our study. These species may have experienced enhanced nesting in the many new snags and gleaned insects from the living junipers in the vicinity.

Brewer's Sparrow (*Spizella brewerii*) and Vesper Sparrow are ground feeders that nest on the ground or in shrubs, and prefer open clearings far from woodland edges (Hardy 1945; O'Meara et al. 1981; Sedgwick 1987). Numbers of both species were negatively influenced by juniper density.

Bird-treatment relationships

Bird species richness, regardless of pinyon-juniper reduction treatment, was lower in treated verses untreated areas, where the former averaged lower pinyon pine densities. At Walnut Canyon, mean pinyon pine density was significantly lower in treated than untreated plots, while at GSENM, mean density of pinyon pines did not vary with treatment, nor did species richness. This matches our findings that species richness was positively related to pinyon pine density in untreated plots across all three sites.

We found few relationships between relative abundance of species and treatment method. Of 16 species considered, there was evidence of significant treatment effects for only two species. Brewer's Sparrow, which prefers the interior of large open brushy areas (O'Meara et al. 1981), was less abundant in chained areas and more abundant in hand cut areas than in untreated woodlands. We found that the Vesper Sparrow, a ground-nester which utilizes only the edges of chained areas (Sedgwick 1987), was less abundant in treated areas, regardless of treatment method. It is possible that our plots, which were a minimum of 100 m from treatment boundaries, did not adequately represent the edges of treatment units.

## CONCLUSIONS

Pinyon-juniper woodlands and their avian communities vary considerably across the landscape (Balda and Masters 1980) and increasing use of fuels reduction treatments by managers makes it vital to identify indicators of treatment impacts to bird populations. Across a wide range of pinyon-juniper forest structure variation, we found consistent relationships among avian pinyon-juniper specialists and their responses.   We found that patches of mature pinyon pine forest are important for maintaining species diversity and number of birds. This is generally in agreement with

| Vegetation variables | GSENM | | | Walnut Canyon | | |
|---|---|---|---|---|---|---|
| | Range | Mean | SE | Range | Mean | SE |
| Canopy cover (%) | 0-11 | 1 | 1.2 | 7-42 | 24 | 4.2 |
| Pinyon pine density (trees/ha) | 0-32 | 4 | 3.6 | 0-1250 | 670 | 169.9 |
| Juniper density (trees/ha) | 0-139 | 16 | 15.4 | 0-88 | 47 | 11.9 |
| PJ height mean | 9-9.5 | 9 | 0.1 | 4-9 | 6.5 | 0.8 |
| Snag density (snags/ha) | 0-22 | 4 | 2.5 | 0-23 | 10 | 2.9 |
| Shrub density | 216-1419 | 618 | 117.6 | 21-439 | 164 | 46.7 |
| Shrub species richness (# species) | 1-3 | 2 | 0.3 | 1 | 1 | 0.0 |
| Grass (% cover) | 0-51 | 17 | 5.3 | 13-41 | 21 | 9.0 |
| Forbs (% cover) | 0-9 | 3 | 0.9 | 0-20 | 6 | 2.4 |
| Litter (% cover) | 7-81 | 45 | 6.7 | 5-45 | 5.1 | 20.0 |
| Bare ground (% cover) | 15-93 | 47 | 7.5 | 37-86 | 54 | 5.6 |

Table 7.7 Handcut plots: ranges and means of vegetation variables in plots cut with chainsaws at Walnut Canyon and GSENM.

the findings of previous studies and what is known about the nesting and feeding habits of pinyon-juniper obligate species.

We modeled relationships between vegetation characteristics of habitat and bird species frequency and richness at a scale that we feel represents proximal cues used in habitat selection by avian species. We did not locate nests in this study, but instead focused on detections of birds at each count station during the breeding season. We felt that this would provide us a robust picture of breeding season habitat requirements.

Although this study focused on bird species that used pinyon-juniper woodlands during the breeding season, several bird species are known to depend on pinyon-juniper woodlands in the fall and winter. Flocks of Pinyon Jays (*Gymnorhinus cyanocephalus*) were observed in our study

but not included in analysis because they were detected only in flight. Townsend's Solitaire feeds nearly exclusively on juniper fruits in winter (Poddar and Lederer 1982) and Juniper Titmouse is known to eat juniper fruits throughout the fall (Latta et al.1999). In addition to breeding birds, managers should also consider how fuels reduction treatments might impact the ability of woodlands to also support the fall and winter requirements of birds that rely on these habitats.

Management implications

Avian communities in pinyon-juniper woodlands include some species that prefer mature woodlands, areas of high tree density, and shrubby openings. These woodlands were historically maintained by fire, but with the recent suppression of natural fires the pinyon-juniper ecosystem requires active management in order to sustain the variety of successional stages that are necessary to support a full assemblage of avian species. We found that a reduction of pinyon pine density will negatively influence many bird species due to positive relationships of pinyon-juniper specialists with pinyon abundance. The density of pinyon also influences overall avian species richness. Previous research has shown that the distribution of pinyon-juniper specialists was limited by the presence of pinyon pine near the boundary of the geographic range of these bird species (Pavlacky and Anderson 2001), but our results indicate that the influence occurs over the entire geographic range of the avian species. To enhance treatments for avian pinyon-juniper specialists, planning and precision implementation of selective handcutting can ensure that a larger proportion of pinyon pines are left standing. However, this is also labor intensive and expensive. Application of prescribed fire in a mosaic of small treatments would also leave pockets of high density pinyon pine.

Unlike mechanical treatments, fire plays an additional role in nutrient cycling, and stimulates seed sprouting (Bock and Block 2005). If selective cutting is used initially to reduce fuel density and ladder fuels, the risk of stand-replacing crown fires will be reduced. This may require a series of smaller-scale treatments. Prescribed fire can then mimic natural fires which historically maintained healthy woodlands with openings that contribute to the diversity of avian habitat provided by pinyon-juniper woodlands. Once restoration goals are obtained, there will no longer be a need to suppress natural fires within the woodland. This process reduces risk to adjacent properties and resources, and will eventually restore a natural fire regime that maintains space between mature trees and preserves woodland openings.

Application of fuels reduction in multiple small-scale treatments could effectively rotate treatment units through successional stages, maximizing the number of species for which habitat is provided. It also needs to be recognized that ecological restoration of pinyon-juniper woodlands may cause declines of some species, while enhancing habitat for other species. The relationships that we found can hopefully better assist land managers in making management decisions regarding consequences to avian community structure when they undertake fuels reduction treatments in pinyon-juniper woodlands with similar vegetation characteristics.

## ACKNOWLEDGEMENTS

We thank the Bureau of Land Management, National Park Service, USDA Forest Service and United States Geological Survey and the staffs of the specific units involved for providing the opportunity to conduct this research. In particular we are indebted to Holly Beck, Jeff Bradybaugh, Kelly Fuhrmann, Kristin Legg, Jan Passek, Melissa Siders and Cynthia Wanschura. We thank the University of Arizona's Advanced Resource Technology (ART) lab for technical assistance. We also thank Alissa Fogg, Sarah Gaines, Becca Lieberg and Donna Shorrock for invaluable assistance with field work

## LITERATURE CITED

Balda, R.P. 1987. Avian impacts on pinyon-juniper woodlands. Pp 525-533 in R.L Everett, ed. Proceedings–Pinyon Juniper Conference. General Technical Report INT-125. USDA Forest Service Intermountain Research Station, Ogden, Utah, USA.

Balda, R.P., and N. Masters. 1980. Avian communities in the pinyon-juniper woodland: a descriptive analysis. Pp 146-169 in General Technical Report INT-86, USDA Forest Service Intermountain Research Station, Ogden, Utah, USA.

Bergtold, W. H. 1913. A study of the House Finch. Auk 30:40-73.

Bock, C.E., and W.M. Block. 2005. Fire and birds in the southwestern United States. Studies in Avian Biology 30:14-32.

Bradshaw, K.E., and J.L. Reveal. 1943. Tree classifications for pinus monophylla and juniperus utahensis. Journal of Forestry 41:100-104.

Brown, B.T. 1992. Nesting chronology, density and habitat use of Black-chinned Hummingbirds along the Colorado River, Arizona. Journal of Field Ornithology 63: 393-400.

Brunson, M.W. and B.A. Shindler. 2004. Geographic variation in social acceptability of wildland fuels management in the western United States. Society and Natural Resources 17:661-678.

Clark, J.R. 1999. The ecosystem approach from a practical point of view. Conservation Biology 13:679-681.

Crow, C., and C. van Riper III. 2010. Avian community responses to mechanical thinning of a pinyon-juniper woodland: Specialist sensitivity to tree reduction. Natural Areas Journal 30:191-201.

deHoop, C.F., A.H. Reddy and R. Smith. 2006. Forest fuel reduction survey analysis: forest administrators. Louisiana Forest Products Development Center Working Paper 73. Louisiana Agricultural Experiment Station, Baton Rouge, Louisiana, USA.

Dombeck, M. 1996. Thinking like a mountain: BLM's approach to ecosystem management. Ecological Applications 6:699-702.

Ehrlich, P.R., D.S. Dobkin, and D. Wheye. 1988. The birder's handbook : A field guide to the natural history of North American birds. Simon and Schuster, New York, New York, USA.

Goguen, C.B., and N.E. Mathews. 2000. Local gradients of cowbird abundance and parasitism relative to livestock grazing in a western landscape. Conservation Biology 14:1862-1869.

Hardy, R. 1945. Breeding birds of the pigmy conifers in the Book Cliff region of eastern Utah. Auk 62: 523-542.

Harris, A. T., G. P. Asner, and M. E. Miller. 2003. Changes in vegetation structure after long-term grazing in pinyon-juniper ecosystems: integrating imaging spectroscopy and field studies. Ecosystems 6:368-383.

Knick, S.T., A.L. Holmes, and R. F. Miller. 2005. The role of fire in structuring sagebrush habitats and bird communities. Studies in Avian Biology 30:63-75.

Latta, M.J., C.J. Beardmore, and T.E. Corman. 1999. Arizona Partners in Flight Bird Conservation Plan, Version 1.0. Nongame and Endangered Wildlife Program Technical Report 142. Arizona Game and Fish Department, Phoenix, Arizona, USA.

Laudenslayer, W.F., Jr., and R.P. Balda. 1976. Breeding bird use of a pinyon-juniper-ponderosa pine ecotone. Auk 93: 571-586.

Martin, T.E., C.R. Paine, C.J. Conway, W.M. Hochachka, P. Allen, and W. Jenkins. 1997. BBIRD Field Protocol. Montana Cooperative Wildlife Research Unit, University of Montana, Missoula, Montana, USA.

Medin, D.E., B.L. Welch, and W.P. Clary. 2000. Bird habitat relationships along a Great Basin elevational gradient. Research Paper RMRS-RP-23. USDA Forest Service Rocky Mountain Research Station, Fort Collins, Colorado, USA.

Miller, R.F., and J.A. Rose. 1999. Fire history and western juniper encroachment in sagebrush steppe. Journal of Range Management 52:550-559.

Moir, W.H. and J.D. Carton. 1987. Classification of pinyon-juniper (P-J) sites on national forests in the southwest. Pages 216-225 in R.L Everett, ed. Proceedings –Pinyon Juniper Conference. General Technical Report INT-125. USDA Forest Service Intermountain Research Station, Ogden, Utah, USA.

Muller-Dombois, D., and H. Ellenberg. 1974. Aims and methods of vegetation ecology. John Wiley & Sons, New York, 547 p.

O'Meara, T.E., J.B. Haufler, L.H. Stelter, and J.G. Nagy. 1981. Nongame wildlife responses to chaining of pinyon-juniper woodlands. Journal of Wildlife Management 46:381-389.

Parrish, J.R., F.P. Howe, and R.E. Norvell. 2002. Utah Partners in Flight Avian Conservation Strategy Version 2.0. UDWR Publication 20-27. Utah Partners in Flight Program, Utah Division of Wildlife Resources, Salt Lake City, Utah, USA. 302 p.

Pavlacky, D.C., and S.H. Anderson. 2001. Habitat preferences of pinyon-juniper specialists near the limit of their geographic range. Condor 103:322-331.

Pavlacky, D.C., and S.H. Anderson. 2004. Comparative habitat use in a juniper woodland bird community. Western North American Naturalist: 64:376-384.

Poddar, S. and R.J. Lederer. 1982. Juniper berries as an exclusive winter forage for Townsend's Solitaires. American Midland Naturalist 108:34-40.

Ramsey, F.L. and D.W. Schafer. 2002. The Statistical Sleuth, 2nd ed. Duxbury Press, Pacific Grove, California, USA.

Reynolds, R.T., J.M. Scott, and R.A. Nussbaum. 1980. A variable circular-plot method for estimating bird numbers. Condor 82:309-313.

Rosenstock, S.S., and C. van Riper III. 2001. Breeding bird responses to juniper woodland expansion. Journal of Range Management 54:226-232.

Samuels, M.L., and J.L. Betancourt. 1982. Modeling the long-term effects of fuelwood harvests on pinyon-juniper woodlands. Environmental Management 6:505-515.

Sedgwick, J.A. 1987. Avian habitat relationships in pinyon-juniper woodland. Wilson Bulletin 99:413-431.

Shindler, B. 2003. Public acceptance of wildland fire conditions and fuel reduction practices: challenges for federal forest managers. Chapter in H.J. Cortner, D.R. Field, P. Jakes, and J.D. Buthman, eds. Humans, Fires and Forests: Social Science Applied to Fire Management. USDA Forest Service North Central Research Station/ Ecological Restoration Institute Northern Arizona University, Flagstaff, Arizona, USA.

Stauffer, D.F. and L.B. Best. 1980. Habitat selection by birds of riparian communities: evaluating effects of habitat alterations. Journal of Wildlife Management 44:1-15.

Tausch, R.J and Hood, S.M. 2007. Pinyon-juniper woodlands. Chapter 4 in Hood, S.M. and M. Miller, eds. Fire Ecology and Management of the Major Ecosystems of Southern Utah. General Technical Report RMRS-GTR-202. USDA Forest Service Rocky Mountain Research Station, Fort Collins, Colorado, USA.

USDAFS. 1994. Neotropical Migratory Bird Reference Book. USDA Forest Service Region 5, San Francisco, California, USA.

USDAFS. 1996. Landbird Monitoring Implementation Plan. USDA Forest Service Region 5, San Francisco, California, USA.

USDAFS and USDIFWS. 2001. Memorandum of understanding between USDA Forest Service and USDI Fish and Wildlife Service. 01-MU-11130117-028.

USDAFS and USDIBLM. 2004. The Healthy Forests Initiative and Healthy Forests Restoration Act Interim Field Guide, FS-799. USDA Forest Service, Washington, DC, USA.

USDI and USDAFS. 2006. Protecting People and Natural Resources: a cohesive fuels treatment strategy. US Department of Interior and USDA Forest Service. Washington, DC, USA.

van Riper, C., III. 1976. Aspects of House Finch breeding biology in Hawaii. Condor 78:224 229.

West, N.E. 1984. Successional patterns and productivity potentials of pinyon-juniper ecosystems. Pp 1301-1322 in Developing strategies for rangeland management. Westview Press, Boulder, Colorado, USA.

West, N.E., R.J. Tausch, P.T Tueller. 1998. A management oriented classification of pinyon-juniper woodlands of the Great Basin. General Technical Report RMRS-GTR-12. USDA Forest Service Rocky Mountain Research Station, Ogden, Utah, USA.

Wethington, S.M. and S.M. Russell. 2003. The seasonal distribution and abundance of hummingbirds in oak woodland and riparian communities in southeastern Arizona. Condor 105: 484-495.

Willson, M.F. 1974. Avian community organization and habitat structure. Ecology 55: 1017-1029.

| Common name | Scientific name | Code | GSENM | Walnut |
|---|---|---|---|---|
| American Robin | *Turdus migratorius* | AMRO | 0 | 4 |
| Ash-throated Flycatcher | *Myiarchus cinerascens* | ATFL | 62 | 93 |
| Black-chinned Hummingbird | *Archilochus alexandri* | BCHU | 9 | 29 |
| Bewick's Wren | *Thyomanes bewickii* | BEWR | 0 | 93 |
| Blue-gray Gnatcatcher | *Polioptila caerulea* | BGGN | 64 | 4 |
| Black-headed Grosbeak | *Pheucticus melanocephalus* | BHGR | 42 | 46 |
| Black-throated Gray Warbler | *Dendroica nigrescens* | BTYW | 27 | 25 |
| Black-throated Sparrow | *Amphispiza bilineata* | BTSP | 31 | 0 |
| Brewer's Sparrow | *Spizella brewerii* | BRSP | 76 | 0 |
| Brown Creeper | *Certhia americana* | BRCR | 0 | * |
| Brown-headed Cowbird | *Molothrus ater* | BHCO | 7 | 50 |
| Bushtit | *Psaltiparus minimus* | BUSH | 16 | 25 |
| Cassin's Kingbird | *Tyrannus vociferans* | CAKI | 4 | 0 |
| Canyon Wren | *Catherpes mexicanus* | CANW | 0 | * |
| Chipping Sparrow | *Spizella passerina* | CHSP | 78 | 89 |
| Common Nighthawk | *Cordeiles minor* | CONI | 7 | 11 |
| Dusky Flycatcher | *Empidonax oberholseri* | DUFL | 2 | 0 |
| Gray Flycatcher | *Empidonax wrightii* | GRFL | 7 | 82 |
| Gray Vireo | *Vireo vicinior* | GRVI | 0 | 0 |
| Green-tailed Towhee | *Pipilo chlorurus* | GRRO | 2 | 0 |
| Hairy Woodpecker | *Picoides villosus* | HAWO | 0 | 32 |
| Hepatic Tanager | *Piranga flava* | HETA | 0 | 3 |
| House Finch | *Carpodacus mexicanus* | HOFI | 4 | 61 |
| Juniper Titmouse | *Baeolophus ridgwayi* | JUTI | 0 | 86 |
| Lark Sparrow | *Chondestes grammacus* | LASP | 42 | 57 |
| Lazuli Bunting | *Passerina amoena* | LAZB | 2 | 0 |
| Lesser Goldfinch | *Carduelis psaltria* | LEGO | 0 | 18 |
| Loggerhead Shrike | *Lanius ludovicianus* | LOSH | 2 | 4 |
| Mountain Bluebird | *Sialia currucoides* | MOBL | 11 | 36 |
| Mountain Chickadee | *Poecile gambeli* | MOCH | 0 | 11 |

| Common name | Scientific name | Code | GSENM | Walnut |
|---|---|---|---|---|
| Mourning Dove | *Zenaida macroura* | MODO | 56 | 64 |
| Northern Flicker | *Colaptes auratus* | NOFL | 7 | 14 |
| Northern Mockingbird | *Mimus polyglottos* | NOMO | 13 | 61 |
| Olive-sided Flycatcher | *Contopus cooperi* | OSFL | 2 | 4 |
| Phainopepla | *Phainopepla nitens* | PHAI | 4 | 0 |
| Plumbeous Vireo | *Vireo plumbeus* | PLVI | 0 | 7 |
| Pygmy Nuthatch | *Sitta pygmaea* | PYNU | 0 | 7 |
| Red-tailed Hawk | *Buteo jamaicensis* | RTHA | 4 | 0 |
| Rock Wren | *Salpinctes obsoletus* | ROWR | 0 | * |
| Sage Sparrow | *Amphispiza belli* | SAGS | 18 | 0 |
| Say's Phoebe | *Sayornis saya* | SAPH | 7 | 7 |
| Spotted Towhee | *Pipilo maculatus* | SPTO | 87 | 93 |
| Steller's Jay | *Cyanocitta stelleri* | STJA | 0 | 14 |
| Townsend's Solitaire | *Myadestes townsendi* | TOSO | 13 | 0 |
| Turkey Vulture | *Cathartes aura* | TUVU | 0 | 4 |
| Vesper Sparrow | *Pooecetes gramineu* | VESP | 71 | 50 |
| Virginia's Warbler | *Vermivora virginiae* | VIWA | 9 | 0 |
| Western Bluebird | *Salia mexicana* | WEBL | 29 | 18 |
| Western Kingbird | *Tyrannus verticalis* | WEKI | 0 | 14 |
| Western Meadowlark | *Sturnella neglecta* | WEME | 22 | 7 |
| Western Tanager | *Piranga ludoviciana* | WETA | 11 | 4 |
| Western  Scrub Jay | *Aphelocoma californica* | WSJA | 18 | 43 |
| Western Wood-Pewee | *Contopus sordidulus* | WEWP | 2 | 11 |
| White-breasted Nuthatch | *Sitta carolinensis* | WBNU | 2 | 46 |
| Wild Turkey | *Meleagris gallopavo* | WITU | 0 | * |
| Yellow Warbler | *Meleagris gallopavo* | YEWA | 0 | 7 |
| Yellow-rumped Warbler | *Dendroica coronata* | YRWA | 2 | 11 |

Appendix 7  Bird species and percent of plots on which species was detected at two sites on the Colorado Plateau during 2005 and 2006 breeding seasons. GSENM=Grand Staircase Escalante National Monument, Walnut=Walnut Canyon National Monument and surrounding Coconino National Forest. *Observed outside of count period.

# LANDSCAPE SCALE FEATURES PREDICT PREDATION AND PARASITISM ON PASSERINE NESTS: A LITERATURE REVIEW

*Katie J. Stumpf*

## ABSTRACT

Loss and degradation of breeding habitat may force breeding birds to nest in habitats with sub-optimal habitat characteristics, potentially exposing them to unnaturally high rates of nest predation and brood parasitism. Species may respond differently to characteristics at different scales, especially in riparian habitats where linear patches may alter predator and/or parasite diversity and abundance. I reviewed the literature to understand how habitat characteristics affect predation and parasitism, because habitat modifications may be a cost-effective method of reducing these pressures. Landscape scale characteristics, including patch size and surrounding landscape were most often associated with nest predation. Factors at all three spatial scales (landscape, territory, and nest site) were equally likely to be associated with parasitism including surrounding patch size, habitat type, surrounding landscape, canopy cover, distance from the nest to the nearest tree, ground cover, tree size, and foliage density. The importance of landscape scale characteristics for predation and parasitism rates may be driven by the effect of large-scale features on abundance and distribution patterns of predators and parasites. Alternatively, an association between smaller spatial scales and parasitism but not predation may reflect the fact that cowbirds rely mainly on visual cues to locate nests, while a diverse suite of predators may mask any association with characteristics at these smaller scales. Given the importance of patch size for both predation and parasitism, I recommend that future studies carefully quantify predation and parasitism effects across a range of patch sizes to elucidate the patch sizes at which these rates are most strongly reduced.

## INTRODUCTION

Riparian forests in the southwestern United States are at risk due to conversion to non-native habitat (e.g., tamarisk, *Tamarix spp.*), increasing urbanization, livestock grazing, and water management practices (Ohmart 1994). Although riparian areas comprise less than 1% of the total land area in the American southwest, they provide breeding habitat for more avian species than any other land type (Anderson and Ohmart 1977), including endangered species (e.g., Southwestern Willow Flycatcher, *Empidonax traillii extimus*) and species of special concern (e.g., Yellow-billed Cuckoo, *Coccyzus americanus*). Additionally, riparian areas provide stopover habitat, shelter, and critical food resources for numerous Neotropical migratory bird species during both spring and fall migration. Riparian habitats are areas of high species diversity, density, and productivity (Hoover et al. 1995) and, therefore, have high conservation value.

Loss or degradation of riparian habitat may force nesting birds into sub-optimal habitats, exposing them to increased risk of brood parasitism by brown-headed cowbirds (*Molothrus ater*) and nest predation, the two main causes of nest failure for most passerine birds (Martin 1992). High rates of predation

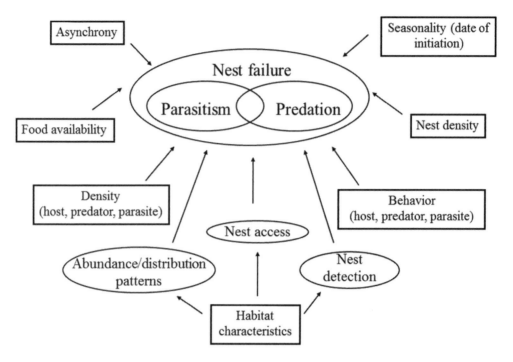

Figure 8.1 Factors that have been proposed as potential effects on the probability of nest failure by predation and/ or parasitism on passerine nests.

and parasitism can negatively affect reproductive success and may contribute to further population declines (Brittingham and Temple 1983; Heske et al. 2001). Understanding the underlying mechanisms associated with the probability of parasitism and predation is critical for maintaining viable populations for many species, and is especially important for endangered or threatened species. Many mechanisms potentially affecting the probability of nest predation and brood parasitism have been proposed (Figure 8.1), including habitat characteristics (e.g., Davidson and Knight 2001; Helzel and Earnst 2006), behavioral adaptations (e.g., nest defense; Briskie et al. 1990; Ward and Smith 2000a), predator/ parasite and host species distribution and abundance patterns (e.g., Verner and Ritter 1983; Robinson et al. 2000; Gustafson et al. 2002), seasonality (date during the breeding season; Finch 1983; Parker 1999), nest density (Best 1978), food availability

(e.g., Verner and Ritter 1983), and nesting asynchrony (e.g., Briskie et al. 1990; Hoover and Robinson 2007). Of these, I focus on how habitat characteristics affect predation and parasitism rates, because habitat modification may be the most cost-effective method for enhancing reproductive success (Robinson et al. 2000).

The spatial scale at which habitat characteristics are measured and how different species respond to those characteristics are two important considerations when assessing the link between habitat characteristics and the risk of predation and parasitism. Multiple species may nest in a given habitat patch, often within meters of one another, and yet may respond differently to habitat characteristics. In spite of this, many studies focus on a single species and results of the few multiple-species studies are often difficult to interpret. For example, Davidson and Knight (2001) found nest success in only one of three species (American

| Predation study | Region | Scale(s) measured | # Species |
|---|---|---|---|
| Best and Stauffer 1980 | Midwest | LS, NS | 15 |
| Borgmann and Rodewald 2002 | Midwest | LS, Terr, NS | 2 |
| Budnik et al. 2002 | Midwest/Central | LS, Terr, NS | 1 |
| Cain et al. 2003 | Mountain West | LS, Terr | 2 |
| Chase 2002 | Coastal West | LS, Terr, NS | 1 |
| Davidson and Knight 2001 | Mountain West | LS | 12 |
| Howe and Knopf 2000 | Mountain West | LS | 1 |
| Kus et al. 2008 | Southwest | LS, NS | 1 |
| Larison 1996 | Southwest | Terr,NS | 1 |
| Larison et al. 2001 | Southwest | Terr, NS | 1 |
| Parker 1999 | Midwest/Central | LS, Terr, NS | 1 |
| Saab and Vierling 2001 | Mountain West | LS | 1 |
| Small et al. 2007 | Mountain West | LS, NS | 1 |
| Tewksbury et al. 1998 | Mountain West | LS | 20 |
| Tewksbury et al. 2006 | Mountain West | LS | 4 |
| Twedt 2001 | Southeast | LS | 5 |
| Wilson and Cooper  1998 | Southeast | NS | 1 |

Table 8.1   Studies of predation region of study, scale(s) at which measurements were taken (NS = nest-site, T= territory, LS = landscape), and number of focal species measured in studies of habitat effects in riparian areas throughout the US.

Robin, *Turdus migratorius*) was negatively correlated with distance to edge, while two other species (Mourning Dove, *Zenaida macroura* and Yellow Warbler, *Dendroica petechia*) showed no such association. As a result, determining general processes that underlie habitat-nest predation/parasitism interactions has been elusive.

In addition to inter-specific variation in how habitat characteristics affect predation/parasitism, variation in how habitat characteristics are measured has made comparison across studies difficult. Habitat characteristics can be measured at multiple spatial scales, and the spatial scale at which habitat characteristics are measured can affect the likelihood of detecting a significant relationship between nest predation/parasitism and habitat characteristics (e.g., Stephens et al. 2004; Tewksbury et al. 2006). Many studies examine characteristics at a single spatial scale (e.g., Davidson and Knight 2001; Stoleson and Finch 2001; Howe and Knopf 2002; Table 8.1) and may miss important, unmeasured characteristics operating at different spatial scales (Bakermans and Rodewald 2006). Likewise, the often large and varied suite of potential nest predators, like that found in riparian areas, may respond differently to habitat characteristics at various spatial scales (Andren 1992; Chalfoun et al. 2002). Understanding how habitat characteristics measured at multiple scales affect nest predation and parasitism is critical for developing more general models of these interactions.

| Parasitism study | Region | Scale(s) measured | # Species |
|---|---|---|---|
| Airola 1986 | Mountain West | LS | 36 |
| Averill 1998 | Southwest | Terr, NS | 2 |
| Best and Stauffer 1980 | Midwest | LS, NS | 15 |
| Briskie et al. 1990 | Mountain West | NS | 2 |
| Brodhead et al. 2007 | Southwest | LS, NS | 1 |
| Budnik et al. 2002 | Midwest/Central | LS, Terr, NS | 1 |
| Cain et al. 2003 | Mountain West | LS, Terr | 2 |
| Davidson and Knight 2001 | Mountain West | LS | 12 |
| Helzel and Earnst 2006 | Mountain West | LS | 15 |
| Hochchka et al. 1999 | US (1 in SW) | LS | multiple |
| Howe and Knopf 2000 | Mountain West | LS | 1 |
| Larison 1996 | Southwest | LS, Terr, NS | 1 |
| Larison et al. 1998 | Southwest | LS, Terr, NS | 1 |
| Parker 1999 | Midwest/Central | LS, Terr, NS | 1 |
| Robinson et al. 2000 | Midwest | LS, NS | 15 |
| Sharp and Kus 2006 | Southwest | LS, Terr, NS | 11 |
| Staab and Morrison 1999 | Southwest | Terr, NS | 4 |
| Stoleson and Finch 2001 | Southwest | NS | 5 |
| Tewksbury et al. 1998 | Mountain West | LS | 20 |
| Tewksbury et al. 2006 | Mountain West | LS | 4 |
| Twedt 2001 | Southeast | LS | 5 |
| Uyehara and Whitfield 2000 | Southwest | Terr, NS | 1 |
| Ward and Smith 2000a | Mountain West | NS | 14 |
| Ward and Smith 2000b | Mountain West | LS | 1 |
| Whitfield 1990 | Southwest | LS, Terr, NS | 1 |
| Wilson and Cooper 1998 | Southeast | NS | 1 |

Table 8.2   Studies of parasitism region of study, scale(s) at which measurements were taken (NS = nest-site, T= territory, LS = landscape), and number of focal species measured in studies of habitat effects in riparian areas throughout the US.

Numerous studies have examined habitat characteristics associated with predation and parasitism (Table 8.2). They range from small-scale, microhabitat features immediately around the nest to large-scale landscape features, but these studies are scattered throughout many journals, and span many geographical regions and passerine species. Few studies have been conducted in riparian habitats, which are inherently different than other habitats due to their linear nature and the presence of water, resulting in higher densities and/or a greater diversity of predators and parasites. As a first step toward elucidating general patterns of the effect of habitat characteristics on the likelihood of predation and parasitism for riparian passerines across spatial scales, I reviewed the current literature to address two major questions: 1) at what spatial scale is the impact of habitat characteristics on nest predation and brood parasitism rates the greatest? 2) What habitat characteristics are most often associated with nest predation and brood parasitism? Answering these questions will facilitate the management of landscapes and future restoration plans for riparian habitats in ways that may enhance passerine reproductive success.

## METHODS

Because I wanted to understand how habitat characteristics affect nest predation and brood parasitism separately, I only included studies of real (versus artificial nest) open-cup nests in riparian areas that separated causes of nest failure (as opposed to those using "nest success" as the response variable). I defined riparian broadly to include as many studies as possible, using the presence of a body of water (lake, ephemeral/perennial streams, rivers, etc.) adjacent to the nesting habitat as the critical parameter. I included multiple-site studies if at least one of the sites was riparian habitat.

I conducted a literature search on JSTOR and CSA using the keywords parasitism or predation, riparian, and habitat. I also included additional studies cited within studies returned from these results if they satisfied the above criteria. I compiled a list of each habitat parameter and the direction of the observed effect from each study. I combined similar characteristics and excluded those that were only measured in one study in order to increase my power of inference. For the resulting characteristics, I ranked each based on the frequency of an observed effect, whether positive or negative.

I further divided vegetation and habitat parameters into three spatial scales: nest (microhabitat), territory (macrohabitat), and landscape, as defined by Robinson et al. (2000). Habitat characteristics at the nest-site scale were those within immediate proximity (less than 5 m) of the nest or those for which measurements were taken at the nest (e.g., nest height, species, concealment). The territory scale included parameters measured within 5-10 m of the nest (e.g., distance to gaps, number and size of surrounding vegetation). The landscape scale included parameters that were indicative of the entire patch (e.g., surrounding landscape type, patch size and shape). For characteristics that can be measured at multiple scales (e.g., canopy cover directly over a nest versus average canopy cover of the territory), I based my categorization on the author's description. Final categories for predation were; perimeter to area ratio, dominant habitat type (e.g., riparian, upland, restored), distance to edge, patch size, type of surrounding landscape (e.g., human-dominated, agriculture, forested), amount of canopy cover, ground cover, and understory vegetation, nest tree height, nest tree species, nest concealment, nest height, and distance from the nest to the edge of the nest patch (Table 8.3). For parasitism studies, the final categories were; patch shape (width and length), distance to the patch edge, patch size, dominant habitat type (same as for

| Habitat characteristic | Predation studies | | | |
|---|---|---|---|---|
| | # studies with + effect | # studies with - effect | # studies with no effect | % studies that found effect |
| **Landscape scale (12 studies)** | | | | 75.0 |
| Perimeter to area ratio | | | 3 | 0.0 |
| Dominant habitat type | 1 | | 3 | 25.0 |
| Distance to habitat edge | | 2 | 5 | 28.6 |
| Patch size | 2 | 1 | 3 | 50.0 |
| Type of surrounding area | 5 | | 3 | 62.5 |
| **Territory scale (8 studies)** | | | | 50.0 |
| Canopy cover | | | 4 | 0.0 |
| Amount of ground cover | | 1 | 6 | 14.3 |
| Amount of understory vegetation | 1 | 3 | 7 | 36.4 |
| **Nest-site scale (10 studies)** | | | | 50.0 |
| Nest tree height | | | 3 | 0.0 |
| Nest tree species | | | 7 | 0.0 |
| Concealment | | 1 | 8 | 11.1 |
| Nest height | | 2 | 8 | 20.0 |
| Distance to edge | | 1 | 3 | 25.0 |

Table 8.3 Number of studies that reported positive, negative and no effects of habitat characteristics on predation. When categorical characteristics (e.g., type of surrounding landscape, dominant habitat type) were reported to affect predation/parasitism, I included that study in the positive effect column.

predation), type of surrounding landscape (same as for predation), canopy height (at territory and nest-site scales), canopy cover (at territory and nest-site scales), type of ground cover (e.g., woody, shrub), amount of understory cover, distance to the nearest tree, nest tree height, distance to the edge of the nest patch, nest height, concealment, nest tree species, ground cover, nest tree dbh (diameter at breast height), and vertical foliage density (Table 8.4).

## RESULTS

I found 17 studies that met my criteria for the effects of habitat on predation (Table 8.1) and 26 studies that related the effects

of habitat on parasitism (Table 8.2). Of these, 12 studies examined both predation and parasitism concurrently. Six of the 12 studies examined characteristics at a single spatial scale (five at the landscape scale and one at the nest-site scale) and six examined characteristics at multiple scales. Characteristics at larger spatial scales were measured more often in both predation and parasitism studies. Of predation studies, 82% included at least one landscape characteristic, 41% at least one territory characteristic, and 59% at least one nest-site characteristic. Of parasitism studies, 73% included at least one landscape characteristic, 38% at least one territory characteristic, and 65% at least one nest-site

| Habitat characteristic | Parasitism studies | | | |
|---|---|---|---|---|
| | # studies with + effect | # studies with - effect | # studies with no effect | % studies that found effect |
| **Landscape scale (18 studies)** | | | | 72.2 |
| Patch shape | | 1 | 2 | 33.3 |
| Distance to habitat edge | 1 | 2 | 5 | 37.5 |
| Patch size | 1 | | 1 | 50.0 |
| Dominant habitat type | 4 | | 3 | 57.1 |
| Type of surrounding area | 3 | 3 | 2 | 75.0 |
| **Territory scale (10 studies)** | | | | 80.0 |
| Canopy height | | | 5 | 0.0 |
| Type of ground cover | 1 | 1 | 5 | 28.6 |
| Understory cover | 3 | 3 | 9 | 40.0 |
| Canopy cover | 2 | 1 | 3 | 50.0 |
| Distance to nearest tree | | 2 | 2 | 50.0 |
| **Nest-site scale (16 studies)** | | | | 68.8 |
| Canopy height | | | 3 | 0.0 |
| Nest tree height | | 1 | 8 | 11.1 |
| Distance to edge | | 1 | 4 | 20.0 |
| Canopy cover | | 1 | 4 | 20.0 |
| Nest height | | 4 | 12 | 25.0 |
| Concealment | 1 | 2 | 8 | 27.3 |
| Nest tree species | 2 | 1 | 6 | 33.3 |
| Ground cover | | 2 | 2 | 50.0 |
| Nest tree dbh | | 3 | 1 | 75.0 |
| Vertical foliage density | 2 | 1 | 1 | 75.0 |

Table 8.4 Number of studies that reported positive, negative and no effects of habitat characteristics on parasitism. When categorical characteristics (e.g., type of surrounding landscape, dominant habitat type) were reported to affect predation/parasitism, I included that study in the positive effect column.

characteristic. The majority of predation (71%) and parasitism (54%) studies included only one or two focal nesting species (Tables 8.1 and 8.2). Only one predation and four parasitism studies were conducted in low elevation southwestern ecosystems.

## Predation

Habitat characteristics measured at the landscape scale were most often associated with predation; overall, 75.0% of studies measuring landscape scale characteristics observed an effect (Table 8.3). Territory and nest-site characteristics were associated with predation less often (50.0% of studies at each scale). Four of five (80.0%) landscape-scale characteristics measured in multiple studies had an effect on predation, while two of three (66.7%) and three of five (60%) characteristics measured at the territory and nest-site scales, respectively, were associated with predation. All factors with an observed effect in more than 50% of studies were landscape scale characteristics, although fewer characteristics were measured at the territory scale, versus the landscape or nest-site scales.

The factors most often associated with predation included patch size/dimension and type of surrounding landscape (Table 8.3). The direction of the effect of patch size on predation risk showed conflicting results. Patch width was positively correlated with predation in two studies (Budnik et al. 2002; Chase et al. 2002), patch length was negatively correlated with predation in one study (Budnik et al. 2002), and overall patch size showed no correlation with predation in three studies (Tewksbury et al. 1998; Davidson and Knight 2001; Cain et al. 2003). Surrounding landscapes consisting of human-modified (rural vs. urban, (Borgmann and Rodewald 2004), golf course/parks, (Kus et al. 2008), agriculture (Tewksbury et al. 2006), and forest (e.g., forest buffer; Tewksbury et al. 2006) were positively associated with predation.

## Parasitism

Characteristics at the landscape and nest-site scales were comparably correlated with parasitism (72.2 and 68.8%, respectively; Table 8.4), and characteristics at the territory scale were slightly more often correlated with parasitism (80.0%; Table 8.4). Of characteristics measured in multiple studies, five of five (100%) landscape scale characteristics, four of five (80%) territory characteristics, and nine of ten (90%) nest-site characteristics were associated with parasitism (Table 8.4). Characteristics that were correlated with parasitism in more than 50% of studies were found at all three spatial scales.

The factors most often associated with parasitism included patch size, dominant habitat type, type of surrounding landscape, canopy cover (at the territory scale), distance from the nest to the nearest tree, ground cover, nest tree size, and vertical foliage density (Table 8.4). Larger patches were positively correlated with parasitism, while distance to the nearest tree, amount of ground cover, and nest tree size showed negative correlations with parasitism. Dominant habitat type, the amount and type of surrounding landscape, canopy cover, and vertical foliage density showed conflicting results. Parasitism rates were higher in habitats dominated by riparian vegetation than other habitat types in three studies (Ward and Smith 2000b; Twedt et al. 2001; Budnik et al. 2002), and in mature riparian patches versus restored patches in one study (Larison et al. 1998). Parasitism rates were higher in habitats surrounded by more human-modified or agricultural landscapes in three studies (Best and Stauffer 1980; Sharp and Kus 2006; Tewksbury et al. 2006), and were lower in habitats surrounded by more forest (including habitat buffers) in three studies (Tewksbury et al. 1998; Hochachka et al. 1999; Tewksbury et al. 2006).

## DISCUSSION

### General Patterns of Spatial Scale

Habitat characteristics measured at the landscape scale were associated with both predation and parasitism. This relationship may have been driven by the fact that abundance and distribution patterns of brood parasites and generalist nest predators are affected by landscape scale habitat characteristics. Cowbirds are more abundant in riparian sites surrounded by agriculture (Thompson 1994; Rodewald and Yahner 2001), and human-dominated landscapes (Tewksbury et al. 1998). Other landscape-scale habitat characteristics such as the dominant vegetation in the breeding habitat (Tewksbury et al. 1998) and the amount and distribution of feeding areas (Verner and Ritter 1983; Tewksbury et al. 1998; Chace 2004) can also affect cowbird distribution patterns. Likewise, abundance of some potential nest predators is positively related to the amount of urbanization (Bakermans 2003) and may account for increased nest predation in more human-dominated landscapes (Bakermans and Rodewald 2006). Predator species composition may likewise be related to large-scale habitat characteristics (Thompson and Burhans 2003) and may thus respond differently to characteristics at the landscape scale.

Unlike predation, brood parasitism rates were associated with characteristics measured at smaller spatial scales, in addition to those at the landscape scale. The association of characteristics such as ground cover and concealment with the likelihood of parasitism may reflect the fact that cowbirds rely mainly on visual cues to locate nests. Small-scale factors (canopy cover, distance to the nearest tree, ground cover, and vertical foliage density) may conceal host species movements and visually oriented cowbirds may be less likely to detect these nests. Conversely, these same characteristics would be less likely to affect the ability of predators using olfactory cues or other search methods to detect nests. The greater diversity of predators, each using different nest-search methods, likely confounds any general relationship between predation and habitat characteristics at smaller spatial scales, since not all predator types would be expected to respond to the same smaller scale characteristics.

### Habitat characteristics associated with predation and parasitism

Nest predation was most often associated with the type of surrounding landscape and the size (and dimensions) of the breeding patch. Predation rates were positively correlated with the amount of agriculture (Tewksbury et al. 2006), urbanization (Borgmann and Rodewald 2004; Kus et al. 2008), and forest cover (Tewksbury et al. 2006) of the land surrounding the breeding patch. Similarly, parasitism rates were positively correlated with amount of agriculture (Sharp and Kus 2006), human presence (e.g., degree of urbanization surrounding the habitat, Tewksbury et al. 1998), and negatively correlated with the amount of forest cover (Tewksbury et al. 1998; Tewksbury et al. 2006) in the landscape surrounding the breeding patch. As discussed above, these patterns likely reflect general distribution and abundance patterns of cowbirds and a diverse suite of predators; habitats that support more cowbirds and predators likely exhibit higher rates of parasitism and predation. The most striking difference was in the response to surrounding forest cover, with cowbirds showing a negative response not unexpected given their evolutionary association with open habitats, while predation was positively associated with forest cover.

Larger patches, and particularly wider patches, were positively correlated with predation rates in some studies (Budnik et al. 2002; Chase 2002) but other studies failed to detect an association (Tewksbury et al. 1998;

Davidson and Knight 2001; Cain et al. 2003). Patch size was also positively correlated with parasitism rates in one study (Brodhead et al. 2007), yet another (Davidson and Knight 2001) failed to detect an association. Cowbirds are a highly mobile species, travelling as much as 6.7 (Rothstein et al. 1984) or 14 km (Curson et al. 2000) in one day, thus the size of the patch may not be a limiting factor. However, larger patches may support a more diverse predator community due to a greater availability of food, resulting in an increase in predation rates. Patch size and the landscape surrounding the patch may also be connected. Bakermans and Rodewald (2006) found that Acadian Flycatchers (Empidonax virescens) needed wider patches and successful nests were located farther from the habitat edge when surrounded by more urban environments than when surrounded by other habitats. Therefore, patches that are surrounded by more urban environments may need to be bigger than those surrounded by more rural or natural landscapes.

The amount of core area, as determined by both shape and size of the habitat, may be related to the predation and parasitism risk at a given patch. Since riparian habitats tend to be linear, the width of a patch may be of particular importance in these areas. Of the studies I reviewed, only one examined the association of the width of a patch with parasitism (Budnik et al. 2002) and only two examined the effect of patch width on predation rates (Budnik et al. 2002; Chase 2002). Three additional parasitism studies (Tewksbury et al. 1998; Robinson et al. 2000; Brodhead et al. 2007) and three predation studies (Tewksbury et al. 1998; Davidson and Knight 2001; Cain et al. 2003) examined the effect of overall patch size. Patch sizes ranged from 0.1 ha to 167 ha in these studies, and did not seem to affect the likelihood of detecting an effect. However, because the size of the patch is clearly important, future analyses should investigate at which patch

sizes predation and parasitism rates are no longer affected.

Canopy cover density measured at the territory scale (Averill 1998) and vertical foliage density (Larison et al. 1998) were positively correlated with parasitism rates, while distance from the nest to the nearest tree (Averill 1998) and ground cover (Averill 1998; Parker 1999) were negatively correlated with parasitism rates. Each of these characteristics is a measure of the degree of concealment surrounding a nest. More cover above and around the nest and less distance between trees would likely conceal host species movements to and from the nest, making nest detection difficult. Although the positive correlation between canopy cover at the territory scale seemingly implies otherwise, the ability of cowbirds to use multiple nest-detection methods may be masking this effect and may account for the lack of a measureable effect when canopy cover is measured at the nest-site scale (though one study did detect a negative relationship between cover and parasitism rates; Averill 1998). Ground cover is unlikely to impact cowbird detection of nests, and the conflicting results here (a negative correlation and no correlation) reflect this fact. Furthermore, studies have revealed that host species density may impact likelihood of parasitism (Arcese et al. 1996, Gates and Giffen 1991, Freeman et al. 1990), and the studies in this review likely had varying host to cowbird ratios, which may impact the detected correlations.

## CONCLUSION

Passerine birds nesting in riparian habitats are faced with dual risks of predation and parasitism. Where possible, habitat restoration and modification aimed at increasing passerine nest success will be most effective if it can minimize both of these pressures. The results of this literature review suggest that landscape scale habitat characteristics are associated with both predation and parasitism, a finding that is corroborated in recent work examining predation (Robinson et al. 1995; Thompson et al. 2003) and parasitism (Tewksbury et al. 2006). Although these studies were conducted in eastern landscapes, management efforts should use information from all available studies until enough research is conducted in arid southwestern habitats. Patch size and the type of surrounding landscape were the two characteristics most often associated with both parasitism and predation. Unfortunately, inconsistent methods of measuring patch size and the lack of any measurement of patch size in the majority of the studies I examined make interpretations across studies difficult. I recommend that future studies seek to determine the effect of patch size on predation and parasitism by more carefully quantifying predation and parasitism effects across a range of patch sizes to elucidate the patch sizes at which these rates are most strongly reduced. Specifically, measuring overall size (area), width, and length of the patch will be more informative for managers seeking to design and implement restoration plans that minimize the risk of predation and parasitism to songbirds. More urbanized landscapes are associated with higher predation and parasitism rates; therefore the surrounding landscape is also an important characteristic to incorporate into future restoration and management plans.

While a diverse suite of predators and host species and their corresponding responses to habitat characteristics make conclusions about specific habitat characteristics difficult, this could be improved by a unified approach to measuring habitat characteristics. Several characteristics that seem logically important, but were rarely measured include distance to edge, canopy cover (at the nest-site scale), and nest concealment. All of these characteristics are likely to impact the ability of cowbirds and nest predators to locate or access nests, but the different methods researchers employ to measure them (if they measure them at all) may be masking the effect that these habitat characteristics have on predation and parasitism rates. Habitat restoration and management of landscapes will benefit from more rigorous, long-term studies incorporating measurements of these characteristics, as well as patch size and surrounding landscape, especially in areas that experience high rates of both predation and parasitism.

## ACKNOWLEDGMENTS

This manuscript was greatly improved by helpful suggestions and comments by T. Theimer, M.A. McLeod, and E. Paxton and reviewer comments from B. Kus and M.J. Whitfield. K. Stumpf was supported by a National Science Foundation IGERT fellowship and National Science Foundation BIOTEC fellowship.

## LITERATURE CITED

Airola, D. 1986. Brown-headed Cowbird parasitism and habitat disturbance in the Sierra Nevada. The Journal of Wildlife Management 50(4):571-575.

Anderson, B.W., and R.D. Ohmart. 1977. Vegetation structure and bird use in the lower Colorado River valley. Pages 23-34 in R.R. Johnson and D.A. Jones, technical coordinators. Important, preservation and management of riparian habitat: a symposium. United States Forest Service, General Technical Report RM-GTR-43.

Andren, H. 1992. Corvid density and nest predation in relation to forest fragmentation:

A landscape perspective. Ecology 73(3):794-804.

Arcese, P., J. N. M. Smith, and M. I. Hatch. 1996. Nest predation by cowbirds and its consequences for passerine demography. Proceedings of the National Academy of Sciences of the United States of America 93(10):4608-4611.

Averill, A. 1998. Brown-headed Cowbird parasitism of Neotropical migratory songbirds in riparian areas along the lower Colorado River. M.S. Thesis, University of Arizona.

Bakermans, M.H. 2003. Hierarchical habitat selection by the Acadian Flycatcher: Implications for conservation of riparian forests. M.S. Thesis, The Ohio State University, Columbus, Ohio.

Bakermans, M. H., and A. Rodewald. 2006. Scale-dependent habitat use of Acadian Flycatcher (Empidonax virescens) in central Ohio. The Auk 123(2):368-382.

Best, L. B. 1978. Field sparrow reproductive success and nesting ecology. The Auk 95:9-22.

Best, L. B., and F. Stauffer. 1980. Factors affecting nesting success in riparian bird communities. The Condor 82:149-158.

Borgmann, K. L., and A. D. Rodewald. 2004. Nest predation in an urbanizing landscape: The role of exotic shrubs. Ecological Applications 14(6):1757-1765.

Briskie, J. V., S. G. Sealy, and K. A. Hobson. 1990. Differential parasitism of Least Flycatchers and Yellow Warblers by the Brown-headed Cowbird. Behavioral Ecology and Sociobiology 27(6):403-410.

Brittingham, M. C., and S. A. Temple. 1983. Have cowbirds caused forest songbirds to decline? BioScience 33(1):31-35.

Brodhead, K. M., S. H. Stoleson, and D. M. Finch. 2007. Southwestern Willow Flycatchers (Empidonax traillii extimus) in a grazed landscape: Factors influencing brood parasitism. The Auk 124(4):1213-1228.

Budnik, J. M., F. R. Thompson III, and M. R. Ryan. 2002. Effect of habitat characteristics on the probability of parasitism and predation of Bell's Vireo nests. The Journal of Wildlife Management 66(1):232-239.

Cain, J. W., M. L. Morrison, and H. L. Bombay. 2003. Predator activity and nest success of Willow Flycatchers and Yellow Warblers. The Journal of Wildlife Management 67(3):600-610.

Chace, J. F. 2004. Habitat selection by sympatric brood parasites in southeastern Arizona: The influence of landscape, vegetation, and species richness. The Southwestern Naturalist 49(1):24-32.

Chalfoun, A., F. Thompson, and M. Ratnaswamy. 2002. Nest predators and fragmentation: A review and meta-analysis. Conservation Biology 16(2):306-318.

Chase, M. K. 2002. Nest site selection and nest success in a Song Sparrow population: The significance of spatial Variation. The Condor 104(1):103-116.

Curson, D.R., C.B. Goguen, and N.E. Mathews. 2000. Long distance commuting by Brown-headed Cowbirds in New Mexico. Auk 117(3):795-799.

Davidson, A. S., and R. L. Knight. 2001. Avian nest success and community composition in a western riparian forest. The Journal of Wildlife Management 65(2):334-344.

Finch, D. M. 1983. Brood parasitism of the Abert's Towhee: Timing, frequency, and effects. The Condor 85(3):355-359.

Freeman, S., D. F. Gori, and S. Rohwer. 1990. Red-winged Blackbirds and Brown-headed Cowbirds: Some aspects of a host-parasite relationship. The Condor 92(2):336-340.

Gates, J. E., and N. R. Giffen. 1991. Neotropical migrant birds and edge effects at a forest-stream ecotone. The Wilson Bulletin 103(2):204-217.

Gustafson, E. J., M. G. Knutson, G. J. Niemi, and M. Friberg. 2002. Evaluation of spatial models to predict vulnerability of forest birds to brood parasitism by cowbirds.

Ecological Applications 12(2):412-426.

Heltzel, J. M., and S. L. Earnst. 2006. Factors influencing nest success of songbirds in aspen and willow riparian areas in the Great Basin. The Condor 108:842-855.

Heske, E. J., S. K. Robinson, and J. D. Brawn. 2001. Nest predation and Neotropical migrant songbirds: Piecing together the fragments. Wildlife Society Bulletin 29(1):52-61.

Hochachka, W. M., T. E. Martin, V. Artman, C. R. Smith, S. J. Hejl, D. E. Andersen, D. Curson, L. Petit, N. E. Mathews, T. M. Donovan, E. E. Klaas, P. B. Wood, J. C. Manolis, K. P. McFarland, J. V. Nichols, J. C. Bednarz, D. M. Evans, J. P. Duguay, S. Garner, J. J. Tewksbury, K. L. Purcell, J. Faaborg, C. B. Goguen, C. Rimmer, R. Dettmers, M. G. Knutson, J. A. Collazo, L. Garner, D. Whitehead, and G. Geupel. 1999. Scale dependence in the effects of forest coverage on parasitism by Brown-headed Cowbirds. Studies in Avian Biology 18:80-88.

Hoover, J. P., M. C. Brittingham, and L. J. Goodrich. 1995. Effects of forest patch size on nesting success of Wood Thrushes. The Auk 112(1):146-155.

Hoover, J. P., and S. K. Robinson. 2007. Retaliatory mafia behavior by a parasitic cowbird favors host acceptance of parasitic eggs. Proceedings of the National Academy of Sciences 104(11):4479-4483.

Howe, W. H., and F. L. Knopf. 2000. The role of vegetation in cowbird parasitism of Yellow Warblers. Pages 200-203 in Smith, J. N. M., T. L. Cook, S. I. Rothstein, S. K. Robinson, and S. G. Sealy, editors. Ecology and Management of Cowbirds and Their Hosts: Studies in the Conservation of North American Passerine Birds. University of Texas Press, Austin, TX.

Kus, B., B. Peterson, and D. Deutschman. 2008. A multiscale analysis of nest predation on Least Bell's Vireos (*Vireo bellii pusillus*). The Auk 125(2):277-284.

Larison, B., S. A. Laymon, P. L. Williams, and T. B. Smith. 1998. Song Sparrows vs. cowbird brood parasites: Impacts of forest structure and nest-site selection. The Condor 100(1):93-101.

Martin, T. E. 1992. Breeding season productivity: What are the appropriate habitat features for management? Pages 455-473 in Hagan, J. M. a. J., D.W., editors. Ecology and Conservation of Neotropical migrant landbirds. Smithsonian Institution Press, Washington, D.C.

Ohmart, R. D. 1994. The effects of human-induced changes in the avifauna of western riparian habitats. Studies in Avian Biology 15:273-285.

Parker, T. H. 1999. Responses of Bell's Vireos to brood parasitism by the Brown-headed Cowbirds in Kansas. The Wilson Bulletin 111(4):499-504.

Robinson, S., F. Thompson, T. M. Donovan, D. Whitehead, and J. Faaborg. 1995. Regional forest fragmentation and the nesting success of migratory birds. Science 267(5206):1987-1990.

Robinson, S., J. P. Hoover, and J. R. Herkert. 2000. Cowbird parasitism in a fragmented landscape: Effects of tract size, habitat, and abundance of cowbirds and hosts. Pages 280-297 in Smith, J. N. M., T. L. Cook, S. I. Rothstein, S. K. Robinson, and S. G. Sealy, editors. Ecology and Management of Cowbirds and Their Hosts: Studies in the Conservation of North American Passerine Birds. University of Texas Press, Austin, TX.

Rodewald, A. Y., and R. H. Yahner. 2001. Avian nesting success in forested landscapes: Influence of landscape composition, stand and nest-patch microhabitat, and biotic interactions. The Auk 118(4):1018-1028.

Rothstein, S. I., J. Verner, and E. Steven. 1984. Radio-tracking confirms a unique diurnal pattern of spatial occurrence in the parasitic Brown-headed Cowbird. Ecology 65(1):77-88.

Saab, V. A., and K. T. Vierling. 2001. Reproductive success of Lewis's Woodpecker in burned pine and cottonwood riparian forests. The Condor 103:491-501.

Sharp, B. L., and B. E. Kus. 2006. Factors influencing the incidence of cowbird parasitism of Least Bell's Vireos. Journal of Wildlife Management 70(3):682-690.

Small, S. L., F. R. Thompson III, G. R. Geupel, and J. Faaborg. 2007. Spotted Towhee population dynamics in a riparian restoration context. 109:721-733.

Stephens, S., D. N. Koons, J. J. Rotella, and D. W. Willey. 2004. Effects of habitat fragmentation on avian nesting success: A review of the evidence at multiple spatial scales. Biological Conservation 115:101-110.

Stoleson, S. H., and D. M. Finch. 2001. Breeding bird use of and nesting success in exotic Russian Olive in New Mexico. The Wilson Bulletin 113(4):452-455.

Tewksbury, J. J., L. Garner, S. Garner, J. D. Lloyd, and V. Saab, and T.E. Martin. 2006. Tests of landscape influence: Nest predation and brood parasitism in fragmented ecosystems. Ecology 87(3):759-768.

Tewksbury, J. J., S. J. Hejl, and T. E. Martin. 1998. Breeding productivity does not decline with increasing fragmentation in a western landscape. Ecology 79(8):2890-2903.

Thompson, F.R. III. 1994. Temporal and spatial patterns of breeding Brown-headed Cowbirds in the midwestern United States. The Auk 111(4):979-990.

Thompson, F. R., III and D. E. Burhans. 2003. Predation of songbird nests differs by predator and between field and forest habitats. The Journal of Wildlife Management 67(2):408-416.

Twedt, D. J., R. R. Wilson, J. Henne-Kerr, and R. B. Hamilton. 2001. Nest survival of forest birds in the Mississippi alluvial valley. Journal of Wildlife Management 65(3):450-460.

Uyehara, J. C., and M. J. Whitfield. 2000. Association of cowbird parasitism and vegetative cover in territories of Southwestern Willow Flycatchers. Pages 204-209 in Smith, J. N. M., T. L. Cook, S. I. Rothstein, S. K. Robinson, and S. G. Sealy, editors. Ecology and Management of Cowbirds and Their Hosts: Studies in the Conservation of North American Passerine Birds. University of Texas Press, Austin, TX.

Verner, J., and L. V. Ritter. 1983. Current status of the Brown-headed Cowbird in the Sierra National Forest. The Auk 100(2):355-368.

Ward, D., and J. N. M. Smith. 2000a. Brown-headed Cowbird parasitism results in a sink population in Warbling Vireos. The Auk 117(2):337-344.

Ward, D., and J. N. M. Smith. 2000b. Interhabitat differences in parasitism frequencies by Brown-headed Cowbirds in the Okanagan Valley, British Columbia. Pages 210-219 in Smith, J. N. M., T. L. Cook, S. I. Rothstein, S. K. Robinson, and S. G. Sealy, editors. Ecology and Management of Cowbirds and Their Hosts: Studies in the Conservation of North American Passerine Birds. University of Texas Press, Austin, TX.

Wilson, M. F. 1998. Nest predation and avian species diversity in northwestern forest understory. Ecology 79(7):2391-2402.

# Synergy Between Human and Environmental Systems: Planning and Managing Frameworks

# KNOWING THE COGS AND WHEELS: USING BIOBLITZ METHODS TO RAPIDLY ASSESS FLORAS AND FAUNAS

*Walter Fertig, Linda Whitham and John Spence*

## ABSTRACT

Although protecting representative examples of the full array of native species is the objective of many conservation projects, circumstances often dictate that actions can be taken before the biodiversity of a site is fully documented. Some approaches can be inefficient, as sites may contribute few new species to the protected area network, while other lands with high species richness or complementarity are bypassed. Bioblitzes are a potential first step in acquiring species presence information for an area of potential conservation interest. Conducted by teams of taxonomic experts and interested members of the public, bioblitzes are intensive 24-48 hour surveys of the entire biota (plants, vertebrates, invertebrates) of a specific area with the goal of documenting as many species as possible. In July 2007 and May 2008, The Utah Nature Conservancy sponsored two bioblitzes in the Deer Creek drainage, east of Boulder, Utah and adjacent to Grand Staircase-Escalante National Monument and Dixie National Forest. A team of agency and university botanists, bryologists, biologists, entomologists, ecologists, and interested landowners participated in the bioblitz and documented 588 taxa in the area and discovered more than two dozen significant new distribution records. Although comparable numbers of species were observed each year, there was nearly 50% turnover in recorded species in the second year. The two bioblitzes increased the known vascular plant and vertebrate richness of the drainage by nearly 20%. The bioblitzes demonstrated the high biological significance of the Deer Creek area and fostered increased interest in conservation in the local community and communication among biologists from multiple disciplines.

## INTRODUCTION

The world's biological diversity is increasingly under threat from a multitude of forces, not limited to habitat destruction, over-harvest, competition by invading species, pollution, and climate change. Conservation biology is a multidisciplinary field of applied science dedicated to stemming the loss of native biological diversity. One of the primary strategies of conservation biology is to establish networks of protected areas that capture adequate samples of the full array of native species and ecosystems of an area (Margules and Sarkar 2007; Stein et al. 2000). Developing such a network requires information on the ecology, distribution, and identity of plant, vertebrate, invertebrate, fungal, and microbe species of a region. More importantly, this information is needed quickly, while the window of opportunity to protect vanishing species and wild lands is still open.

The idea of assembling teams of taxonomic specialists to quickly establish baseline conditions of biological diversity in threatened habitats took hold in the late 1980s with the development of the Rapid Assessment Program (RAP) by several international conservation groups (Wilson 1992). Initially, RAP was used primarily in areas of known or suspected high species richness in the tropics., In the mid 1990s similar methods began to be applied to less exotic landscapes such as suburban and city parks in North America, Europe, and Australia. Now commonly called bioblitzes, these efforts differ from traditional RAPs by involving both skilled and untrained amateurs along with professional taxonomists, focusing on specific target areas for a 24-48 hour period, and in placing greater emphasis on education. Bioblitzes have become a popular tool for gathering baseline data on local biodiversity, encouraging collaboration between professional biologists and the public, and promoting interest in nature (Conniff 2000; Lundmark 2003).

In July 2007 and May 2008, The Utah Nature Conservancy (TNC) conducted a bioblitz along the middle reach of the Deer Creek watershed near Boulder, Utah (see Figure 9.1). Deer Creek is a perennial tributary of the Escalante River and is bordered by private inholdings within the Grand Staircase-Escalante National Monument and Dixie National Forest. The general area had previously been identified as a landscape of conservation significance by TNC (Tuhy et al. 2002) because of the presence of a perennial water feature for wildlife, the occurrence of several rare plant and animal species, and significant threats from residential development around Boulder. Although some baseline data existed on rare species and the general flora and fauna of neighboring Grand Staircase-Escalante National Monument, relatively little information was known from the middle reach of Deer Creek itself. The Nature Conservancy was interested in employing bioblitz methods to document the biodiversity of the middle reach of the watershed, verify the significance of the area, identify potential management concerns, and build interest and support for conservation efforts in the Boulder community.

## STUDY AREA

Deer Creek originates below the south summit of Boulder Mountain at 3170 m in Dixie National Forest and meanders for nearly 27 km to its confluence with Boulder Creek in Grand Staircase-Escalante National Monument. The upper third of the drainage lies above 2400 m and is dominated by subalpine conifer and Ponderosa pine forests on volcanic-derived soils. The lower two-thirds of the watershed is mostly contained within a deep canyon that cuts more than 800 m into sandstone slickrock. The 2007-2008 bioblitz focused primarily on a 1620 ha area centered on private lands along the middle reach of Deer Creek above its confluence with a small, perennial tributary called Nazer Draw.

The mesas and canyons enclosing the Deer Creek study area are comprised mostly of Jurassic-age Navajo Sandstone cliffs that may be as thick as 4270 m (Doelling et al. 2003). Mesas on the west side of Deer Creek are capped by a thin, limey layer of Jurassic Page Sandstone. Both sandstone formations may be covered by patches of Oligocene or Miocene age basaltic boulders derived from alluvial outwash from Boulder Mountain. The bottom of Deer Creek Canyon consists of Quaternary sand and clay alluvium derived from local sandstone and volcanic deposits (Doelling et al. 2003).

Six main vegetation types are recognized in the study area (Fertig et al. 2009) derived from the classification of Schulz (2007) and Anderson et al. (1998). Alluvial bottomlands in Deer Creek Canyon are dominated by riparian forests of narrowleaf cottonwood (*Populus angustifolia*), Fremont cottonwood (*P. fremontii*), water birch (*Betula occidentalis*), and silver buffaloberry (*Shepherdia argentea*) interrupted by wet meadows and willow thickets of Nebraska sedge (*Carex nebrascensis*), Baltic rush (*Juncus arcticus*), coyote willow (*Salix exigua*), and yellow willow (*S. eriocephala var. watsonii*). Canyon bottoms above the floodplain of Deer Creek are vegetated by big sagebrush (*Artemisia tridentata*)/fourwing saltbush (*Atriplex canescens*) shrublands with a sparse understory of Indian ricegrass (*Stipa hymenoides*), galleta (*Hilaria jamesii*), and blue grama (*Bouteloua gracilis*). Some floodplain sites were cleared in the past for pasture lands and are now dominated by weedy annuals and introduced grasses, including smooth brome (*Bromus inermis*), orchardgrass (*Dactylis glomerata*), and crested wheatgrass (*Agropyron cristatum*). Navajo sandstone-derived dunes on slopes and mesa tops above the canyon contain open woodlands of ponderosa pine (*Pinus ponderosa*), two-needle pinyon (*P. edulis*), Utah juniper (*Juniperus osteosperma*), and single-leaf ash (*Fraxinus anomala*). Exposures of Navajo Sandstone slickrock are unvegetated or sparsely covered by mountain mahogany (*Cercocarpus intricatus*), rubber rabbitbrush (*Chrysothamnus nauseosus var. leiospermus*), or isolated ponderosa pine (Fertig et al. 2009; Schulz 2007).

## METHODS

Prior to the first bioblitz, potential species lists were assembled for vascular plants and vertebrates based on a review of existing published and unpublished checklists and inventory work elsewhere in the Deer Creek watershed or on the adjacent Grand Staircase-Escalante National Monument (Bosworth 2003; Evenden 1999; Fertig 2005; Flinders et al. 2002; Oliver 2003; Welsh and Atwood 2002). Additional information was compiled on vertebrates observed by local Boulder residents (summarized in Fertig et al. 2009). These draft lists (consisting of nearly 1600 taxa) were used to assist survey teams in quickly tallying their results and for highlighting new or unusual discoveries.

The Deer Creek bioblitzes occurred over two 48-hour periods on 20-22 July 2007 and 4-6 May 2008. Survey teams were organized by major biological discipline (birds, mammals, herptiles and fish, invertebrates, bryophytes and lichens, and vascular plants) and included at least one subject expert and one or more assistants and note-takers. Team leaders used a baseline vegetation map (Fertig et al. 2009) and a reconnaissance field trip to identify survey sites from a cross-section of all major habitat types. Identifications for most taxonomic groups were done in the field. Vascular plant, bryophyte, and lichen specimens that could not be reliably identified on-site or which represented noteworthy distribution records were collected for later study and ultimately for deposit in major regional herbaria. Photographs and detailed field notes were prepared at each sample location and for any rare or unusual species.

Members of the invertebrate team captured diurnal insect specimens using hand-held nets and pitfall traps buried in sandy dune and alluvial bottomland sites. At night, insects were attracted to white sheets with a bug zapper and black light. Insect specimens were pinned or preserved in alcohol for later identification. Specimens were identified to order or family in the field when possible and assigned to morpho-species. Many still are in the process of being identified by their collectors or have been sent to specialists for confirmation.

The vertebrate teams used binoculars and their skills at deciphering avian songs to identify bird species. Surveys of fish, amphibians, and reptiles were done by observation rather than trapping. The mammal team identified species by analyzing tracks, scats, and other sign (such as trees cut by beaver) and by direct observation. Sherman traps were employed to capture small rodents. One hundred traps along two transect grids were set up each night and opened the following morning. All identifications of small mammals were done in the field and trapped animals were returned to the wild unharmed.

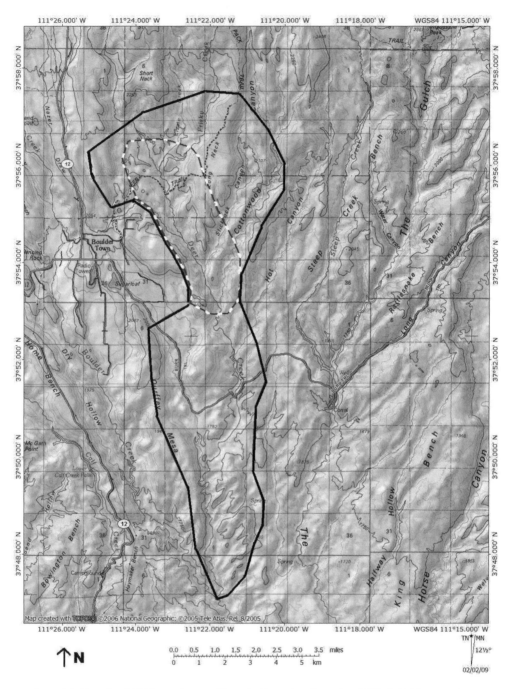

Figure 9.1   Deer Creek bioblitz study area.  Watershed boundary shown in solid black, bioblitz area shown
by dotted line.

## RESULTS

### Vascular Plants (team leader Walter Fertig)

In July 2007, the bioblitz plant team recorded 256 species of vascular plants in the Deer Creek study area (Table 9.1). Fewer plant species were reported in May 2008, in part because many had not emerged or flowered yet and because surveys focused on sites not visited the previous year. Of the 159 vascular plant species observed in 2008, 34 had not been recorded in July 2007 (just over 21%). Between the two bioblitzes, 290 vascular plant species were found in the study area, representing nearly 52% of all plant and animal species recorded in 2007-08. Fertig et al. (2009) compiled a list of 66 additional vascular plant species known from elsewhere in the Deer Creek watershed in Grand Staircase-Escalante National Monument and Dixie National Forest but not found in the 2007-2008 bioblitz. If all of these additional species were present within the study area, the total flora would contain at least 356 vascular plant taxa, and the two bioblitzes would have detected 81.5% of the known species. Based on other vascular plant species known from the broader "Escalante Canyons" subregion of the national monument, the total flora of the Deer Creek study area may be as high as 400-500 taxa (Fertig 2005).

The flora of the Deer Creek study area (as currently known) contains 9.7% of the 3569 native and naturalized vascular plant species known from Utah (Welsh et al. 2008) and 35.6% of the flora of the Grand Staircase-Escalante National Monument (Fertig 2005). At least 20 species from private lands along Deer Creek are not presently known or protected within the monument. Populations of five plant species tracked as "species of concern" by the state natural heritage program were discovered during the 2007-2008 bioblitz (Utah Division of Wildlife Resources 1998). Most notable of the new discoveries was an occurrence of Neese's pepperplant (*Lepidium montanum var. neeseae*) from shallow sand dunes above the confluence of Deer Creek and Nazer Draw. This species was formerly a candidate for potential listing under the Endangered Species Act and was previously known entirely from the Death Hollow and Hell's Backbone areas of northern Garfield County, Utah. The population of 100-150 individuals found in the 2008 bioblitz is the first to be recorded from east of Boulder and is only the fifth extant population currently known.

### Bryophytes and Lichens (team leader John Spence)

Mosses and liverworts were not included in the 2007 bioblitz. In May 2008 the bryophyte team identified 35 species of bryophytes from the Deer Creek area representing 16 families (Table 9.1). Based on habitat availability and the known distribution of bryophytes elsewhere in southwestern Utah, as many as 60 bryophyte taxa may occur in the watershed (Fertig et al. 2009). Among the more noteworthy species discovered in 2008 were *Crumia latifolia* (a rare moss in Utah), found along basalt boulders in Deer Creek, and the first record of *Anomobryum julaceum var. mexicanum* for the state of Utah from a small hanging garden.

Lichens were also not surveyed in the 2007 bioblitz, but 16 species were documented in 2008 (Table 9.1). Over 90% of the sampled lichens were species with crustose or foliose growth forms. Only one squamulose species and no fruticose taxa were observed. Team leader John Spence estimates that the total number of lichens in the Deer Creek drainage may be as high as 130 based on the availability of suitable habitat (Fertig et al. 2009).

## Insects (team leaders Evelyn Cheng, Tim Graham, and Larry Stevens)

Between the two bioblitzes, the invertebrate teams documented at least 122 taxa of insects in 68 families and 13 orders (Fertig et al. 2009). Most samples were assigned to morpho-species in the field (many still need to be identified to species). Based on bioblitz data, insect species richness at Deer Creek is second only to that of vascular plants, with insects accounting for nearly 22% of all identified taxa (Table 9.1). Insect diversity, however, is probably significantly under-estimated. Only eight species were apparently sampled in both years of the bioblitz, suggesting a significant turnover in insect species richness between years and seasons. Using a ratio of ten insect species for every vascular plant species (Ricketts et al. 1999), the total insect diversity in the Deer Creek watershed may be between 4000 to 5000 species. Additional sampling, especially in late June before the annual monsoon and in late summer and fall for pollinators of flowering Asteraceae, will likely increase the number of documented insect species (Larry Stevens, personal communication).

The most notable insect species reported so far is an unidentified specimen of Ammobaenetes (sand treader cricket) that team co-leader Tim Graham suspects may be a new species to science.

## Fish, Amphibians, and Reptiles (team leaders Rhett Boswell, Neil Perry, and Kevin Wheeler)

Cold-blooded vertebrates are the least species-rich of all the taxonomic groups studied in the 2007-2008 bioblitzes. Between the two years, nine species of reptiles, three amphibians, and one fish were encountered in the Deer Creek study area.

Based on surveys of other portions of the watershed, a total of twelve reptile, four amphibian, and five fish species are known from the Deer Creek drainage (Bosworth 2003; Oliver 2003) and perhaps as many as 50 species of herptiles and fish could occur based on inventories of the entire Escalante River system (Fertig et al. 2009). Several uncommon fish and reptile species are known from elsewhere in the Deer Creek watershed, including the roundtail chub *(Gila robusta)*, bluehead sucker *(Catostomus discobolus)*, flannelmouth sucker *(C. latipinnis)* and plateau striped whiptail *(Cnemidophorus velox)*, but none were relocated during the bioblitzes.

## Birds (team leaders Rhett Boswell and John Spence)

Birds are the most species-rich of the vertebrate taxa studied at Deer Creek, with 88 species from 33 families reported by local residents and the bioblitz teams (Table 9.1). Forty bird species were found in the July 2007 bioblitz and 58 observed in May 2008 (18 of which represented new records for the study area). Based on available habitat and known distribution in Grand Staircase-Escalante National Monument, 150-190 bird species might occur in the Deer Creek area (Atwood et al. 1980; Jensen et al. no date). Most of these additional potential species are non-resident migratory waterfowl and Neotropical songbirds.

Among the more notable birds observed in the 2007-08 bioblitzes were five species of concern tracked by the Utah Division of Wildlife, state natural heritage program, and Utah Partners in Flight (Bosworth 2003; Parrish et al. 2002). These species included the broad-tailed hummingbird *(Selasphorus platycercus)*, cordilleran flycatcher *(Empidonax occidentalis)*, Virginia's warbler *(Vermivora virginiae)*, black-throated gray warbler *(Dendroica nigrescens)*, and Brewer's sparrow *(Spizella breweri)*. Other uncommon birds previously recorded in the vicinity are the bald eagle *(Haliaeetus leucocephalus)*, peregrine falcon *(Falco peregrinus)*, and Lewis' woodpecker *(Melanerpes lewis)*.

Table 9.1.  Summary of the Biodiversity of the Deer Creek Watershed

| Taxonomic group | # of Taxa found during 2007-08 Bioblitz Study Area | | | # of Taxa known from entire Middle Reach Deer Creek Watershed* | Potential # of Taxa in entire Middle Reach Deer Creek Watershed** |
| | 2007 | 2008 (# new taxa) | Cumulative Total | | |
| --- | --- | --- | --- | --- | --- |
| Lichens | 0 | 16 (16) | 16 | 16 | 130 |
| Bryophytes | 0 | 35 (35) | 35 | 35 | 60 |
| Vascular Plants | 256 | 159 (34) | 290 | 356 | 400-500 |
| Insects | 42 | 88 (80) | 122 | 122 | 4000-5000 |
| Fish | 0 | 1 (1) | 1 | 5 | 20 |
| Amphibians | 3 | 1 (0) | 3 | 4 | 10 |
| Reptiles | 7 | 6 (2) | 9 | 12 | 20 |
| Birds | 40 | 58 (18) | 58 | 88 | 150-190 |
| Mammals | 18 | 24 (6) | 24 | 37 | 50 |
| **TOTAL** | **366** | **388 (192)** | **558** | **675** | **4840-5980** |

*Additional species known from the watershed based on earlier studies (Bosworth 2003, Evenden 1999, Fertig 2005, Fertig et al. 2009, Flinders et al. 2002, Oliver 2003, Welsh and Atwood 2002), but not documented in the 2007-2008 bioblitzes.

** Estimates based on number of species known from adjacent areas of Grand Staircase-Escalante NM that might be expected (Fertig 2005, Flinders et al. 2002, Jensen et al. no date) and extrapolations by taxon experts leading each team (Dr. John Spence for bryophytes and lichens, Dr. Larry Stevens and Dr. Tim Graham for insects).

### Mammals (team leaders Rhett Boswell and Neil Perry)

The mammal teams identified 18 species in July 2007 and 24 species in May 2008 (Table 9.1). Combined with additional reports from local residents and previous research, a total of 37 mammal species is known from the Deer Creek watershed (Fertig et al. 2009). Based on suitable habitat and records from the Grand Staircase-Escalante National Monument, at least 50 mammal species may occur in the study area (Flinders et al. 2002).

Several notable mammal species were detected in the 2008 bioblitz. Neil Perry and Rhett Boswell live-trapped specimens of long-tailed pocket mouse (*Chaeteodipus formosus*) and little pocket mouse (*Perognathus longimembris*), neither of which were previously known from the vicinity. Recently cut tree stems and a fresh dam provided evidence of beaver (*Castor canadensis*) in the area, a species that was previously thought to be extirpated in the lower Deer Creek drainage (Fertig et al. 2009).

## Plant Communities (team leaders Keith Schulz and Walter Fertig)

Although not officially targeted in the 2007 bioblitz, ecologists identified a potentially new and undescribed plant community in the Deer Creek area. Small but extensive patches of Bigelow sagebrush (*Artemisia bigelovii*), blue grama (*Bouteloua gracilis*), and black grama (*B. eriopoda*) were observed in thin, sandy soils amid volcanic basalt boulders and rubble overlying Navajo Sandstone slickrock mesa tops and slopes. Other Bigelow sagebrush/blue grama community types described in the literature occur on shale or limestone breaks or have associated species more typical of the Chihuahuan Desert (Keith Schulz, personal communication; Anderson et al. 1998). Similar communities probably occur elsewhere on Navajo Sandstone outcrops between Boulder and Escalante, Utah, and possibly in the Kolob area of Zion National Park, but are otherwise apparently not known (Fertig et al. 2009).

## Overall Biodiversity

In all, 558 species of plants and animals were observed during the 2007-2008 bioblitzes at Deer Creek. Comparable numbers of species were documented in July 2007 (366) and May 2008 (388). The second bioblitz, however, included 192 species not observed the first year, a turnover rate of nearly 50% (Table 9.1). Conducting the second bioblitz increased the known flora and fauna of the area by 34%. Another 117 species have been detected outside the immediate study area in the Deer Creek watershed, but were not found in either bioblitz (Fertig et al. 2009). If all of these

taxa are present in the study area, the total known biota would increase to 675 taxa (of which the two bioblitzes documented 82.7%).

Of course these figures represent just a fraction of the total species richness of the area. When total insect diversity is better quantified, the estimated number of plant and animal species in the study area may be as great as 5980 species (Table 9.1).

## DISCUSSION

If, as Aldo Leopold famously wrote "to keep every cog and wheel is the first precaution of intelligent tinkering" (Leopold 1966), the second ought to be to know the identity of all the cogs and wheels. Because resources of time, money, and opportunity are always limited, conservationists need to know which specific areas and plant and animal species are under-represented in the protected area network and which are in greatest need of protection and management (Margules and Sarkar 2007). Unfortunately, decisions to protect or not protect an area often must be made quickly and with incomplete information. This can result in redundant conservation actions, or omission of biologically important (but poorly known) areas. Bioblitzes can be an inexpensive tool for rapidly learning the cogs and wheels of a specific site, and thus aid conservation practitioners in wisely allocating scarce resources.

Deer Creek is a case study of the potential value of bioblitzes. The Nature Conservancy was initially interested in the study area because of its proximity to a protected landscape (Grand Staircase-Escalante National Monument), rare species were known from the vicinity, and the Boulder region was known to be vulnerable to rampant growth and change in existing land uses. Because the site contained perennial wetlands in an otherwise arid landscape, it was also assumed to be rich in native biological diversity, though this was initially unproven due to a lack of field data. Sponsoring a bioblitz was a relatively low cost and rapid way for TNC to acquire a significant pool of location data on various taxonomic groups from recognized regional experts. The resulting data confirmed that Deer Creek had significant biological values including 558 species, an undescribed and rare plant community type, more than one dozen rare plant and animal taxa, at least one state distribution record, and one possibly undescribed insect.

Bioblitzes are not without limitations. No single weekend-long event will ever capture all the species present in an area over the course of an entire year. No more than 58% of the known vascular flora and vertebrate fauna of the study area were observed in any single Deer Creek bioblitz event. In addition, nearly 20% of the vascular plant and vertebrate species known from elsewhere in the watershed were not encountered during either bioblitz. These findings suggest that additional bioblitzes, scheduled during other seasons of the year, would very likely document a number of new species.

Another drawback is that bioblitzes focus mostly on presence/absence of species rather than on their population size and ecological or life-history needs (although mammal and insect trapping can include an abundance component). Merely identifying that a species is present in an area is no guarantee that a viable population exists, or that the site is significant for the survival of the species. Additional monitoring, surveys, or research are needed to answer these questions, especially for rare species or others of high management interest. More rigorous probabilistic sampling would improve the repeatability of bioblitz inventories and provide more defensible parameters to extrapolate potential numbers of taxa, but would require considerably greater time and expense to complete.

The number of species encountered in a bioblitz can vary depending on the skill of the assembled team and the diversity of taxonomic groups studied (Karns et al. 2006). Overall species richness is probably significantly higher at the Deer Creek site than presently known due to incomplete sampling of insects (perhaps as few as 3% of all insect species are presently known from the area if our current estimates of invertebrate diversity are correct). Diversity would also have been higher if our team had included experts on algae, soil microbes, fungi, or other invertebrate groups besides insects.

Despite these limitations, the 2007-2008 Deer Creek bioblitzes successfully demonstrated the biological richness and significance of the study area. TNC has used the information from the bioblitzes to help justify establishment of several conservation easements on private properties within the watershed. Involving adjacent landowners and other stakeholders from the local community in the bioblitz events has also created greater support and enthusiasm for TNC's conservation efforts in the Boulder area. Most importantly, the bioblitzes brought academic professionals, amateur naturalists, and the lay public together to share information across disciplines and to have fun. Hopefully the lessons learned from the Deer Creek bioblitzes of 2007 and 2008 will be more widely applied to other conservation lands or areas proposed for protection in Utah and the Colorado Plateau.

## APPENDIX

Flora and Fauna of the Deer Creek Watershed, Garfield County, Utah

This species list is based on observations made during the July 2007 and May 2008 Deer Creek bioblitzes and on distribution data from previous studies (Fertig et al. 2009). Species are organized by major taxonomic group, then by family and scientific name. Plant families are arranged alphabetically, while animal families are organized phylogenetically. Non-native (introduced) species are indicated by * preceding the scientific name. Columns are checked with an "x" if a species was documented during the 2007-2008 bioblitzes or was known previously from the vicinity of the study area based on a review of pertinent literature (cited in Fertig et al. 2009).

## Bryophytes

| Family | Scientific Name | Known Prev. | 2007-08 Bioblitz |
|---|---|---|---|
| Amblystegiaceae | *Amblystegium juratzkanum* | | x |
| | Species 2 | | x |
| Bartramiaceae | *Philonotos marchica* | | x |
| Brachytheciaceae | *Brachythecium rivulare* | | x |
| | Species 2 | | x |
| | Species 3 | | x |
| Bryaceae | *Anomobryum julaceum* | | x |
| | *Bryum lanatum* | | x |
| | *Gemmabryum caespiticium* | | x |
| | *Gemmabryum kunzei* | | x |
| | *Rosulabryum flaccidum* | | x |
| Dicranaceae | Species 1 | | x |
| Ditrichaceae | *Ceratodon purpureus* | | x |
| Encalyptaceae | *Encalypta rhaptocarpa* | | x |
| Fontinalaceae | *Fontinalis sp.* | | x |
| Funariaceae | *Funaria hygrometrica* | | x |
| Grimmiaceae | *Grimmia anodon* | | x |
| | *Grimmia longirostris* | | x |
| | Species 3 | | x |
| | Species 4 | | x |
| Hypnaceae | *Hypnum revolutum* | | x |
| Leskeaceae | *Pseudoleskeella tectorum* | | x |
| Mielichhoferiaceae | Species 1 | | x |
| Mniaceae | *Plagiomnium medianum* | | x |

| Pottiaceae | *Crumia latifolia* | | x |
| | *Didymodon brachyphyllus* | | x |
| | *Didymodon tophaceus* | | x |
| | *Syntrichia ruralis* | | x |
| | *Tortula sp.* | | x |
| Other Families | Species 1 | | x |
| | Species 2 | | x |
| | Species 3 | | x |
| | Species 4 | | x |
| | Species 5 | | x |
| | Species 6 | | x |

## Lichens

| Family | Scientific Name | Known Prev. | 2007-08 Bioblitz |
|---|---|---|---|
| Acarosporaceae | *Acarospora stapfiana* | | x |
| Cladoniaceae | *Cladonia sp.* | | x |
| Hymeneliaceae | *Aspicilia sp.* | | x |
| Lecanoraceae | *Candelariella citrina* | | x |
| | *Lecanora sp.* | | x |
| Lecideaceae | *Lecidea sp.* | | x |
| | *Pleopsidium flavum* | | x |
| Parmeliaceae | *Flavopunctelia soredica* | | x |
| | *Xanthoparmelia coloradoënsis* | | x |
| Peltigeraceae | *Peltigera sp.* | | x |
| Physiaceae | *Physcia stellaris* | | x |
| | *Physcia sp. 2* | | x |
| Rhizocarpaceae | *Rhizocarpon sp.* | | x |
| Teloschistaceae | *Caloplaca trachyphylla* | | x |
| | *Xanthomendoza fallax* | | x |
| | *Xanthoria elegans* | | x |

## Vascular Plants

| Family | Scientific Name | Common Name | Known Prev. | 2007-08 Bioblitz |
|---|---|---|---|---|
| Aceraceae (Maples) | *Acer negundo* | Box-elder | x | |
| Agavaceae (Agaves) | *Yucca harrimaniae var. harrimaniae* | Harriman's yucca | x | x |
| Amaranthaceae (Pigweeds) | *Amaranthus blitoides* | Prostrate pigweed | | x |
| | *\*Amaranthus retroflexus* | Redroot pigweed | x | x |
| Anacardiaceae (Sumacs) | *Rhus aromatica var. trilobata* | Skunkbush | x | x |
| | *Toxicodendron rydbergii* | Poison ivy | | x |
| Apiaceae or Umbelliferae (Carrots) | *Angelica pinnata* | Small-leaf angelica | | x |
| | *Conioselinum scopulorum* | Rock lovage | | x |
| | *\*Conium maculatum* | Poison-hemlock | x | |
| | *Cymopterus newberryi* | Sweetroot spring-parsley | | x |
| | *Cymopterus purpurascens* | Wide-wing spring-parsley | | x |
| Apocynaceae (Dogbanes) | *Apocynum cannabinum* | Dogbane | | x |
| | *Apocynum x floribundum* | Hybrid dogbane | | x |
| Asclepiadaceae (Milkweeds) | *Asclepias speciosa* | Showy milkweed | x | x |
| Asteraceae or Compositae (Sunflowers) | *Achillea millefolium var. lanulosa* | Yarrow | | x |
| | *Ambrosia acanthicarpa* | Bur ragweed | x | x |
| | *Antennaria marginata* | Sandstone pussytoes | | x |
| | *Antennaria microphylla* | Rosy pussytoes | | x |
| | *Artemisia bigelovii* | Bigelow's sagebrush | x | x |
| | *Artemisia campestris var. scouleriana* | Sand wormwood | x | x |
| | *Artemisia carruthii* | Carruth's wormwood | | x |
| | *Artemisia dracunculus ssp. glauca* | Tarragon | | x |
| | *Artemisia filifolia* | Sand sagebrush | | x |

| Family | Scientific Name | Common Name | Known Prev. | 2007-08 Bioblitz |
|--------|-----------------|-------------|-------------|------------------|
| Asteraceae or Compositae (Sunflowers) con't | *Artemisia frigida* | Fringed sagebrush | | x |
| | *Artemisia ludoviciana var. albula* | Louisiana wormwood | x | x |
| | *Artemisia ludoviciana var. mexicana* | Louisiana wormwood | x | x |
| | *Artemisia nova var. nova* | Black sagebrush | | x |
| | *Artemisia tridentata var. tridentata* | Basin big sagebrush | x | x |
| | *Aster ascendens* | Chilean aster | | x |
| | *Aster eatonii* | Eaton's aster | x | x |
| | *Aster foliaceus* | Leafy aster | | x |
| | *Aster hesperius* | Siskiyou aster | | x |
| | *Aster spathulatus* | Western aster | | x |
| | *Aster welshii* | Welsh's aster | | x |
| | *Brickellia grandiflora* | Tasselflower | | x |
| | *Brickellia microphylla var. scabra* | Rough brickellbush | | x |
| | *Brickellia oblongifolia var. linifolia* | Mohave brickellbush | | x |
| | *Chaenactis douglasii* | Douglas' dusty-maiden | | x |
| | *Chrysopsis villosa var. minor* | Hispid goldenaster | | x |
| | *Chrysopsis zionensis* | Zion golden-aster | x | x |
| | *Chrysothamnus linifolius* | Spreading rabbitbrush | | x |
| | *Chrysothamnus nauseosus var. arenarius* | Sand rabbitbrush | x | |
| | *Chrysothamnus nauseosus var. hololeucus* | Graystem rabbitbrush | x | |
| | *Chrysothamnus nauseosus var. leiospermus* | Smoothseed rabbitbrush | | x |
| | *Chrysothamnus nauseosus var. oreophilus* | Rubber rabbitbrush | | x |
| | *Chrysothamnus viscid-iflorus var. viscidiflorus* | Lanceleaf rabbitbrush | x | x |
| | *Cirsium arizonicum* | Arizona thistle | | x |
| | *\*Cirsium vulgare* | Bull thistle | x | x |
| | *Conyza canadensis var. glabrata* | Horseweed | x | x |

| Family | Scientific Name | Common Name | Known Prev. | 2007-08 Bioblitz |
|--------|-----------------|-------------|:-----------:|:----------------:|
| Asteraceae or Compositae (Sunflowers) con't | *Erigeron divergens var. divergens* | Spreading daisy | x | x |
| | *Erigeron eatonii* | Eaton's daisy | | x |
| | *Erigeron lonchophyllus* | Longleaf daisy | | x |
| | *Erigeron pumilus var. concinnus* | Vernal daisy | | x |
| | *Erigeron religiosus* | Religious daisy | x | |
| | *Erigeron utahensis var. sparsifolius* | Slenderleaf daisy | | x |
| | *Erigeron utahensis var. utahensis* | Utah daisy | | x |
| | *Euthamia occidentalis* | Western goldenrod | x | x |
| | *\*Gaillardia pulchella* | Firewheel blanket-flower | | x |
| | *Gnaphalium canescens* | Wright's cudweed | | x |
| | *Gnaphalium stramineum* | Cottonbatting cudweed | x | |
| | *Grindelia squarrosa var. serrulata* | Curly gumweed | | x |
| | *Gutierrezia microcephala* | Thread snakeweed | x | |
| | *Gutierrezia sarothrae* | Broom snakeweed | x | x |
| | *Helianthella microcephala* | Smallhead sunflower | | x |
| | *Helianthus annuus var. lenticularis* | Common sunflower | x | |
| | *Helianthus petiolaris var. fallax* | Prairie sunflower | | x |
| | *Hymenopappus filifolius var. cinereus* | Hyalineherb | | x |
| | *Hymenoxys acaulis var. arizonica* | Arizona woollybase | | x |
| | *Hymenoxys acaulis var. ivesiana* | Ives' woollybase | x | x |
| | *\*Lactuca serriola* | prickly lettuce | x | |
| | *Lactuca tatarica var. pulchella* | prickly lettuce | x | x |
| | *Lygodesmia spinosa* | thorny wirelettuce | | x |
| | *Machaeranthera canescens var. aristata* | hoary aster | x | x |
| | *Machaeranthera tanacetifolia* | tansyleaf aster | x | x |
| | *Petradoria pumila var. pumila* | Rock goldenrod | | x |
| | *Psilostrophe sparsiflora* | Greenstem paper-flower | | x |
| | *Ratibida columnifera* | Prairie coneflower | | x |
| | *\*Rudbeckia hirta* | Black-eyed susan | | x |
| | *Senecio multilobatus* | Basin groundsel | x | x |
| | *Senecio spartioides var. spartioides* | Broom groundsel | x | x |

| Family | Scientific Name | Common Name | Known Prev. | 2007-08 Bioblitz |
|---|---|---|---|---|
| Asteraceae or Compositae (Sunflowers) con't | *Solidago canadensis var. salebrosa* | Canada goldenrod | x | |
| | *Solidago velutina* | Alcove goldenrod | | x |
| | *Stephanomeria exigua* | Annual wirelettuce | | x |
| | *Stephanomeria tenuifolia var. tenuifolia* | Slender wirelettuce | x | x |
| | *\*Taraxacum officinale* | Common dandelion | | x |
| | *Tetradymia canescens* | Gray horsebrush | | x |
| | *Townsendia incana* | Hoary Easter-daisy | x | x |
| | *\*Tragopogon dubius* | Yellow salsify | x | x |
| | *Xanthium strumarium var. canadense* | Cocklebur | | x |
| Berberidaceae (Barberries) | *Mahonia fremontii* | Fremont's mahonia | | x |
| Betulaceae (Birches) | *Betula occidentalis* | Water birch | x | x |
| Boraginaceae (Borages) | *Cryptantha cinerea var. jamesii* | Head cryptanth | x | |
| | *Cryptantha cinerea var. pustulosa* | Pustular cryptanth | | x |
| | *Cryptantha confertiflora* | Golden cryptanth | | x |
| | *Cryptantha fendleri* | Fendler's cryptanth | | x |
| | *Cryptantha flavoculata* | Yellow-eye cryptanth | | x |
| | *Cryptantha gracilis* | Slender cryptanth | | x |
| | *Cryptantha pterocarya* | Wing-nut cryptanth | x | x |
| | *Lappula occidentalis var. cupulata* | Western stickseed | | x |
| | *Lithospermum incisum* | Showy stoneseed | | x |
| Brassicaceae or Cruciferae (Mustards) | *Arabis perennans var. perennans* | Common rockcress | | x |
| | *Arabis selbyi* | Selby's rockcress | | x |
| | *\*Camelina microcarpa* | Littlepod false flax | | x |
| | *Descurainia pinnata var. osmiarum* | Pinnate tansy-mustard | x | x |
| | *Dithyrea wislizenii* | Spectacle-pod | x | x |
| | *Erysimum asperum var. purshii* | Wallflower | | x |
| | *\*Lepidium campestre* | Field pepperwort | | x |
| | *Lepidium densiflorum var. densiflorum* | Prairie pepperwort | | x |
| | *Lepidium montanum var. neeseae* | Neese's pepperplant | | x |
| | *\*Nasturtium officinale* | Water-cress | | x |
| | *Physaria chambersii var. chambersii* | Chambers' twindpod | | x |
| | *Physaria rectipes* | Colorado bladder-pod | | x |

| Family | Scientific Name | Common Name | Known Prev. | 2007-08 Bioblitz |
|---|---|---|---|---|
| Brassicaceae or Cruciferae (Mustards) con't | *Sisymbrium altissimum* | Tumbling mustard | | x |
| | *Streptanthella longirostris* | Long-beak fiddle-mustard | x | x |
| | *Thelypodium integri-folium var. integrifolium* | Tall thelypody | | x |
| Cactaceae (Cacti) | *Coryphantha vivipara var. arizonica* | Arizona pincushion | | x |
| | *Echinocereus triglochid-iatus var. melanacanthus* | Claretcup | x | x |
| | *Opuntia phaeacantha var. major* | Large pricklypear | x | |
| | *Opuntia polyacantha* | Plains pricklypear | x | x |
| | *Sclerocactus whipplei var. roseus* | Whipple's fishhook cactus | x | x |
| Campanulaceae (Bellflowers) | *Campanula parryi* | Parry's bellflower | | x |
| Capparaceae (Capers) | *Cleome lutea* | Yellow beeplant | | x |
| Caryophyllaceae (Pinks) | *Arenaria fendleri var. eastwoodiae* | Fendler's sandwort | x | x |
| | *Gypsophila paniculata* | Tall baby's-breath | x | x |
| Chenopodiaceae (Goosefoots) | *Atriplex canescens var. canescens* | Four-wing saltbush | | x |
| | *Atriplex rosea* | Tumbling orach | | x |
| | *Bassia scoparia* | Summer-cypress | x | x |
| | *Chenopodium album* | Lambsquarter | x | x |
| | *Chenopodium atrovirens* | Mountain goosefoot | x | |
| | *Chenopodium berlandieri var. zschackei* | Berlandier's goose-foot | x | |
| | *Chenopodium desiccatum* | Desert goosefoot | | x |
| | *Chenopodium fremontii var. fremontii* | Fremont goosefoot | | x |
| | *Chenopodium glaucum var. salinum* | Oakleaf goosefoot | | x |
| | *Chenopodium leptophyllum* | Narrowleaf goosefoot | x | |
| | *Corispermum americanum* | American bugseed | | x |
| | *Corispermum welshii* | Welsh's bugseed | x | |
| | *Salsola tragus* | Russian-thistle | x | x |
| Commelinaceae (Spiderworts) | *Tradescantia occidentalis* | Western spiderwort | | x |

| Family | Scientific Name | Common Name | Known Prev. | 2007-08 Bioblitz |
|---|---|---|---|---|
| Cupressaceae (Cypresses) | *Juniperus osteosperma* | Utah juniper | x | x |
| | *Juniperus scopulorum* | Rocky Mountain juniper | x | |
| Cyperaceae (Sedges) | *Carex athrostachya* | Slender-beak sedge | x | |
| | *Carex aurea [C. hassei form]* | Golden sedge | | x |
| | *Carex nebrascensis* | Nebraska sedge | x | x |
| | *Carex pellita* | Woolly sedge | x | x |
| | *Carex praegracilis* | Blackcreeper sedge | | x |
| | *Carex rossii* | Ross' sedge | | x |
| | *Eleocharis palustris* | Common spikerush | x | x |
| | *Eleocharis rostellata* | Torrey's spikerush | x | |
| | *Scirpus acutus* | Hardstem bulrush | | x |
| | *Scirpus microcarpus* | Panicled bulrush | x | |
| | *Scirpus pungens var. longispicatus* | Common threesquare | | x |
| Elaeagnaceae (Oleasters) | *Shepherdia argentea* | Silver buffaloberry | x | x |
| | *Shepherdia rotundifolia* | Roundleaf buffaloberry | x | x |
| Ephedraceae (Mormon-teas) | *Ephedra viridis var. viridis* | Green Mormon-tea | x | x |
| Equisetaceae (Horsetails) | *Equisetum arvense* | Meadow horsetail | | x |
| | *Equisetum laevigatum* | Smooth scouring-rush | x | x |
| Ericaceae (Heaths) | *Arctostaphylos patula* | Greenleaf manzanita | | x |
| Euphorbiaceae (Spurges) | *Chamaesyce glyptosperma* | Ridge-seeded spurge | | x |
| | *Euphorbia brachycera* | Shorthorn spurge | | x |
| Fabaceae or Leguminosae (Peas) | *Astragalus amphioxys* | Crescent milkvetch | x | |
| | *Astragalus ceramicus var. ceramicus* | Painted milkvetch | | x |
| | *Astragalus mollissimus var. thompsoniae* | Woolly milkvetch | x | x |
| | *Astragalus nuttallianus var. micranthiformis* | Small-flowered milkvetch | x | |
| | *Glycyrrhiza lepidota* | Licorice | x | x |
| | *Lathyrus brachycalyx var. zionis* | Zion sweetpea | | x |
| | *Lupinus pusillus var. intermontanus* | Great Basin lupine | | x |
| | *Medicago lupulina* | Black medick | | x |
| | *Medicago sativa* | Alfalfa | x | x |

| Family | Scientific Name | Common Name | Known Prev. | 2007-08 Bioblitz |
|---|---|---|---|---|
| Fabaceae or Leguminosae (Peas) con't | *Melilotus alba | White sweet-clover | x | x |
| | *Melilotus officinalis | Yellow sweet-clover | x | x |
| | Thermopsis montana | Golden pea | | x |
| | *Trifolium fragiferum | Strawberry clover | x | |
| | *Trifolium pratensered | Clover | | x |
| | *Trifolium repens | White clover | | x |
| | Vicia americana var. americana | American vetch | | x |
| Fagaceae (Oaks) | Quercus gambelii var. gambelii | Gambel's oak | x | x |
| Fumariaceae (Fumitories) | Corydalis aurea | Golden corydalis | | x |
| Gentianaceae (Gentians) | Centaurium exaltatum | Great Basin centaury | x | |
| Geraniaceae (Geraniums) | *Erodium cicutarium | Storksbill | | x |
| Hydrophyllaceae (Waterleafs) | Phacelia ivesiana | Ives' phacelia | | x |
| Iridaceae (Irises) | Iris missouriensis | Missouri iris | | x |
| | Sisyrinchium demissum | Blue-eyed grass | | x |
| Juncaceae (Rushes) | Juncus alpinus | Northern rush | | x |
| | Juncus arcticus | Baltic rush | x | x |
| | Juncus ensifolius var. brunnescens | Tracy's rush | | x |
| | Juncus ensifolius var. montanus | Sword-leaf rush | x | |
| | Juncus longistylis | Longstyle rush | x | x |
| | Juncus tenuis | Poverty rush | x | |
| | Juncus torreyi | Torrey's rush | x | x |
| Juncaginaceae (Arrow grasses) | Triglochin maritima | Maritime arrowgrass | | x |
| Lamiaceae or Labiatae (Mints) | Mentha arvensis var. glabrata | Field mint | x | x |
| Lemnaceae (Duckweeds) | Lemna minor | Lesser duckweed | x | |
| Liliaceae (Lilies) | Allium cernuum | Nodding onion | | x |
| | Allium macropetalum | San Juan onion | | x |
| | Androstephium brevifolium | Funnel lily | | x |
| | *Asparagus officinalis | Asparagus | x | |
| | Fritillaria atropurpurea | Leopard-lily | | x |

| Family | Scientific Name | Common Name | Known Prev. | 2007-08 Bioblitz |
|---|---|---|---|---|
| Liliaceae (Lilies) con't | *Smilacina stellata* | false Solomon's-seal | | x |
| | *Zigadenus paniculatus* | foothills death camas | | x |
| Linaceae (Flaxes) | *Linum aristatum* | Broom flax | x | |
| | *Linum perenne ssp. lewisii* | Blue flax | x | x |
| Loasaceae (Stickleafs) | *Mentzelia multiflora* | Desert stickleaf | | x |
| Malvaceae (Mallows) | *Sidalcea candida* | White checkermallow | | x |
| | *Sphaeralcea coccinea* | Scarlet globe mallow | x | x |
| | *Sphaeralcea parvifolia* | Small-leaf globe mallow | | x |
| Monotropaceae (Indian pipes) | *Pterospora andromedea* | Pinedrops | | x |
| Nyctaginaceae (Four o'clocks) | *Abronia fragrans var. fragrans* | Fragrant sand-verbena | | x |
| | *Mirabilis linearis var. linearis* | Narrowleaf four'-o'clock | x | x |
| | *Mirabilis oxybaphoides* | Spreading four'-o'clock | | x |
| Oleaceae (Olives) | *Pterospora andromedea* | Fraxinus anomala | x | x |
| Onagraceae (Evening-primroses) | *Epilobium ciliatum* | Northern willowherb | x | x |
| | *Epilobium halleanum* | Hall's willowherb | | x |
| | *Epilobium hornemannii var. hornemannii* | Hornemann's willowherb | x | |
| | *Epilobium saximontanum* | Rocky Mountain willowherb | x | |
| | *Oenothera caespitosa var. navajoensis* | Tufted evening-primrose | x | x |
| | *Oenothera elata* | Hooker's evening-primrose | x | |
| | *Oenothera longissima* | Bridges' evening-primrose | x | |
| | *Oenothera pallida var. pallida* | Pale evening-primrose | | x |
| Orchidaceae (Orchids) | *Corallorrhiza maculata* | Spotted coralroot | | x |
| | *Epipactis gigantea* | Helleborine | | x |
| | *Habenaria hyperborea* | Northern bog orchid | | x |
| | *Habenaria sparsiflora var. laxiflora* | Few-flower bog orchid | | x |
| | *Spiranthes diluvialis* | Ute ladies-tresses | x | |

| Family | Scientific Name | Common Name | Known Prev. | 2007-08 Bioblitz |
|---|---|---|---|---|
| Pinaceae (Pines) | *Picea pungens* | Blue spruce | | x |
| | *Pinus edulis* | Two-needle pinyon | x | x |
| | *Pinus ponderosa var. scopulorum* | Ponderosa pine | x | x |
| Plantaginaceae (Plantains) | *Plantago eriopoda* | Woolly-foot plantain | x | |
| | *Plantago lanceolata* | English plaintain | x | x |
| | *Plantago major* | Common plantain | x | x |
| | *Plantago patagonica* | Pursh's plantain | x | x |
| Poaceae or Gramineae (Grasses) | *Agropyron cristatum* | Crested wheatgrass | x | x |
| | *Agrostis stolonifera* | Redtop | x | x |
| | *Aristida adscensionis* | Sixweeks threeawn | | x |
| | *Aristida purpurea* | Purple threeawn | | x |
| | *Bouteloua curtipendula* | Sideoats grama | | x |
| | *Bouteloua eriopoda* | Black grama | | x |
| | *Bouteloua gracilis* | Blue grama | x | x |
| | *Bromus carinatus* | California brome | x | |
| | *Bromus inermis* | Smooth brome | x | x |
| | *Bromus japonicus* | Japanese brome | x | |
| | *Bromus tectorum* | Cheatgrass | x | x |
| | *Calamagrostis scopulorum* | Jones' reedgrass | | x |
| | *Calamovilfa gigantea* | Big sandreed | | x |
| | *Cenchrus longispinus* | Field sandbur | x | x |
| | *Dactylis glomerata* | Orchardgrass | x | x |
| | *Distichlis spicata* | Desert saltgrass | x | |
| | *Elymus canadensis* | Canada wildrye | x | x |
| | *Elymus elymoides* | Squirreltail | | x |
| | *Elymus hispidus* | Intermediate wheatgrass | x | |
| | *Elymus repens* | Quackgrass | x | |
| | *Elymus trachycaulus* | Slender wheatgrass | x | x |
| | *Elymus x pseudorepens* | False quackgrass | | x |
| | *Enneapogon desvauxii* | Spike pappusgrass | x | |
| | *Festuca octoflora* | Sixweeks fescue | x | x |
| | *Festuca ovina var. saximontana* | Sheep fescue | x | |
| | *Festuca pratensis* | Meadow fescue | | x |
| | *Festuca rubra* | Red fescue | | x |

| Family | Scientific Name | Common Name | Known Prev. | 2007-08 Bioblitz |
|---|---|---|---|---|
| Poaceae or Gramineae (Grasses) con't | Glyceria striata | Fowl mannagrass | | x |
| | Hilaria jamesii | Galleta | x | x |
| | Hordeum jubatum | Foxtail barley | | x |
| | Lycurus phleoides | Wolftail | | x |
| | Muhlenbergia andina | Foxtail muhly | | x |
| | Muhlenbergia asperifolia | Scratchgrass | x | x |
| | Muhlenbergia pauciflora | New Mexican muhly | | x |
| | Muhlenbergia pungens | Sandhill muhly | x | x |
| | Muhlenbergia thurberi | Thurber's muhly | | x |
| | Muhlenbergia wrightii | Spike muhly | x | |
| | Munroa squarrosa | False buffalograss | x | |
| | Panicum capillare | Witchgrass | | x |
| | *Phleum pratense | Timothy | | x |
| | Phragmites australis | Common reed | | x |
| | Poa fendleriana | Muttongrass | x | x |
| | *Poa pratensis | Kentucky bluegrass | x | |
| | Poa secunda | Sandberg bluegrass | x | x |
| | *Polypogon monspeliensis | Rabbit-foot grass | x | |
| | Schizachyrium scoparium | Little bluestem | | x |
| | *Setaria viridis | Green bristlegrass | | x |
| | Sphenopholis obtusata | Prairie wedgegrass | | x |
| | Sporobolus contractus | Spike dropseed | x | x |
| | Sporobolus cryptandrus | Sand dropseed | x | x |
| | Stipa comata var. comata | Needle-and-thread | | x |
| | Stipa hymenoides | Indian ricegrass | x | x |
| | *Triticum aestivum | Wheat | x | |
| Polemoniaceae (Phloxes) | Gilia subnuda | Carmine gilia | | x |
| | Ipomopsis aggregata | Scarlet gilia | | x |
| | Ipomopsis congesta var. frutescens | Shrubby gilia | x | x |
| | Ipomopsis longiflora | Longflower gilia | | x |
| | Leptodactylon pungens | Granite prickly-phlox | | x |
| | Leptodactylon watsonii | Watson's prickly-phlox | | x |
| | Phlox hoodii var. canescens | Hood's phlox | | x |

| Family | Scientific Name | Common Name | Known Prev. | 2007-08 Bioblitz |
|---|---|---|---|---|
| Polygonaceae (Buckwheats) | *Eriogonum alatum* | Winged buckwheat | x | x |
| | *Eriogonum cernuum var. cernuum* | Nodding buckwheat | | x |
| | *Eriogonum corymbosum var. corymbosum* | Crispleaf buckwheat | x | |
| | *Eriogonum corymbosum var. orbiculatum* | Rimrock buckwheat | | x |
| | *Eriogonum leptocladon var. ramosissimum* | Eastwood's buckwheat | | x |
| | *Eriogonum microthecum var. simpsonii* | Slender buckwheat | | x |
| | *Eriogonum umbellatum var. subaridum* | Arid buckwheat | x | x |
| | *\*Polygonum aviculare* | Knotweed | x | |
| | *Polygonum douglasii var. utahense* | Utah knotweed | | x |
| | *\*Polygonum lapathifolium* | Willow-weed | x | |
| | *\*Polygonum persicaria* | Ladysthumb | x | |
| | *\*Rumex crispus* | Curly dock | x | x |
| Polypodiaceae (Ferns) | *Adiantum capillus-veneris* | Maidenhair fern | x | |
| | *Adiantum pedatum var. aleuticum* | Northern maidenhair fern | x | |
| | *Cystopteris fragilis* | Brittle fern | | x |
| | *Pellaea glabella var. simplex* | Smooth cliffbrake | x | |
| Portulacaceae (Purslanes) | *Portulaca oleracea* | Purslane | x | x |
| | *Talinum brevifolium* | Sausageleaf talinum | | x |
| | *Talinum parviflorum* | Scapose talinum | | x |
| Potamogeton-aceae (Pondweeds) | *Potamogeton filiformis var. occidentalis* | Fineleaf pondweed | x | |
| Ranunculaceae (Buttercups) | *Clematis ligusticifolia* | White virgin's-bower | x | x |
| | *Delphinium scaposum var. andersonii* | Anderson's larkspur | | x |
| | *Ranunculus cymbalaria* | Marsh buttercup | | x |
| Rosaceae (Roses) | *Amelanchier utahensis* | Utah serviceberry | x | x |
| | *Cercocarpus intricatus* | Dwarf mountain mahogany | x | x |
| | *Holodiscus dumosus* | Mountain spray | | x |
| | *Petrophyton caespitosum* | Rock spiraea | | x |

| Family | Scientific Name | Common Name | Known Prev. | 2007-08 Bioblitz |
|---|---|---|---|---|
| Rosaceae (Roses) con't | *Potentilla gracilis var. elmeri* | Elmer's beautiful cinquefoil | | x |
| | *\*Prunus persica* | Peach | x | x |
| | *Purshia mexicana var. stansburyana* | Cliff-rose | | x |
| | *Purshia tridentata* | Bitterbrush | x | x |
| | *Rosa woodsii* | Woods' rose | x | x |
| Salicaceae (Willows) | *Populus angustifolia* | Narrowleaf cottonwood | x | x |
| | *Populus fremontii var. fremontii* | Fremont cottonwood | x | x |
| | *Populus x intercurrens* | lanceleaf cottonwood | x | x |
| | *Salix boothii* | Booth's willow | x | |
| | *Salix eriocephala var. watsonii* | Yellow willow | x | x |
| | *Salix exigua var. stenophylla* | Coyote willow | x | x |
| | *Salix lucida ssp. caudata* | Whiplash willow | x | x |
| Santalaceae (Sandalwoods) | *Comandra umbellata var. pallida* | Bastard toadflax | x | x |
| Scrophulariaceae (Figworts) | *Castilleja chromosa* | Common paintbrush | | x |
| | *Castilleja exilis* | Annual paintbrush | | x |
| | *Castilleja linariifolia* | Wyoming paintbrush | | x |
| | *Castilleja scabrida var. scabrida* | Eastwood's paintbrush | | x |
| | *Cordylanthus wrightii* | Wright's bird-beak | x | x |
| | *Mimulus glabratus ssp. fremontii* | Glabrous monkeyflower | x | |
| | *Mimulus guttatus* | Yellow monkeyflower | | x |
| | *Penstemon angustifolius var. venosus* | Veined penstemon | x | x |
| | *Penstemon comarrhenus* | Dusty penstemon | | x |
| | *Penstemon eatonii var. undosus* | Firecracker penstemon | x | x |
| | *Penstemon pachyphyllus var. congestus* | Rockville penstemon | | x |
| | *Penstemon palmeri var. eglandulosus* | Palmer's penstemon | | x |
| | *Penstemon rostriflorus* | Beaked penstemon | | x |
| | *Penstemon strictus* | Rocky mountain penstemon | | x |
| | *Penstemon utahensis* | Utah penstemon | | x |
| | *\*Veronica anagallis-aquatica* | Water speedwell | | x |

| Family | Scientific Name | Common Name | Known Prev. | 2007-08 Bioblitz |
|---|---|---|---|---|
| Selaginellaceae (Spikemosses) | *Selaginella mutica* | Awnless spikemoss | | x |
| Solanaceae (Nightshades) | *Solanum triflorum* | Cut-leaf nightshade | x | |
| Tamaricaceae (Salt-cedars) | *Tamarix chinensis salt-cedar* | Salt-cedar | | x |
| Typhaceae (Cattails) | *Typha domingensis* | Southern cattail | x | x |
| | *Typha latifolia* | Broad-leaved cattail | | x |
| Urticaceae (Nettles) | *Urtica dioica var. occidentalis* | Stinging nettle | | x |
| Violaceae (Violets) | *Viola nephrophylla* | Bog violet | | x |
| Viscaceae (Mistletoes) | *Arceuthobium divaricatum* | Pinyon dwarf-mistletoe | | x |
| | *Phoradendron juniperinum* | Juniper mistletoe | x | x |
| Zygophyllaceae (Caltrops) | *Tribulus terrestris* | Puncture vine | x | |

Invertebrates

| Order | Family | Scientific Name | Known Prev. | 2007-08 Bioblitz |
|---|---|---|---|---|
| **Coleoptera** | Cerambycidae (Longhorn beetles) | Species 1 | | x |
| | Chrysomelidae (Leaf beetles) | Species 1 | | x |
| | | Species 2 | | x |
| | Cicindelidae (Tiger beetles) | *Cicindela oregona* | | x |
| | | *Cicindela repanda tanneri* | | x |
| | Curculiondidae (Weevils) | Species 1 | | x |
| | Dytiscidae (Diving beetles) | *Agabus nr. lugens* | | x |
| | Elateridae (Click beetles) | Species 1 | | x |
| | Gyrinidae (Whirligig beetles) | *Gyrinis plicifer* | | x |
| | Histeridae (Hister beetles) | Species 1 | | x |
| | Hydrophilidae (Water scavenger beetles) | Species 1 | | x |
| | Lampyridae (Fireflies) | Species 1 | | x |
| | Nitidulidae (Sap beetles) | Species 1 | | x |

| Order | Family | Scientific Name | Known Prev. | 2007-08 Bioblitz |
|-------|--------|-----------------|-------------|------------------|
| **Coleoptera con't** | Scarabaeidae (Scarab beetles) | *Euphoria inda* | | x |
| | | *Polyphylla decemlineata* | | x |
| | | Species 3 | | x |
| | | Species 4 | | x |
| | | Species 5 | | x |
| | | Species 6 | | x |
| | Staphylinidae (Rove beetles) | Species 1 | | x |
| | Tenebrionidae (Darkling or Stink beetles | *Eleodes obscurus* | | x |
| | | *Embapion sp.* | | x |
| | | *Eusattus nr. muricatus* | | x |
| | | Species 4 | | |
| **Diptera** | Asilidae (Robber flies) | *Efferia sp.* | | x |
| | | *Lestomyia sp. 1* | | x |
| | | *Lestomyia sp. 2* | | x |
| | | Species 4 | | x |
| | Bombyliidae (Bee flies) | Species 1 | | |
| | | Species 2 | | x |
| | Calliphoridae (Bluebottle flies) | Species 1 | | x |
| | Culicidae (Mosquitoes) | *Culiseta incidens* | | x |
| | | Species 2 | | x |
| | Sarcophagidae (Flesh flies) | Species 1 | | x |
| | Stratiomyidae (Soldier flies) | Species 1 | | x |
| | Syrphidae (Hover flies) | Species 1 | | x |
| | | Species 2 | | x |
| | | Species 3 | | x |
| | Tachinidae (Tachinid flies) | Species 1 | | x |
| | Tipulidae (Crane flies) | Species 1 | | x |
| | | Species 2 | | x |
| **Ephemeroptera** | Family unknown (Mayflies) | Species 1 | | x |
| **Hemiptera** | Aphidae (Aphids) | Species 1 | | x |
| | Cercopidae (Froghoppers) | Species 1 | | x |
| | Cicadellidae (Leafhoppers) | Species 1 | | x |
| | | Species 2 | | x |
| | Cicadidae (Cicadas) | Species 1 | | x |
| | Fulgoridae (Plant hoppers) | Species 1 | | x |

| Order | Family | Scientific Name | Known Prev. | 2007-08 Bioblitz |
|---|---|---|---|---|
| **Hemiptera con't** | Gerridae (Water striders) | *Aquarius remigis* | | x |
| | Lygaeidae (Seed bugs) | Species 1 | | x |
| | Miridae (Leaf bugs) | Species 1 | | x |
| | Nabidae (Damsel bugs) | Species 1 | | x |
| | | Species 2 | | x |
| | Notonectidae (Backswimmers) | *Notonecta kirbyi* | | x |
| | Reduviidae (Assassin Bugs) | Species 1 | | x |
| **Hymen-toptera** | Apidae (Bees) | *Anthophorus sp.* | | x |
| | | *Apis mellifera* | | x |
| | Formicidae (Ants) | Species 1 | | x |
| | | Species 2 | | |
| | | Species 3 | | |
| | | Species 4 | | |
| | Halictidae (Halictid bees) | Species 1 | | x |
| | Ichneumonidae (Ichneumon wasps) | *Culiseta incidens* | | x |
| | Megachilidae (leafcutting bees) | Species 1 | | x |
| | Mutillidae (Velvet ants) | Species 1 | | x |
| | Pompilidae (Spider wasps) | Species 1 | | x |
| | Sphecidae (Digger wasps) | *Podalonia sp.* | | x |
| | | Species 2 | | x |
| | | Species 3 | | x |
| | | Species 4 | | x |
| | Tenthridinae (Common sawflies) | Species 1 | | x |
| | Vespidae (Vespid wasps) | *Mischocyattarus flavitarsus* | | x |
| | | *Polistes aurifer* | | x |
| **Lepidotera** | Arctiidae (Tiger moths) | Species 1 | | x |
| | | Species 2 | | x |
| | Geometridae (Geometer moths) | Species 1 | | x |
| | | Species 2 | | x |
| | | Species 3 | | x |
| | | Species 4 | | x |
| | | Species 5 | | x |
| | Hesperiidae (Skippers) | *Erynnis meridianus* | | x |
| | | *Pyrgus communis complex* | | x |
| | | Species 3 | | x |

| Order | Family | Scientific Name | Known Prev. | 2007-08 Bioblitz |
|-------|--------|-----------------|-------------|------------------|
| **Lepidoptera con't** | Lycaenidae (Gossamer-winged butterflies) | Species 1 | | x |
| | Moths – family undetermined | Species 1 | | x |
| | Noctuidae (Noctuid moths) | Species 1 | | x |
| | | Species 2 | | x |
| | | Species 3 | | x |
| | | Species 4 | | x |
| | Nymphalidae (Brush-footed butterflies) | *Euphydryas editha* | | x |
| | | *Phyciodes campestris* | | x |
| | | Species 3 | | x |
| | Pieridae (Whites and Sulfurs) | *Colias eurytheme* | | x |
| | | *Eurema nicippe* | | x |
| | | *Pontia protodice* | | x |
| | Sphingidae (Hawk moths) | *Hyles lineata* | | x |
| **Mantodea** | Mantidae (Preying Mantids) | Species 1 | | x |
| **Megaloptera** | Raphiidae (Snakeflies) | Species 1 | | x |
| **Neuroptera** | Chrysopidae (Green lacewings) | Species 1 | | x |
| | Hemerobiidae (Brown lacewings) | Species 1 | | x |
| | Myrmeleontidae (Antlions) | Species 1 | | x |
| **Odonata** | Aeshnidae (Darners) | *Rhionaeschina?* | | x |
| | Libelluidae (Skimmers) | *Erythemis collocata* | | x |
| **Orthoptera** | Acrididae (Short-horned grasshoppers) | *Melanoplus sp.* | | x |
| | | Species 2 | | x |
| | | Species 3 | | x |
| | | Species 4 | | x |
| | | Species 5 | | x |
| | | Species 6 | | x |
| | | Species 7 | | x |
| | | Species 8 | | x |
| | Gryllidae (Field crickets) | *Gryllus?* | | x |
| | Stenopelmatidae (Jerusalem crickets) | Species 1 | | x |
| | Rhaphidiphoridae (Camel crickets) | *Ceuthophilus sp.* | | x |
| | | *Ammobaenetes sp.* (possibly undescribed) | | x |

| Order | Family | Scientific Name | Known Prev. | 2007-08 Bioblitz |
|---|---|---|---|---|
| **Thysanura** | Family unknown (Silverfish) | Species 1 | | x |
| **Trichoptera** | Limnephilidae (Northern caddisflies) | Species 1 | | x |
| | Trichoptera (family unknown) | Species 1 | | x |
| | | Species 2 | | x |
| | | Species 3 | | x |
| | | Species 4 | | x |
| | | Species 5 | | x |

Fish

| Family | Scientific Name | Common Name | Known Prev. | 2007-08 Bioblitz |
|---|---|---|---|---|
| **Cyprinidae (Minnows)** | *Gila robusta* | Roundtail chub | x | |
| **Catostomidae (Suckers)** | *Catostomus discobolus* | Bluehead sucker | x | |
| | *Catostomus latipinnis* | Flannelmouth sucker | x | |
| **Salmonidae (Trout)** | *Oncorhynchus clarki pleuriticus* | Colorado River cutthroat trout | x | |
| | *\*Salmo trutta* | Brown trout | | x |

Amphibians

| Family | Scientific Name | Common Name | Known Prev. | 2007-08 Bioblitz |
|---|---|---|---|---|
| **Ambystoma-tidae (Mole salamanders)** | *Ambystoma tigrinum* | Tiger Salamander | x | |
| **Pelobatidae (Spadefoot toads)** | *Spea intermontana* | Great Basin spadefoot | x | x |
| **Bufonidae (True toads)** | *Bufo woodhousii* | Woodhouse's toad | x | x |
| | *Bufo punctatus* | Red-spotted toad | | x |

| Family | Scientific Name | Common Name | Known Prev. | 2007-08 Bioblitz |
|---|---|---|---|---|
| **Phrynosoma-tidae (Spiny lizards)** | *Sceloporus magister* | Desert spiny lizard | | X |
| | *Sceloporus undulatus tristichus* | Southern plateau lizard | | x |
| | *Sceloporus graciosus* | Sagebrush lizard | | x |
| | *Uta stansburiana* | Common side-blotched lizard | | x |
| | *Urosaurus ornatus* | Ornate tree lizard | | x |
| **Teiidae (Whiptails)** | *Cnemidophorus tigris* | Western (Tiger) whiptail | x | |
| | *Cnemidophorus velox (includes Aspidoscelis innotata)* | Plateau striped whiptail | x | |
| **Colubridae (Colubrid snakes)** | *Masticophis taeniatus* | Striped whipsnake | | x |
| | *Pituophis catenifer* | Gopher snake | x | x |
| | *Hypsiglena torquata* | Night snake | x | |
| | *Thamnophis elegans* | Western terrestrial gartersnake | | x |
| **Viperidae (Rattle-snakes)** | *Crotalus viridis* | Prairie rattlesnake | x | x |

Birds

| Family | Scientific Name | Common Name | Known Prev. | 2007-08 Bioblitz |
|---|---|---|---|---|
| **Cathartidae (Vultures)** | *Cathartes aura* | Turkey vulture | x | X |
| **Anatidae (Ducks)** | *Anas platyrhynchos* | Mallard | x | |
| **Accipitridae (Hawks)** | *Haliaeetus leucocephalus* | Bald eagle | x | |
| | *Circus cyaneus* | Northern harrier | x | |
| | *Accipiter striatus* | Sharp-shinned hawk | x | |
| | *Accipiter cooperii* | Cooper's hawk | | x |
| | *Buteo jamaicensis* | Red-tailed hawk | x | |
| | *Buteo lagopus* | Rough-legged hawk | x | |
| | *Aquila chrysaetos* | Golden eagle | x | x |
| **Falconidae (Falcons)** | *Falco sparverius* | American kestrel | x | x |
| | *Falco peregrinus* | peregrine falcon | x | |
| | *Falco mexicanus* | prairie falcon | x | |

| Family | Scientific Name | Common Name | Known Prev. | 2007-08 Bioblitz |
|---|---|---|---|---|
| **Phasianidae (Pheasants)** | *Alectoris chukar | Chukar | x | |
| | Meleagris gallopavo | Wild turkey | x | x |
| **Columbidae (Pigeons)** | Zenaida macroura | Mourning dove | x | x |
| **Strigidae (Owls)** | Otus kennicottii | Western screech-owl | x | |
| | Bubo virginianus | Great horned owl | x | x |
| **Caprimulgidae (Goatsuckers)** | Chordeiles minor | Common nighthawk | x | |
| | Phalaenoptilus nuttallii | Common poorwill | x | x |
| **Apodidae (Swifts)** | Aeronautes saxatilis | White-throated swift | | x |
| **Trochilidae (Hummingbirds)** | Archilochus alexandri | Black-chinned hummingbird | x | x |
| | Selasphorus platycercus | Broad-tailed hummingbird | x | x |
| | Selasphorus rufus | Rufous hummingbird | x | |
| **Picidae (Woodpeckers)** | Melanerpes lewis | Lewis' woodpecker | x | |
| | Sphyrapicus nuchalis | Red-naped sapsucker | x | |
| | Picoides pubescens | Downy woodpecker | x | |
| | Picoides villosus | Hairy woodpecker | x | x |
| | Colaptes auratus | Northern flicker | x | x |
| **Tyrannidae (Tyrant flycatchers)** | Contopus sordidulus | Western wood-pewee | x | |
| | Empidonax oberholseri | Dusky flycatcher | x | |
| | Empidonax wrightii | Gray flycatcher | x | |
| | Empidonax occidentalis | Cordilleran flycatcher | | x |
| | Sayornis saya | Say's phoebe | x | x |
| | Myiarchus cinerascens | Ash-throated flycatcher | | x |
| | Tyrannus verticalis | Western kingbird | x | |
| **Laniidae (Shrikes)** | Lanius ludovicianus | Loggerhead shrike | x | |
| **Vireonidae (Vireos)** | Vireo plumbeus | Plumbeous vireo | | x |
| | Vireo gilvus | Warbling vireo | | x |
| **Corvidae (Jays)** | Aphelocoma californica | Western scrub jay | x | x |
| | Gymnorhinus cyanocephalus | Pinyon jay | x | x |
| | Nucifraga columbiana | Clark's nutcracker | x | |
| | Pica hudsonia | Black-billed magpie | x | |
| | Corvus corax | Common raven | x | x |

| Family | Scientific Name | Common Name | Known Prev. | 2007-08 Bioblitz |
|---|---|---|---|---|
| **Hirundinidae (Swallows)** | *Tachycineta thalassina* | Violet-green swallow | x | x |
| | *Stelgidopteryx serripennis* | Northern rough-winged swallow | | x |
| **Paridae (Chickadees)** | *Poecile atricapilla* | Black-capped chickadee | x | x |
| | *Poecile gambeli* | Mountain chickadee | x | x |
| | *Baelophus ridgwayi* | Juniper titmouse | x | x |
| **Aegithalidae (Bushtits)** | *Psaltriparus minimus* | Bushtit | x | x |
| **Certhiidae (Creepers)** | *Certhia americana* | Brown creeper | | x |
| **Sittidae (Nuthatches)** | *Sitta carolinensis* | White-breasted nuthatch | x | x |
| **Troglodytidae (Wrens)** | *Troglodytes aedon* | House wren | | x |
| | *Thryomanes bewickii* | Bewick's wren | | x |
| | *Salpinctes obsoletus* | Rock wren | | x |
| | *Catherpes mexicanus* | Canyon wren | x | x |
| **Regulidae (Kinglets)** | *Regulus calendula* | Ruby-crowned kinglet | | x |
| **Sylviidae (Gnatcatchers)** | *Polioptila caerulea* | Blue-gray gnatcatcher | | x |
| **Turdidae (Thrushes)** | *Sialia mexicana* | Western bluebird | x | |
| | *Sialia currucoides* | Mountain bluebird | x | |
| | *Turdus migratorius* | American robin | x | x |
| **Mimidae (Thrashers)** | *Mimus polyglottos* | Northern mockingbird | x | x |
| **Parulidae (Wood warblers)** | *Vermivora celata* | Orange-crowned warbler | | x |
| | *Vermivora virginiae* | Virginia's warbler | | x |
| | *Dendroica coronata* | Yellow-rumped warbler | | x |
| | *Dendroica nigrescens* | Black-throated gray warbler | | x |
| | *Dendroica petechia* | Yellow warbler | | x |
| | *Oporornis tolmiei* | MacGillivray's warbler | x | |
| | *Icteria virens* | Yellow-breasted chat | | x |

| Family | Scientific Name | Common Name | Known Prev. | 2007-08 Bioblitz |
|---|---|---|---|---|
| **Thraupidae (Tanagers)** | *Piranga ludoviciana* | Western tanager | x | |
| **Emberizidae (Sparrows)** | *Pipilo chlorurus* | Green-tailed towhee | | x |
| | *Pipilo maculatus* | Spotted towhee | x | x |
| | *Spizella passerina* | Chipping sparrow | x | x |
| | *Spizella breweri* | Brewer's sparrow | | x |
| | *Chondestes grammacus* | Lark sparrow | x | x |
| | *Zonotrichia leucophrys* | White-crowned sparrow | x | x |
| | *Junco hyemalis* | Dark-eyed junco | x | x |
| **Cardinalidae (Buntings)** | *Pheuticus melanocephalus* | Black-headed grosbeak | | x |
| | *Passerina amoena* | Lazuli bunting | | x |
| **Icteridae (Blackbirds)** | *Sturnella neglecta* | Western meadowlark | x | |
| | *Euphagus cyanocephalus* | Brewer's blackbird | | x |
| | *Molothrus ater* | Brown-headed cowbird | x | x |
| | *Icterus bullockii* | Bullock's oriole | | x |
| | *Icterus parisorum* | Scott's oriole | x | |
| **Fringillidae (Finches)** | *Carpodacus mexicanus* | House finch | x | x |
| | *Carduelis pinus* | Pine siskin | x | x |
| | *Carduelis psaltria* | Lesser goldfinch | x | x |
| | *Carduelis tristis* | American goldfinch | | x |
| **Passeridae (Weaver finches)** | *\*Passer domesticus* | House sparrow | x | |

Mammals

| Family | Scientific Name | Common Name | Known Prev. | 2007-08 Bioblitz |
|---|---|---|---|---|
| **Soricidae (Shrews)** | *Sorex palustris* | American water shrew | x | |
| **Vespertilionidae (Bats)** | *Eptesicus fuscus* | Big brown bat | | x |
| | *Pipistrellus hesperus* | Western pipistrelle | x | |
| **Canidae (Dogs)** | *Canis latrans* | Coyote | x | |
| | *Urocyon cinereoargenteus* | Common gray fox | | x |
| | *Vulpes vulpes* | Red fox | x | |
| **Ursidae (Bears)** | *Ursus americanus* | American black bear | | x |

| Family | Scientific Name | Common Name | Known Prev. | 2007-08 Bioblitz |
|---|---|---|---|---|
| **Procyonidae (Raccoons)** | *Bassariscus astutus* | Ringtail | | x |
| **Felidae (Cats)** | *Lynx rufus* | Bobcat | x | |
| | *Puma concolor* | Mountain lion | x | x |
| **Mephitidae (Skunks)** | *Mephitis mephitis* | Striped skunk | | x |
| **Mustelidae (Weasels)** | *Mustela frenata* | Long-tailed weasel | | x |
| | *Mustela erminea* | Ermine (Short-tailed weasel) | | x |
| **Cervidae (Deer)** | *Odocoileus hemionus* | Mule deer | x | x |
| | *Cervus elaphus* | Elk | x | x |
| **Sciuridae (Squirrels)** | *Ammospermophilus leucurus* | White-tailed antelope squirrel | | x |
| | *Spermophilus variegatus* | Rock squirrel | | x |
| | *Tamias dorsalis* | Cliff chipmunk | x | |
| | *Tamias minimus* | Least chipmunk | x | |
| | *Marmota flaviventris* | Yellow-bellied marmot | | x |
| **Geomyidae (Gophers)** | *Thomomys bottae* | Botta's pocket gopher | x | x |
| **Heteromyidae (Pocket mice)** | *Dipodomys ordii* | Ord's kangaroo rat | x | x |
| | *Perognathus parvus* | Great Basin pocket mouse | | x |
| | *Chaeteodipus formosus* | Long-tailed pocket mouse | | x |
| **Castoridae (Beavers)** | *Castor canadensis* | Beaver | | x |
| **Muridae (Mice)** | *Microtus montanus* | Montane vole | | x |
| | *Onychomys leucogaster* | Northern grasshopper mouse | x | |
| | *Neotoma cinerea* | Bushy-tailed woodrat | x | |
| | *Neotoma lepida* | Desert woodrat | x | |
| | *Peromyscus crinitus* | Canyon mouse | x | |
| | *Peromyscus boylii* | Brush mouse | x | |
| | *Peromyscus maniculatus* | Deer mouse | | x |
| | *Peromyscus truei* | Pinyon mouse | | x |
| | *Reithrodontomys megalotis* | Western harvest mouse | | x |
| | *\*Mus musculus* | House mouse | x | |

| Family | Scientific Name | Common Name | Known Prev. | 2007-08 Bioblitz |
|--------|-----------------|-------------|-------------|------------------|
| **Leporidae (Rabbits)** | *Sylvilagus audubonii* | Desert cottontail | x | x |
| | *Lepus californicus* | Black-tailed jackrabbit | x | x |

## ACKNOWLEDGMENTS

The 2007 and 2008 Deer Creek bioblitzes were generously sponsored by landowners Tom and Caroline Hoyt, who provided camping sites, lavish meals, ample liquid refreshment, and wonderful hospitality. The events were organized by The Nature Conservancy's Moab office with the assistance of Gen Green, Chris Pague, and Nicol Gagstetter. Local landowners in the Deer Creek watershed provided access to their properties for field research. Eric Feiler provided important logistic support. Local residents Nina and Walt Gove, Angie Evenden, and Carl Dede shared information from their wildlife observations in the watershed. Thanks to all of our fellow bioblitz teammates for donating their time and skills to cataloguing the local biota: Rhett Boswell, Scott Brodie, Jim Catlin, Evelyn Cheng, Eric Feiler, Tim Graham, Gen Green, Tom and Caroline Hoyt, Allison Jones, Jeri Ledbetter, Max Licher, Mary O'Brien, Chris Pague, Neil Perry, Keith Schulz, David Smuin, Larry Stevens, and Kevin Wheeler.

## LITERATURE CITED

Anderson, M., P. Bourgeron, M.T. Bryer, R. Crawford, L. Engelking, D. Faber-Langendoen, M. Gallyoun, K. Goodin, D.H. Grossman, S. Landaal, K. Metzler, K.D. Patterson, M. Pyne, M. Reid, L. Sneddon, and A.S. Weakley. 1998. International classification of ecological communities: terrestrial vegetation of the United States. Volume II. The National Vegetation Classification System: list of types. The Nature Conservancy, Arlington, VA.

Atwood, N.D., C.L. Pritchett, R.D. Porter, and B.W. Wood. 1980. Terrestrial vertebrate fauna of the Kaiparowits Basin. Great Basin Naturalist 40(4):303-350.

Bosworth, W.B., III. 2003. Vertebrate information compiled by the Utah Natural Heritage Program: A progress report. Report prepared for the Utah Reclamation Mitigation and Conservation Commission and US Department of the Interior. Cooperative Agreement # 1-FC-UT-00840.

Conniff, R. 2000. Wanted, dead or alive. Smithsonian 31(1):21-24.

Doelling, H.H., R.E. Blackett, A.H. Hamblin, J.D. Powell, and G.L. Pollock. 2003. Geology of Grand Staircase-Escalante National Monument, Utah. Pp 188-231. In: D.A. Sprinkel, T.C. Chidsey, Jr., and P.B. Anderson, eds. Geology of Utah's Parks and Monuments. Utah Geological Association Publication 28 (second edition).

Evenden, A.G. 1999. Inventory and monitoring for Ute ladies'-tresses orchid (Spiranthes diluvialis Sheviak) – Grand Staircase-Escalante National Monument. Unpublished report prepared for Grand Staircase-Escalante National Monument, Kanab, UT.

Fertig, W. 2005. Annotated checklist of the flora of Grand Staircase-Escalante National Monument. Moenave Botanical Consulting, Kanab, UT. 54 pp.

Fertig, W., J. Spence, L. Stevens, J. Ledbetter, N. Perry, R. Boswell, and T. Graham. 2009. The biota of the Deer Creek watershed, Garfield County, Utah: Summary of the 2007-2008 bio-blitz. Report prepared for the Utah Nature Conservancy. Moenave Botanical Consulting, Kanab, UT. 54 pp.

Flinders, J.T., D.S. Rogers, J.L. Webber-Alston, and H.A. Barber. 2002. Mammals of the Grand Staircase-Escalante National Monument: a literature and museum survey. Monographs of the Western North American Naturalist 1:1-64.

Jensen, F.C., H. Barber, and S. Hedges. No date. Field checklist of the birds of Grand Staircase-Escalante National Monument. Bureau of Land Management and Utah Division of Wildlife Resources. 3 pp.

Karns, D.R., D.G. Ruch, R.D. Brodman, M.T. Jackson, P.E. Rothrock, P.E. Scott, T.P. Simon, and J.O. Whitaker Jr. 2006. Results of a short-term bioblitz of the aquatic and terrestrial habitats of Otter Creek, Vigo County, Indiana. Proceedings of the Indiana Academy of Science 115(2):82-88.

Leopold, A. 1966. A Sand County Almanac with Essays on Conservation from Round River. Sierra Club/Ballantine Books, New York. 295 pp.

Lundmark, C. 2003. Bioblitz: Getting into backyard biodiversity. Bioscience 53(4):329.

Margules, C.R. and S. Sarkar. 2007. Systematic Conservation Planning. Cambridge University Press, Cambridge, UK. 270 pp.

Oliver, G.V. 2003. Amphibians and reptiles of the Grand Staircase-Escalante National Monument: Distribution, abundance, and taxonomy. Final report under Cooperative Agreement # JSA990010 between the USDI Bureau of Land Management and the Utah Division of Wildlife Resources. 102 pp.

Parrish, J.R., F.P. Howe, and R.E. Norvell. 2002. Utah Partners in Flight Avian Conservation Strategy version 2.0. Utah Partners in Flight, Utah Division of Wildlife Resources, Salt Lake City, UT. 302 pp.

Ricketts, T.H., E. Dinerstein, D.M. Olson, C.J. Loucks, W. Eichbaum, D. DellaSala, K. Kavanagh, P. Hedao, P.T. Hurley, K.M. Carney, R. Abell, and S. Walters. 1999. Terrestrial Ecoregions of North America: A Conservation Assessment. Island Press, Washington, DC. 485 pp.

Schulz, K. 2007. Plant community survey of the Hoyt property, Garfield County, Utah. Report prepared for The Nature Conservancy. 4 pp.

Stein, B.A., L.S. Kutner, and J.S. Adams. 2000. Precious Heritage, the Status of Biodiversity in the United States. The Nature Conservancy and Association for Biodiversity Information, Oxford University Press, New York. 399 pp.

Tuhy, J.S., P. Comer, G. Bell, D. Dorfman, B. Neeley, M. Lammert, S. Silbert, J. Humke, L. Whitham, B. Cholvin, and B. Baker. 2002. A conservation assessment of the Colorado Plateau ecoregion. The Nature Conservancy.

Utah Division of Wildlife Resources. 1998. Inventory of sensitive species and ecosystems in Utah. Endemic and rare plants of Utah: An overview of their distribution and status. Report prepared for the Utah Reclamation Mitigation and Conservation Commission and US Department of the Interior. 566 pp + app.

Welsh, S.L. and N.D. Atwood. 2002. Flora of Bureau of Land Management Grand Staircase-Escalante National Monument and Kane County, Utah. Brigham Young University, Provo, UT. 166 pp.

Welsh, S.L., N.D. Atwood, S. Goodrich, and L.C. Higgins. 2008. A Utah Flora, fourth edition, revised. Brigham Young University, Provo, UT. 1019 pp.

Wilson, E.O. 1992. The Diversity of Life. Belknap Pres of Harvard University Press. Cambridge, MA. 424 pp.

# VALUING THE VERDE RIVER WATERSHED: AN ASSESSMENT

*Patricia West, Dean Howard Smith and William Auberle*

## ABSTRACT

This valuation study is designed to be the first phase of a larger series of studies to value the ecosystem services of the Verde River and its watershed. Interviews were conducted with 35 anonymous community leaders who live in, work with, or manage some aspect of the watershed (or a combination of the three). The interviews resulted in a large list of values for the watershed and provide a starting point for more studies. This report includes preliminary analysis of the data collected from these interviews, a brief literature review on ecosystem services, and recommendations for future research.

This study found that the most valued aspect of the river is not as a place to get things from, but as an entity that is valued for its very existence for a wide variety of reasons, most of which are categorized as "cultural" using the Millennium Ecosystem Assessment (2003, 2005 a-e) criteria. These cultural values contribute to the quality of life of residents of the area and the entire State of Arizona, as well as visitors who provide the area with much of its economic base.

## INTRODUCTION

The Verde River creates a vibrant and verdant series of ecosystems in central Arizona, supports numerous communities, and faces rapidly growing threats. The river is the last remaining mostly free-flowing river in otherwise arid Arizona. The challenges facing the stakeholders of the river and its environs are numerous and complex. The communities throughout the Verde Valley (Cottonwood, Jerome, Clarkdale, Cornville, Sedona, Rim Rock, and Camp Verde), Chino Valley, Prescott, and Prescott Valley are growing rapidly. The impacts of population growth are of great concern to residents. The watershed drains approximately 6,600 square miles and runs 140 miles (Arizona NEMO 2005, Arizona Department of Water Resources 2008). The Big Chino and Little Chino Aquifers initiate the majority of the flow of the river and combine with tributaries to form the discharge and recharge system for the watershed system. This system provides drinking and irrigation water for the local communities and the Phoenix area and supports various distinct ecosystems that include endangered species. As pressures on the watershed increase, there is an increasingly active public debate as to the management processes required to maintain the competing water flows, the water quality, and the habitat. Investigating these competing and in many instances mutually exclusive ends is the purpose of this study.

In what we proposed as a preliminary study of the ecosystem services of the Verde River and its watershed, we have conducted numerous interviews, based on a questionnaire and a semi-structured interview with a variety of community leaders. From the interviews we have developed specific recommendations for further research in valuing the watershed and educating its stakeholders. The interviews and the resulting list of values herein are the first step in a valuation study of the Verde River watershed. Next steps include determining metrics for subsequent valuation studies and determining what specific studies will aid decision-makers in the region. The next step would include securing funding and developing a research team to complete specific valuation studies aimed at providing information to decision-makers for more informed decisions about the Verde River and its watershed.

We first present a brief description of the watershed for readers not familiar with the geography of central Arizona; however, this is not intended to be a complete description of the watershed, the relevant ecosystems and/or geological formations. In order to frame the results presented herein, a brief review of the literature is provided, followed by a discussion of the survey instrument and the methodology of the analysis. Subsequent to the interviews, the data and an analysis were reviewed. Finally, a series of recommendations is presented regarding important areas for further research.

### Description of the Verde River Watershed

The Verde River Watershed (watershed), located in the heart of Arizona is a complex and dynamic system. The watershed ranges from 1,323 to 12,617 feet above sea level over its 6,622 square miles (Arizona NEMO 2005). The watershed has over 9,037 miles of streams, but "only 6% (578 miles) of streams are perennial, and are mostly restricted to the main stem of the Verde River" (Arizona NEMO 2005, 2-4).

From the headwaters in the Chino Valley to Horseshoe and Bartlett Reservoirs, the Verde River is free-flowing and unlike most large rivers in the west, most of the watershed is unregulated with no significant dams and thus retains a natural cyclical flood regime (Pearthree 2008). The 40.5-mile section of the Verde River between Beasley Flats and Sheep Bridge is designated as "Wild and Scenic." This stretch of river is the only designated Wild and Scenic River in Arizona. Before a river corridor can be considered for designation as a Recreation, Scenic, or Wild River Area, the Wild and Scenic Rivers Act (P.L. 90-542) requires that the river and its immediate environments possess one or more specific outstandingly remarkable values. The Environmental Impact Statement approved in 1981 for the Verde River found that this portion of the river corridor contained outstandingly remarkable scenic, fish and wildlife, as well as historic and cultural values (National Wildland and Scenic Rivers System 2007).

### Hydrology and Watershed Health

The Verde River Watershed is made up of three subbasins, including the Upper, Middle and Lower Verde Watersheds. The discharge at the Verde River springs that feed the perennial start of the river comes from the Big Chino (80%) and the Little Chino (14%) aquifers, as well as a small percentage from surface runoff (Wirt et al. 2005; Springer and Haney 2008). "The Big Chino subbasin, in that upper Verde River Watershed is 1850 square miles in area. The Little Chino subbasin in the upper Verde River watershed is the smallest of the three subbasins in the study area and has had the greatest groundwater development" (Blasch et al. 2006, pp 1). Additional groundwater, tributaries, and surface water contribute to the base flow, especially after the first 26 miles (Wirt et al. 2005). Human uses and climate conditions affect the base flow and have both contributed to declines in the base flow since 1994 (Springer and Haney 2008). Decreases in base flows can produce many effects on people, flora and fauna, and ecological processes in the Verde River Watershed. One of the changes predicted is a decline in cottonwood and willow abundance (Haney et al. 2008).

Arizona Nonpoint Education for Municipal Officials (NEMO) (2005) conducted a watershed-based plan that concluded that "the primary sources for nonpoint source pollutants concerns in the Verde Watershed include abandoned mine sites, new development and increased urbanization, and new road construction." Other threats to the watershed included livestock grazing (Fossil Creek and Cherry Creek), animal wastes and failure of residential septic systems (across the watershed) (Arizona NEMO 2005, Ex-2).

## Social/Economic Characteristics

The Verde Watershed spans four counties (Coconino, Gila, Maricopa and Yavapai), although 50% of the watershed is in Yavapai County (Arizona NEMO 2005). The watershed is primarily rural with several urban areas including Sedona, Prescott, and part of Scottsdale (Arizona NEMO 2005). Land managers of the watershed include Forest Service (64%), private landowners (23%), state trust (9%), tribal lands (2%), military (1%), local and state parks (1%), and other (<1%) (Springer and Haney 2008).

A study conducted from 2006-2007 indicated that at least one third (31%) of the visitors to the Verde Valley were from Arizona (Arizona Office of Tourism 2008). This survey also suggested that visiting state and national parks, as well as historic places were the most popular activities for visitors to the Verde Valley, followed by hiking and shopping, bird watching and observing wildlife, and enjoying area streams and rivers (Arizona Office of Tourism 2008). Aside from shopping, these activities were directly related to the natural beauty of the Verde Valley and were directly linked to the Verde River itself.

## Water Rights

The majority of the rights to the water in the Verde River are held by the Salt River Project (SRP) which consists of the Salt River Valley Water User's Association and the Salt River Agricultural Improvement and Power District (Gooch et al. 2007). The predecessor's of the SRP established flow rights as early as 1869, and since its establishment in 1903, SRP has had an interest in maintaining the flows that the shareholders rely upon. Recently, SRP has attempted to maintain the flow by assisting those extracting water from the river illegally to find water elsewhere. When these negotiations failed, SRP was forced to litigate and defend the rights of its shareholders.

## Biological Attributes

### Flora.

The Verde Valley is home to many unique plant species and ecosystems which have great intrinsic value, but also serve as habitat for a number of animal species (Stevens et al. 2008). Bailey (2002) classified the vegetation cover in the Verde River Watershed in the "Dry Domain" with the most prominent division being the Tropical/Subtropical Steppe Division (70%) (Arizona NEMO 2005). Brown et al. (1979) classified the vegetation in nine different biotic communities, the most common being the Great Basin Conifer Woodland (Brown et al. 1979; Arizona NEMO 2005). Within this watershed, Arizona Game and Fish identified 10 types of riparian areas in the almost 14,000 acres. Cottonwood-willow areas cover only 0.13% of the riparian area of the watershed, but are among the riparian types more widely used in the watershed by non-fish vertebrates, and are the second most sensitive riparian habitat to changes in streamflow (Stromberg 2008; Stevens et al. 2008). These ecosystems provide habitat for the multitude of wildlife found in the watershed including many threatened and endangered species. There are no federally-listed, rare or endemic plant species known to occur in riparian areas in the watershed (Stromberg 2008). The one federally-listed threatened endemic plant in the watershed is the Arizona Cliffrose (Purshia subintegra) that has a range limited to a few small limestone outcrops in the Verde Valley.

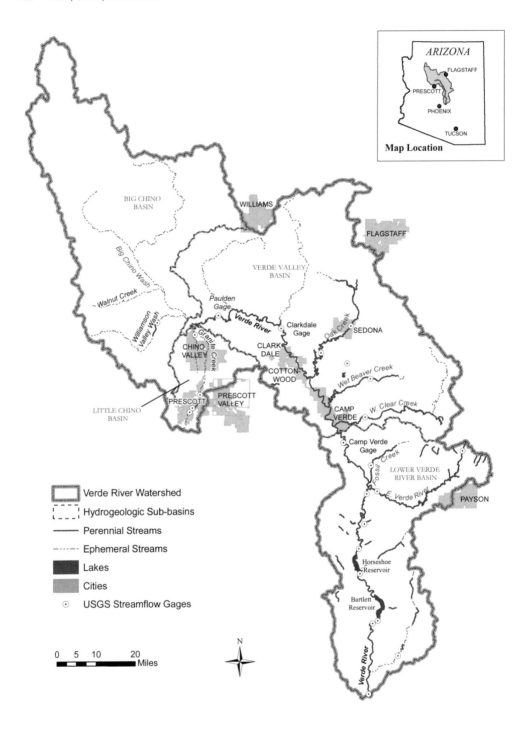

Figure 10.1  Map of the Verde River Watershed printed with permission by The Nature Conservancy.

*Fauna.*

"The Verde River supports an enormous diversity of Arizona's invertebrate and vertebrate species, but anthropogenic activities pose immediate and potentially irrecoverable threats to its aquifers, surface flows, habitat availability and connectivity" (Stevens et al. 2008). Of special concern are the threatened and endangered species that make the watershed their home.

The Verde River Watershed is seasonal home to over 248 species of birds (Schmidt et al. 2005; Stevens et al. 2008), including two endangered bird species: the Desert Nesting Bald Eagle (*Haliaeetus leucocephalus*) and the Southwestern Willow Flycatcher (*Empidonax traillii extimus*). The watershed is also home to the Western Yellow-billed Cuckoo (*Coccyzus americanus occidentalis***) (Schmidt et al. 2005). [We use the symbols* to denote Federally listed species, and ** to denote Candidate for Federal listing.]

Native fish species are threatened by changes to the Verde River. Thirteen native fish species occurred in the Verde River basin, including Gila Trout (*Onchorhynchus gilae*); Desert and Sonora Suckers (*Catognisus clarki* and *C. insignis*, respectively); Speckled Dace (*Rhinichthys osculus*); Razorback Sucker (*Xyrauchen texanus***); Longfin Dace (*Agosia chrysogaster*); Gila, Headwater, and Roundtail Chubs (*Gila intermedia**, G. nigra, and G. robusta***); Spikedace (*Meda fulgida**); Colorado Pikeminnow (*Ptychocheilus lucius*); Loach Minnow (*Tiaroga cobitis**); and Gila Topminnow (*Poeciliopsis occidentalis*) (Stevens et al. 2008).

It was calculated that 92 species of mammals call the Verde River Basin home (Hoffmeister 1986; Feldhamer et al. 2003; Schmidt et al. 2005; Stevens et al. 2008). Some groups of fauna are mostly unnoticed or considered pests by some and as such, their importance is often underestimated. Two of these groups are the invertebrates herpetofauna (reptiles and amphibians). Protected species in these groups are present in the Verde River Watershed (Stevens et al. 2008).

Challenges to The Verde River

In 2006, American Rivers proclaimed the Verde River as the 10th most endangered river in the United States. This was based on the challenges facing the river and the diversity of wildlife. Population growth is also a major challenge, for example, Yavapai County was determined to be the fastest growing rural county in the United States in 1999 (Woods and Poole Economics Inc. 1999). Population predictions have estimated that the population of the county would go from 132,000 in 2000 to over 260,000 in 2050. Simultaneously, regional droughts have created concerns about water sustainability throughout the state and the region. The combination of population growth and drought conditions has led to concerns about where the water will come from to support growing populations.

Another major challenge is climate change. Water shortages are predicted throughout the southwest U.S. Globally, population growth has been predicted to have more negative impacts on water resources than climate change, but arid and semi-arid regions face greater challenges because of the already low water supply (Vörösmarty et al. 2000).

This combination of factors creates an urgent need to provide information to decision-makers about the compromises that will have to be made. The valuation issues in the Verde River Watershed are broad, although many studies have covered aspects of the complex situation in the Verde River Watershed, stakeholders and Native Americans need to be included in the planning process. Non-native species, wildlife habitat and endangered species need to be included in calculations, and shifting urban and rural needs have to be addressed.

Valuation of the Verde River Watershed is in its infancy, and this study is one of the first steps in a complex process that should - according to current literature - include a feedback loop with stakeholders, decision-makers and resource managers.

## Literature Review

Ecosystem services are recognized as important throughout the world for human health and well-being. These services range from providing water and oxygen to providing a feeling of well-being from beautiful places. Many of these are critical to human survival (climate regulation, air purification, crop pollination) while others enhance it (aesthetics) (Kremen 2005). There has been increasing awareness and focus on valuing these services, because losing natural resources leads to losing these services. Valuations also assist decision-makers in designing policies that preserve and/or offset the environmental costs of human changes in the landscape (Bingham et al. 1995; Knetsch 2007).

Ecosystem services have been defined as "the conditions and processes through which natural ecosystems, and the species that make them up, sustain and fulfill human life." (Daily 1997) or "the benefits people obtain from ecosystems" (Millennium Ecosystem Assessment 2003). As the list of ecosystem services has expanded, there have been attempts to categorize the services. For example, Daily (1997) proposed three categories 1) the provision of production inputs, 2) the sustenance of plant and animal life, and 3) the provision of non-use values, which include existence and option values. Kramer (2005) proposed that the total economic value of an environmental resource can be calculated as a sum of four main components: use value, indirect use value, option value and nonuse value.

Much of the valuation literature from 1971 to 1997 and information on ecosystem services was compiled by Wilson & Carpenter (1999). The most recent comprehensive collection of information on ecosystem services is the series published in 2005 by the Millennium Ecosystem Assessment (2005 a-e). The study "was carried out between 2001 and 2005 to assess the consequences of ecosystem change for human well-being and to establish a scientific basis for actions needed to enhance the conservation and sustainable use of ecosystems and their contributions to human well-being" (Millennium Ecosystem Assessment 2005e, p. vii). This five volume series includes information from and references to most previously published information on the subject of ecosystem services (Millennium Ecosystem Assessment 2005a-e). In addition to this, The Proceedings of the National Academy of Sciences compiled a special feature on ecosystem services in July 2008 (Volume 105, number 28) that includes a wide variety of some of the latest research in the field The Millennium Ecosystem Assessment (2003) breaks services into 1) Supporting services, 2) Provisioning services, 3) Regulating services, and 4) Cultural services (Figure 10.2).

The field of Ecological Economics has expanded in many directions over the past twenty years. These shifts include changing scales ranging from very local to global (including multiscale approaches), increasing participation of communities and stakeholders, using models to predict changes in ecosystem services, and including biodiversity and all its services in the calculations of value. Bateman et al. (2004) suggest that scope is also an important factor in determining the effectiveness of valuations, making the process of evaluating ecosystem services both more complex and more effective.

The shift to include community input, participation and collaboration when managing, valuing, and preserving environmental resources, especially Native American communities, has brought

| Provisioning Services: | Regulating Services: | Cultural Services: | Supporting Services: |
|---|---|---|---|
| Products obtained from ecosystems. | Benefits obtained from regulation of ecosystem processes. | Non-material benefits obtained from ecosystems. | Necessary for the production of all other ecosystem services. |
| -Food | -Water purification | -Recreation | -Soil formation |
| -Fresh water | -Disease control | -Ecotourism | -Nutrient cycling |
| -Fuelwood | -Climate regulation | -Spiritual inspiration | -Primary production |
| -Fiber | | -Artistic inspiration | |
| -Biochemicals | | -Education | |
| -Genetic resources | | -Cultural heritage | |

**Security/Health/Work and Social Opportunities:**

Ability to live in a clean and safe environment.

Ability to reduce vulnerability to ecological shocks and stresses.

Ability to access resources to earn income and make choices for your life.

Ability to be nourished and have clean drinking water.

Ability to be free from avoidable disease.

Ability to use energy to keep warm and cool.

Ability to have clean air.

Opportunities to use ecosystems for recreation, expression and education.

Freedom and choice

Figure 10.2  Ecosystem services and their links to human well-being

more informed decisions about resource management (Cronin & Ostergren 2007b; Hein et al. 2006; Terer et al. 2004; Venn & Quiggin 2007). Community participation has always been valued, but techniques for achieving it have been refined over the past 15 years. Part of this shift is to include Traditional Ecological Knowledge (TEK) in valuations and management of natural areas, systems, and resources that are jointly owned, managed and used (Cronin and Ostergren 2007a).

Many recent studies have focused on specific geographic regions because valuation is site specific; however, sites can range from a specific park or lake, to the protected areas in a country (Ingraham & Foster 2008). These sites can cross political boundaries, which raises complex issues. For example, Lange et al. (2007) documented the value of ecosystem services in the Orange River Basin, which spans Botswana, Namibia, Lesotho, and South Africa. Designers of valuation studies need to be aware enough to account for the full extent of beneficiary groups, even for spatially distant groups (Pate and Loomis (1997) and Loomis (2000). In the current context this involves shareholders and customers of SRP and tourists.

These studies have tended to be specific to concerns of the area, and have provided information to resource managers and decision-makers for the site. If research is not focused on the social context in which the values are determined and if studies are not designed to directly supply information needed for management decisions within that context, then the research may have little value (Cowling et al. 2008).

Many studies seek to improve on the original valuation techniques (Norton & Noonan 2007) including contingent valuation (Hanemann 1994; Spash 2000; Holmes et al. 2004; Wiser 2007; Bateman et al. 2006a; Marta-Pedroso et al. 2007) willingness to pay (Bateman et al. 2006b) and cost-benefit

analysis (Kuosmanen & Kortelainen 2007). For example, Kontogianni et al. (2001) used a combination of focus groups and surveys to provide results to decision-makers focusing on willingness to pay for a variety of options and attitudes about the environment. Another shift has been towards modeling. Many studies have used models as a way to value natural resources. Biodiversity has also been valued in many ways for its contribution to human well-being, quality of life, and economic contributions. For example, non-timber forest products are one valued aspect of wild places (Croitoru 2007). Kellermann et al. (2008) documented the specific ecological and economic services that birds provided to coffee farms in the Blue Mountains of Jamaica. Changes in biodiversity are also tracked by looking at non-native invasive species (or "weeds") (Sinden & Griffith 2007) or evaluated as a separately valued part of the ecological system (Eppink & van den Bergh 2007).

Allen & Loomis (2006) determined that valuation of wildlife - particularly estimation of non-use and non-consumptive use values for wildlife - is an important input into various policy decisions. Sensitive species, including federally listed threatened and endangered species need to be accounted for (De Nooij et al. 2006). Numerous studies have been performed to evaluate the value of restoration and habitat preservation projects specifically in riparian areas and wetland ecosystems (Costanza et al 1989; Vicory & Stevenson 1995; Spash 2000; Varady et al. 2001; Amigues et al. 2002; Holmes et al. 2004; Hanley et al. 2006; Ojeda et al. 2008). These studies are especially urgent in the light that "inland water habitats and species are in worse condition than those of forest, grassland, or coastal systems (medium certainty), in that 50% of inland water habitats were lost during the twentieth century. It is well established that for many ecosystem services, the capacity of inland water systems to produce these services is in decline and is as bad or worse than that of other systems" (Millennium Ecosystem Assessment 2005a p.553).

This study was based on the typologies presented by Daily (1997), Millennium Ecosystem Assessment (2003), and Kramer (2005). We selected two systems of typology in order to be able to categorize the areas that are of highest value to stakeholders and use that information to design further studies. This process is best suited to this study because of the wide variety and large number of responses received.

## METHODOLOGY

In order to determine the various uses and values for the Verde River Watershed, a survey was developed, which consisted of two parts . The first part of the survey was a mail-back questionnaire, and the second part was a semi-structured interview. The interviews were completed on an individual basis, recorded and transcribed. A total of 35 individuals knowledgeable of the watershed and current issues were interviewed. Respondents were assured confidentiality, in that all personally identifiable information was maintained by the research team. Due to the small size of the community of people directly involved with the Verde Watershed our Institutional Review Board requires us to maintain the complete anonymity of the respondents. As such, we are unable to provide even the most cursory descriptors for the interviewees. The individuals who were selected represented various sides of the discussion, including elected officials, non-governmental leaders, and professional resource managers. The description of the values rubric follows.

### Values

A two-stage classification of values indicated during the interviews is listed below. For budgetary reasons, only two questions could be analyzed in detail. The replies for Question 3 a-c were included in this portion of the rubric.

Question 6 was also included in the values analysis.

*3.How do you use the river?*
a. What plants and animals that rely on the river are important to you?
b. Do you collect or use any plants or animals that rely on the river? If so, which ones?
c. Do you have a spiritual, religious, or personal connection to the river? If so, could you describe this connection?
*6. What functions, processes or services does the Verde provide that are important to you and the community? (e.g. flooding, filtration, seed dispersal...)*

Each comment as recorded during the data collection process was evaluated based on a rubric comprised of the following categories: use, option, non-use, supporting, provisioning, regulating, and culture. A more detailed description of the rubric categories are explained below.

### Use.

Uses were considered to be goods and services provided by the watershed that the respondents used in some form. Irrigation water and fishing are obvious examples. Enjoying one's five-generation family history with the river is a far less obvious use, because it is closely tied to values the respondents gain.

The discussion of flora and fauna was problematic. Values were placed under the use category if it was not specifically indicated that they fell into the other categories. Species of flora and fauna were viewed as specific current uses as opposed to future values of ecosystem protection.

### Option.

This category involved future use without current use. The primary candidate for this group related to recharging the aquifer. The fewest number of responses fell into this category.

| Primary value | Secondary value | Number of responses | | |
|---|---|---|---|---|
| | | Question 3 | Question 6 | Total |
| Use | | | | |
| | Servicing | 1 | 2 | 3 |
| | Provisioning | 145 | 39 | 184 |
| | Regulating | 6 | 31 | 37 |
| | Cultural | 167 | 48 | 215 |
| | | **319** | **120** | **439** |
| Option | | | | |
| | Servicing | 0 | 0 | 0 |
| | Provisioning | 3 | 0 | 3 |
| | Regulating | 5 | 3 | 8 |
| | Cultural | 2 | 0 | 2 |
| | | **10** | **3** | **13** |
| Non-use | | | | |
| | Servicing | 1 | 0 | 1 |
| | Provisioning | 0 | 0 | 0 |
| | Regulating | 13 | 26 | 39 |
| | Cultural | 4 | 0 | 4 |
| | | **18** | **26** | **44** |

Table 10.1  Ecosystem values as collected in categories.

## Non-Use.

The best example of this category was provided as: "Give value to the person in Wisconsin that may never see it." Indeed, this is a textbook definition of the term non-use. As mentioned above, most items in the non-use list involve ecosystem protection.

Once the comments were evaluated using the use, option, and non-use rubric, a second classification system was used following Figure 10.1 (Millennium Ecosystem Assesment 2003). This includes four different ways of viewing a particular value of the watershed: supporting, provisioning, regulating and cultural.

## Servicing.

These minimally listed items involved supporting the ecosystem from a structural standpoint such as soil formation.

## Provisioning.

The items included in this list involved things that people physically took from the watershed including water, plants, animals, and minerals. Comments concerning property values and economic growth were also included. The list includes all flora and fauna including endangered species.

## Regulating.

Water cleansing and pollution mitigation were examples of regulating services. Several respondents itemized the fact that the greenery and watershed provided for temperature reductions in "microclimates." Transportation was included in this category.

## Cultural.

The most common of values placed on the river fall into the cultural use category: Spiritual, recreational, aesthetic, inspirational, educational and cultural heritage aspects fall into this category.

## Data Processing

Once the rubrics for analysis were determined and the data entered, a two-stage classification of replies to questions three and six was conducted. The responses were initially classified as use, option or non-use, and then further classified as servicing, provisioning, regulating or cultural. Following a sorting of the replies, a second team member validated the classification. As presumed, some of the classifications fell into rather fuzzy zones. For example, one respondent said "recreation" in response to how he/she used the river; however, further discussion indicated that he/she used the river for fishing, swimming, hunting, and kayaking, as well as limited recreational activities, because of inaccessibility to the public. This single response was labeled as a cultural use, but could also fit into providing use because the response indicated that fish and animals were extracted. A second respondent also itemized recreation as hiking, biking, hang-gliding, picnicking, fishing, gardening, and dog-walking.

## Valuation Of The Verde River Watershed

In total, the 35 respondents to the qualitative survey mentioned nearly 500 ways of valuing the river and its watershed. As presented in Table 10.1, responses fell into two primary categories: provisioning use and cultural use.

Perhaps the strongest conclusion from the analysis was that people valued the river as a place and not just a thing. It was not simply a thing where they acquired goods and services; rather it was a place where respondents did activities. The ability to use the watershed as a source of water was vitally important, but this was by no means the only value stakeholders placed on the system.

Less than 40% of the responses would be listed as provisioning, and, of that number, mostly involved aspects and items beyond simple water provision such as ranching and fishing. Indeed, there were more replies that could be categorized as cultural rather than provisioning. Although very broad in scope

– from spiritual to educational – the cultural category included all the reasons people viewed the watershed as a place to interact with, as opposed to a thing from which to take resources.

As previously mentioned, most comments regarding flora and fauna were placed into the use category. To further analyze flora and fauna using the second rubric, each entry was individualized, but this also created numerous fuzzy responses. These were then mostly placed into the provisioning section. Depending on respondents' secondary comment, this methodology was followed even when it was clear that a particular comment fit into several categories. Included in the itemized list of valued things were 79 instances of "flora" and/or "fauna." Specific comments regarding the habitat provisions of a particular animal or plant (or flora and fauna in general) were listed as non-use and regulating. Specific comments regarding protected species (including federally-listed threatened and endangered species) could easily be listed under numerous categories, but were generally placed under use and cultural. The ability to witness a Bald Eagle is a current use that allowed participants to take the experience from the watershed. Many respondents also itemized a spiritual or other cultural value to having Bald Eagles living near the river. Clearly, most respondents also viewed this as an option value, and most, if pressed, would also call it a non-use value.

Although the numerical count of the regulating values was small, respondents were very familiar with the idea of looking at the watershed as a connected system and even a system of systems. Furthermore, the non-use aspects of regulating values showed how people viewed the importance of the watershed as a watershed. The importance of habitat preservation and the biodiversity of the area were highly valued. At both the micro and macro levels, the ecosystems within the watershed were critical. As an area that sustained otters and served as stopover habitat for migrating birds, people wished to protect the watershed.

## Valued Places

The interviews created a geographical lesson for an adventure map of the Verde River Watershed as discussed during Question 3d. There were 226 individual itemized responses in the 35 interviews. Given the intimate working and living relationship most of the interviewees had with the river, and their requisite vested interests, the question concerning the "Valued Places in the Watershed" provided a plethora of information.

The relevant question specifically included the descriptor "watershed" instead of "river." Had the question been limited to areas of the river itself, many of the important realms of interest would have been lost. Indeed many of the interviewees distinguished between the watershed and the river itself. The tributaries could then be distinguished as rivers, springs, creeks, or washes. The former were sources of river water from the aquifer system; whereas, the latter were collectors from rain and snow melt. Several of the respondents included the Big Chino Aquifer as a valued place.

Respondents distinguished between places with easy public access and wild places. The obvious public access places were listed by many of the respondents: Sedona, Oak Creek Canyon, the Beaver Creeks, Montezuma's Castle and Well National Parks Monuments, Arizona State Parks in the region (Verde River Greenway State Natural Area, Dead Horse Ranch State Park, Red Rock State Park, Slide Rock State Park, Fort Verde State Historic Park), Windmill Park, and Tuzigoot National Monument. Many remote and even wilderness places were reported; however, respondents either requested secrecy or refused to indicate their individual special places. Beasley Flats and Beaver creek appear to be popular hiking areas.

Given the concerns about the geological importance and condition of the Big Chino Aquifer, the headwaters of the river were included by several of the respondents. The historical and archeological importance of the Verde Valley and the watershed was considered as valued by many of the respondents. These include battle sites and Indian ruins.

## Threats to the Watershed

Question 4 asked respondents: "What do you feel threatens the river?" The list of the 191 itemized threats to the watershed was grouped into several distinct issues. For the most part, these were not surprising, but the interviews validated the need for further research into the specific issues regarding the future of the river. Although these are discussed in general herein, future research should be based on the very specific issues addressed by the individual respondents.

Of the 35 interviews, 26 individuals were concerned about the amount of pumping taking place from the aquifer and withdrawals from the river. The concerns included the amount of water that was being withdrawn for various reasons including drinking water and irrigation. A consequence of these withdrawals and diversions was the presence of a fragmented river, with sections that were dry during certain times of the year. These dry spots occurred when water was diverted from the river for irrigation, and partially returned to the river downstream.

Closely connected to the concern about pumping was the concern of human development. Twenty-three of the 35 respondents itemized some aspect – in many cases several aspects – of the growth of the Verde Valley and environs. As discussed above, the population of the Verde Valley and the Prescott/Chino Valley area were seeing enormous growth around this time and were expected to continue growing in the foreseeable future. The envisioned impacts of this growth on the Verde River and its watershed have raised urgent concerns among the population of community leaders that were interviewed. The Verde River Watershed has been forced to address the issue of changing from a collection of rural communities to a collection of developed communities. Consequently, this is changing social norms over concerns about water resources in central Arizona.

A number of environmental impacts were mentioned by interviewees. First, human impact on water quality was a serious concern among respondents. A total of 18 of the 35 interviewees included some aspects of pollution as a threat to the river. Of those aspects specifically itemized, five included runoff from septic systems. Other concerns were mining runoff, agricultural runoff, and dumping of solid waste. Second, invasive species were mentioned by eight of the respondents. Finally, climate change was seen as a threat to the river because of the possibility of continuing drought.

Among the interviews, a series of concerns were also expressed with respect to the lack of education and policy concerning the river. The respondents identified a need for increased community policy concerning the management of the watershed resources. Of course, many of these concerns may be mutually exclusive. This requires increased education of the existing residents – as with the presentation at the Arizona Riparian Council meeting in April 2008 and expanded education of new residents and visitors concerning their impact on the watershed.

## DISCUSSION

The principal goal of this project was to collect and analyze data concerning the reasons for valuing the Verde River Watershed. Data were collected using a series of direct interviews with various stakeholders. A two-stage valuing rubric was used to analyze the data. This research was conducted to make a series of recommendations for further research, public comment and education, and comprehensive watershed management.

The interviewees selected for this research reflected a set of values. The series of interviews were targeted toward specific individuals whom the research team knew had professional and personal interactions with the watershed.  As such, they were not representative of the general population of stakeholders living and working within the watershed. Knowingly or unknowingly, anyone living within the watershed or using water from a tap is a current and possibly a future stakeholder in the watershed.  In addition, many stakeholders, current and future, do not actually live within the watershed.

There were important services offered by the Verde Valley. The findings from this research suggested that interviewees placed great value on food and cultural meanings, as well as the health of the aquifer system and access to potable water. A number of threats were also mentioned, including human development and environmental impacts such as invasive species and climate change. As an initial foray into understanding the non-market valuation of the Verde River Watershed, the research team discerned that numerous aspects of the watershed are important as viewed by the respondents. Besides the obvious provisioning of water for myriad uses, the river is a place to interact with showed up extensively in the categorization of the responses.

## RECOMMENDATIONS

There are several recommendations to be gleaned from this research.  First, the survey techniques could be improved.  For example, a less comprehensive survey instrument could be administered at various venues within the geographical area.  In addition to the list of valued aspects on the original survey, demographic and locational information could be collected, which are particularly important for understanding

the spatial dynamics of demand and for generalizing results beyond the Verde Valley. This broader scoped survey could use general areas of interest collected from this initial survey.  Considering that interviews placed great importance on wildlife of the watershed, future research should focus on refining the categorization of wildlife-related values.  Survey efforts should also pay careful attention to phrasing more specific questions concerning how people value the flora and fauna within the watershed.

Millennium Ecosystem Assessment (2003, 2005 a-e) offered a useful conceptualization of ecosystem services and as such, detailed analyses should be undertaken in this context. This approach links ecosystem services, as discussed above with the "Determinants and Constituents of Well-Being" (shown in Figure 10.2), using the categories of security, basic materials for a good life, health and good social relations.  This approach would help to explain not only "what" is important about the watershed, but also "why" it is important.  A detailed analysis of the data described herein has been conducted to link the values of the watershed directly to human well-being.  This more recent research will be presented elsewhere.

The cultural components itemized by the interviewees suggest that educational efforts should be expanded.  A series of public forums and workshops can be developed to educate both stakeholders and decision-makers within the region.  For example, meetings such as the Arizona Riparian Council are available for public outreach purposes. Education is important and could be expected to alter preferences.  Lack of awareness or familiarity with public goods among the general public is a widely recognized problem in nonmarket valuation – one that education might be expected to remedy.

To validate the comments made by our respondents and to increase the overall knowledge of the watershed, ongoing scientific and policy research should be continued, expanded, and coordinated. The importance of understanding both the science and policy issues concerning the watershed cannot be overstated. The Landsward Institute continues to work with local residents on evaluating the Verde River. A new project is being initiated to study the economic impact of the river: The Verde River Economic Development Study. The details of the study are still being developed.

## CONCLUSION

We can conclude that the Verde River and its watershed are valued in a multitude of ways. For example, there was substantial interest among the stakeholders in terms of flora and fauna (e.g., wildlife) within the watershed. Cultural components and water resources were also seen as valuable ecosystem services. There was also a major cultural component itemized by the interviewees that involved education. Putting monetary value on these values may be desired in order to make the value of some ecosystem services clear to all stakeholders, but valuation is not necessary to show that the stakeholders value the river itself, not just what it can give to them. In other words, preferences can be expressed in monetary or nonmonetary forms, both of which are valuable and important pieces of information in management contexts and in evaluating management/economic tradeoffs (i.e., water). Water and fish have meaningful market values, but an eagle or otter does not.

## ACKNOWLEDGMENTS

We owe gratitude primarily to the participants in this study who took the time to be interviewed and gave thoughtful and indispensable information - you know who you are. We would like to thank the many assistants we had on the project, including Shawn Newell and Fred Solop of the NAU Social Research Laboratory for assisting in developing the survey instrument and making this project more effective. Thanks to everyone in the IRB office who guided us including Paula Garcia McAllister, Patrick Schnell, and Tim Ryan. The authors would like to thank students Nick Sheets, Sarah Viglucci, James Worden and Brandt Weathers for work on the data sets developed in this program. We would especially like to thank Karan English for reviewing drafts, and fundraising. We would like to thank Amanda Cronin for her assistance in developing the questions and assisting with the IRB process. We would like to thank Michelle James for helping to develop the original proposal idea and seeking funds. Partial funding for this project was provided by ERDENE (Environmental Research, Development and Education for the New Economy), Ecological Monitoring & Assessment Program, the Sustainable Energy Solutions research group at Northern Arizona University and the Arizona State Parks Foundation, and the Salt River Project. Earlier versions of this research were presented at the following meetings: 2009 Annual Water Symposium, Scottsdale, Arizona; 10th Biennial Conference of Research on the Colorado Plateau, Flagstaff, Arizona; 22nd Meeting of the Arizona Riparian Council, Prescott, Arizona; Verde Watershed Association, Cottonwood, Arizona.

## LITERATURE CITED

Allen, B.P., and J.B. Loomis. 2006. Deriving Values for the Ecological Support Function of Wildlife: An indirect valuation approach. Ecological Economics 56: 49-57.

Amigues, J.-P., C. Boulatoff, B. Desaigues, C. Gauthier and J. E. Keith. 2002. The Benefits and Costs of Riparian Analysis Habitat Preservation: A willingness to accept/ willingness to pay contingent valuation approach. Ecological Economics 43: 17-31.

Arizona Department of Water Resources. 2008. "Verde River Watershed" Http://Www.

Adwr.State.Az.Us/Dwr/Content/Find_By_Category/Abcs_Of_Water/Rural_AZ/Centralhighlands/Verde_River_Watershed. Pdf, accessed November 2, 2008.

Arizona NEMO. 2005. Watershed Based Plan, Verde Watershed. http://www.srnr.arizona.edu/nemo. Accessed December 2008.

Arizona Office of Tourism. 2008. Verde Valley Tourism Survey. Prepared for the Arizona Office of Tourism by Arizona Hospitality Research & Resource Center, the Center for Business Outreach, and the W.A. Franke College of Business, Northern Arizona University. Flagstaff, Arizona.

Bailey, R.G. 2002. Ecoregion-based Design for Sustainability. Springer-Verlag. New York. 222p.

Bateman, I.J., M. Cole, P. Cooper, S. Georgiou, D. Hadley, and G.L. Poe. 2004 On Visible Choice Sets and Scope Sensitivity. Journal of Environmental Economics and Management 47: 71-93.

Bateman, I.J., M.A. Cole, S. Georgiou, and D.J. Hadley. 2006a. Comparing Contingent Valuation and Contingent Ranking: A case study considering the benefits of urban river quality improvements. Journal of Environmental Management 79: 221-231.

Bateman, I. J., B. H. Day, S. Georgiou, and I. Lake. 2006b. The Aggregation of Environmental Benefit Values: Welfare measures, distance decay and total WTP. Ecological Economics 60(2): 450-460.

Bingham, G., R. Bishop, M. Brody, D. Bromley, E. Clark, W. Cooper, R. Costanza, T. Hale, G. Hayden, S. Kellert, R. Norgaard, B. Norton, J. Payne, C. Russell, G. Suter. 1995. Issues in Ecosystem Valuation: Improving information for decision-making. Ecological Economics 14: 73-90.

Blasch, K.W., J.P. Hoffmann, L.F. Graser, J.R. Bryson, and A.L. Flint. 2006. Hydrogeology of the Upper and Middle Verde River Watersheds, Central Arizona. U. S. Geological Survey, Reston, Virginia. http://pubs.usgs.gov/sir/2005/5198/

Brown, D.E., C.H. Lowe, and C.P. Pace. 1979. A Digitized Classification System for the Biotic Communities of North America, with Community (Series) and Association Examples for the Southwest. Journal of the Arizona-Nevada Academy of Sciences 14 (suppl. 1): 1-16.

Costanza, R., S. C. Farber, and J. Maxwell. 1989. Valuation and Management of Wetland Ecosystems. Ecological Economics 1: 335-361.

Cowling, R.M., B. Egoh, A.T. Knight, P.J. O'Farrell, B. Reyers, M. Rouget, D.J. Roux, A. Welz, A. Wilhelm-Rechman. 2008. An Operational Model for Mainstreaming Ecosystem Services for Implementation. Proceedings of the National Academy of Sciences 105(28): 9483-9488.

Croitoru, L. 2007. Valuing the Non-Timber Forest Products in the Mediterranean Region. Ecological Economics 63: 768-775.

Cronin, A.E., and D.M. Ostergren. 2007a. Democracy, Participation, and Native American Tribes in Collaborative Watershed Management. Society and Natural Resources 20: 527-542.

Cronin, A., and D.M. Ostergren. 2007b. Tribal Watershed Management: Culture, science, capacity, and collaboration. American Indian Quarterly 31(1): 87-109.

Daily, G., ed. 1997. Nature's Services: Societal Dependence on Natural Ecosystems. Washington, D.C.: Island Press.

De Nooij, R.J.W., K.M. Lotterman, P.H.J. van de Sande, T. Pelsma, R.S.E.W. Leuven, and H.J.R. Lenders. 2006. Validity and Sensitivity of a Model for Assessment of Impacts of River Floodplain Reconstruction on Protected and Endangered species. Environmental Impact Assessment Review 26: 677-695.

Eppink, F. V., and J.C.J.M. van den Bergh. 2007. Ecological Theories and Indicators in Economic Models of Biodiversity Loss and Conservation: A critical review. Ecological Economics 61: 284-293.

Feldhamer, G.A., B.C. Thompson,

J.A. Chapman. 2003. Wild Mammals of North America: Biology, management and conservation. John's Hopkins University Press. Baltimore, MD.

Gooch, R.S., P.A. Cherington, and Y. Reinink. 2007. Salt River Project in conversion from Agriculture to Urban Water Use. Irrigation and Drainage Systems 21: 145-157.

Hanemann, M.W. 1994. Valuing the Environment through Contingent Valuation. Journal of Economic Perspectives 8(4): 19-43.

Hanley, N., R.E. Wright, and B. Alvarez-Farizo. 2006. Estimating the Economic Value of Improvements in River Ecology Using Choice Experiments: An application to the water framework directive. Journal of Environmental Management 78: 183-193.

Hein, L., K. van Koppen, R.S. de Groot, and E.C. van Ierland. 2006. Spatial Scales, Stakeholders and the Valuation of Ecosystem Services. Ecological Economics 57(2): 209-228.

Hernández, J.M., C.J. León. 2007. The Interactions Between Natural and Physical Capitals in the Tourist Lifecycle Model. Ecological Economics 62: 184-193.

Hoffmeister, D. 1986. Mammals of Arizona. University of Arizona Press, Tucson, AZ.

Holmes, T.P., J.C. Bergstrom, E. Huszar, S.B. Kask and F. Orr III. 2004. Contingent Valuation, Net Marginal Benefits, and the Scale of Riparian Ecosystem Restoration. Ecological Economics 49: 19-30

Ingraham, M.W., and S. G. Foster. 2008. The Value of Ecosystem Services Provided by the U.S. National Wildlife Refuge System in the Contiguous U.S. Ecological Economics 67: 608-616.

Kellermann, J.L., M.D. Johnson, A.M. Stercho, and S.C. Hackett. 2008. Ecological and Economic Services Provided by Birds on Jamaican Blue Mountain Coffee Farms. Conservation Biology 22(5): 1177-1185.

Knetsch, J.L. 2007. Biased Valuations, Damage Assessments, and Policy Choices: The choice of measure matters. Ecological Economics 63: 684-689.

Kontogianni, A., H.S. Skourtos, I.H. Langford, I.J. Bateman, and S. Georgiou. 2001. Integrating Stakeholder Analysis in Non-market Valuation of Environmental Assets. Ecological Economics 37(1): 123-138.

Kramer, R.A. 2005. Economic Tools for Valuing Freshwater and Estuarine Ecosystem Services. Nicholas School of the Environment and Earth Sciences, Duke University, Durham, NC.

Kremen, C. 2005. Managing Ecosystem Services: What do we need to know about their ecology? Ecology Letters 8: 468-479.

Kuosmanen, T., and M. Kortelainen. 2007. Valuing Environmental Factors in Cost-benefit Analysis Using Data Envelopment Analysis. Ecological Economics 62: 56-65.

Lange, G., E. Mungatana, and R. Hassan. 2007. Water Accounting for the Orange River Basin: An economic perspective on managing a transboundary resource. Ecological Economics 61: 660-670.

Loomis, J.B. 2000. Vertically summing public good demand curves: An empirical comparison of economic versus political jurisdictions. Land Economics 76 (2): 312-321.

Marta-Pedroso, C., H. Freitas, and T. Domingos. 2007. Testing for the Survey Mode Effect on Contingent Valuation Data Quality: A case study of web based versus in-person interviews. Ecological Economics 62: 388-398.

Millennium Ecosystem Assessment. 2003. Ecosystems and Human Well-being: A Framework for Assessment. Washington D.C., Island Press.

Millenium Ecosystem Assessment. 2005a. Ecosystems and Human Well-Being: Policy Responses Vol.1. Washington D. C., Island Press.

Millenium Ecosystem Assessment. 2005b. Ecosystems and Human Well-Being:

Future Scenarios Vol.2. Washington D. C., Island Press.

Millenium Ecosystem Assessment. 2005c. Ecosystems and Human Well-Being: Policy Responses Vol.3. Washington D. C., Island Press.

Millenium Ecosystem Assessment. 2005d. Ecosystems and Human Well-Being: Multiscale Assessments Vol.4. Washington D. C., Island Press.

Millenium Ecosystem Assessment. 2005e. Ecosystems and Human Well-Being: Our Human Planet. Summary for Decision Makers. Washington D. C., Island Press.

National Wildland and Scenic Rivers System. 2007. http://www.rivers.gov/ wsr-verde.html. updated January 1, 2007, accessed December 15, 2008.

Norton, B.G., and D. Noonan. 2007. Ecology and Valuation: Big changes needed. Ecological Economics 63: 664-675.

Ojeda, M.I., A.S. Mayer, and B.D. Solomon. 2008. Economic Valuation of Environmental Services Sustained by Water Flows in the Yaqui River Delta. Ecological Economics 65(1): 155-166

Pate, J. and Loomis, J. 1997. The effect of distance on wiliness to pay values: A case study of wetlands and salmon in California. Ecological Economics 20: 199-207.

Pearthree, P.A. 2008. Chapter 3. Background: Falluvial Geomorphology and Flood History of the Verde River. Pp 15-32 in Haney, J.A., D.S. Turner, A.E. Springer, J.C. Stromberg, L.E. Stevens, P.A. Pearthree, V. Suplee. 2008. Ecological Implications of Verde River Flows. A report by the Arizona Water Institute, The Nature Conservancy, and the Verde River Basin Partnership.

Schmidt, C.A., B.F. Powell, and W.L. Halvorson. 2005. Vascular Plant and Vertebrate Inventory of Tuzigoot National Monument. U.S. Geological Survey Open-File Report 2005-1347.

Sinden, J.A., and G. Griffith. 2007. Combining Economic and Ecological Arguments to Value the Environmental Gains from Control of 35 Weeds in Australia. Ecological Economics 61: 396-408.

Spash, C.L. 2000. Ecosystems, Contingent Valuation and Ethics: The case of wetland re-creation. Ecological Economics 34(2): 195-215.

Springer, A.E., and J.A. Haney. 2008. Chapter 2. Background: Hydrology of the Upper and Middle Verde River. Pp 5-14 Haney, J.A., D.S. Turner, A.E. Springer, J.C. Stromberg, L.E. Stevens, P.A. Pearthree, V. Suplee. 2008. Ecological Implications of Verde River Flows. A report by the Arizona Water Institute, The Nature Conservancy, and the Verde River Basin Partnership.

Stevens. L. E., D. S. Turner and V. Supplee. 2008. Chapter 5. Background: Wildlife and Flow Relationships in the Verde River Watershed. Pp 51-70 in Haney, J.A., D.S. Turner, A.E. Springer, J.C. Stromberg, L.E. Stevens, P.A. Pearthree, V. Suplee. 2008. Ecological Implications of Verde River Flows. A report by the Arizona Water Institute, The Nature Conservancy, and the Verde River Basin Partnership.

Stromberg, J. C. 2008. Background: Stream Flow Regimes and Riparian Vegetation of the Verde River. Pp 33-50 in Haney, J.A., D.S. Turner, A.E. Springer, J.C. Stromberg, L.E. Stevens, P.A. Pearthree, V. Suplee. 2008. Ecological Implications of Verde River Flows. A report by the Arizona Water Institute, The Nature Conservancy, and the Verde River Basin Partnership.

Terer, T., G.G. Ndiritu, and N.N. Gichuki. 2004. Socio-economic Values and Traditional Strategies of Managing Wetland Resources in Lower Tana River, Kenya. Hydrobiologia 527: 3-14.

Varady, R.G., K.B. Hankins, A. Kaus, E. Young, and R. Merideth. 2001. ...to the Sea of Cortes: nature, water, culture, and livelihood in the Lower Colorado River basin and delta – an overview of issues, policies, and approaches to environmental restoration. Journal of Arid Environments 49: 195-209.

Venn, T.J., and J. Quiggin. 2007. Accommodating Indigenous Cultural Heritage Values in Resource Assessment: Cape York Peninsula and the Murray-Darling Basin, Australia. Ecological Economics 61: 334-344.

Vicory, A.H. Jr., and A.K. Stevenson. 1995. What's a River Worth Anyway? A resource valuation survey of the Ohio River. Water Science and Technology.32(5-6): 63-70.

Vörösmarty, C. J., P. Green, J. Salsbury, and R.B. Lammers. 2000. Global Water Resources: Vulnerability from Climate Change and Population Growth. Science 289: 284-288.

Ward, F.A. and M. Pulido-Velázquez. 2008. Efficiency, Equity, and Sustainability in a Holistic Water Quantity-quality Optimization Model in the Rio Grande basin. Ecological Economics 66: 23-37.

Wilson, M. A., and S. R. Carpenter. 1999. Economic Valuation of Freshwater Ecosystem Services in the United States: 1971-1997. Ecological Applications 9(3): 772-783.

Wirt, L., E. DeWitt, and V. E. Langenheim, eds. 2005. Geologic Framework of Aquifer Units and Ground-water Flow Paths, Verde River Headwaters, North-central Arizona. U. S. Geological Survey Open-File Report 204-1411.

Wiser, R. H. 2007. Using Contingent Valuation to Explore Willingness to Pay for Renewable Energy: A comparison of collective and voluntary payment vehicles." Ecological Economics 62: 419-432.

Woods and Poole Economics, Incorporated 1999. 1999 Arizona State Profile Report, January. Washington D.C., 220p.

# VEGETATIVE CAPPING OF ARCHAEOLOGICAL MASONRY WALLS

*Frank G. Matero, Alex B. Lim and Michael Henry*

## ABSTRACT

Exposed ruin walls are typically protected from water in all its forms and invasive vegetation by hard caps of lime, amended soils and lime/cement mortars with or without additional masonry. These "hard" caps often fail when cracks develop and can ultimately lead to masonry collapse. The repeated repairs needed to maintain hard caps have been inadequate in the long term management of ruin sites as they are remedial and often hide and exacerbate water-related damage through entrapment, dissolution, and ice and salt crystallization. "Soft" capping by utilizing vegetation, soil, gavel, or geosynthetics offers an effective, light-weight, and sustainable alternative to traditional hard capping. This paper presents ongoing laboratory and field research in Mancos, CO, at Far View House, Mesa Verde National Park, that addresses the question of how best to protect exposed masonry walls from environmental stresses, and to offer an attractive alternative to hard caps.

## INTRODUCTION

Exposed archaeological ruins are subject to various weathering conditions that accelerate their deterioration. Of particular importance in the deterioration are moisture levels and temperature variance. Combined with intrinsic material characteristics, age, building design, and past restorations, these factors can significantly threaten the durability of masonry ruins. In order to conserve and manage archaeological structures effectively, one needs to understand the sources and effects of these environmental factors and to identify remedial and preventive conservation methods to minimize their damaging impacts upon features and sites.

Compound masonry walls are a common construction feature of many archaeological sites. They are usually constructed of two veneer faces filled in-between with rubble, mortar or soil. In exposed archaeological ruins, the preservation of compound walls and vaults poses a particular challenge due to the lack of protection from fragmentation and direct exposure to the weather. Such exposure leads to severe moisture penetration and thermal movement. Over time, these continued cycles of weathering bring about irreversible damage which causes material attrition and displacement and can ultimately lead to masonry collapse.

In the past, a 'hard capping' of lime, cement, and modified soil mortars has conventionally protected exposed compound walls. This method has been popular due to its minimal intervention to the standing wall and the relative ease and economy of its initial application. In reality however, the procedure requires persistent repairs and maintenance that can increase cost and risk to the wall. Hard caps tend to crack under prolonged compressive and tensile stresses from thermal movement and ground subsidence (Figure 11.1). Cracks allow easy access for water to further penetrate and concentrate inside the cavity (Ashurst 2007: 98). At the same time, the cracked cap retards drying and desorption of moisture

Figure 11.1 Typical damages resulting from hard cap failure. (Counter clockwise; a) Disintegrating hard cap; b) Hard cap crack extending through masonry component; c) Wall bulging through accumulation of water in the wall core among other factors. (Photos by A. Lim)

from the top of the wall, especially if it is cementitious (Torraca 1998: 21). Cracks and voids can provide a foothold for vegetation, whose root systems can physically and chemically further damage the masonry. Increased moisture can cause dissolution of core mortars or soil and break down the masonry through freeze/thaw cycling and salt crystallization, both of which eventually weakens the wall's masonry (Lee et al 2009: 9). In addition, the damaged cap must be periodically replaced or repaired or capped over by a new one, the latter treatment adding unnecessary weight to the wall. In either case, hard capping does not usually adequately address the long-term management of moisture and thermal damage that will continue to stress the structure. Instead of protecting the compound wall as initially designed, hard capping can actually accelerate deterioration over time.

A procedure called 'soft capping' aims to counter such problems posed by hard capping. Introduced in recent years at several archaeological sites in England, Turkey and elsewhere, soft capping replaces hard caps with vegetation planted on top of a layer of soil, with optional layers of gravel, and geo¬synthetics (Ashurst 1998, 2007; Wood 2004, 2005; Sass and Viles 2006; Stokely 2007; Viles and Wood 2007; Matero et al 2008). Taking advantage of plants' transpirative ability to utilize the water, it seeks to prevent water penetration, reduce thermal fluctuations, and provide a protective, aesthetically pleasing barrier on the wall top.

Geosynthetics provide further protection from moisture and temperature control. Often used in landfill waste control for their low permeability of water and toxic solutions, geosynthetics provide a moisture barrier and drainage system, in addition to functioning as filter layers and soil support (Kavazanjian 2001). In recent years, architects have adopted geosynthetics for use in "green" roof technology in architecture (Liu 2005) while archaeologists gave used the material for reburial of sites (Demas 2004; Kavazanjian 2004).

## Site Introduction

Far View House, located at Mesa Verde National Park, Colorado, offers an excellent opportunity to test the application of soft capping in the arid climate of the arid Southwest. When compared to the more protected alcove sites at Mesa Verde National Park, the mesa top sites including Far View House, exhibit serious deterioration from direct exposure to moisture, solar radiation, and wind. Daily, they are subject to highly fluctuating air and surface temperatures due to heating from the sun. Seasonally, they are subject to extreme temperatures as well as dry and wet conditions in summer and heavy snow in winter. At Far View House over forty exposed rooms divided by single and compound masonry walls experience such weathering conditions. Over the years, these walls have undergone numerous stabilization campaigns utilizing a variety of mortar caps and repointing. While temporarily effective against precipitation, they do not benefit the walls in the long term as they require frequent repair and can give the false illusion of wall stability. Unintended consequences such as cracks lead to serious destabilization of the wall by allowing water to enter the interior earthen mortar core resulting in structural displacement (bulging and leaning) and in some cases, collapse.

Candidates for likely mechanisms of stone and mortar deterioration include hygric expansion, subsurface salt crystallization (cryptofluorescence), and freeze-thaw cycling. All these mechanisms all require moisture. At the material level they cause stone and mortar erosion; at the macroscale,

they can destabilize the wall causing structural displacement and wall collapse.

Despite the high vulnerability to environmental weathering, there is no easy solution for proper protection at Far View House. With the collapse of the original roofs after an extended abandonment of the site, the pueblo would have been slowly buried with aeolian deposition, construction debris, and vegetation. While the resultant mounds would have offered some protection to direct exposure, the ground moisture in an aerobic burial environment would have continued to deteriorate both organic and inorganic materials. Excavation in the early twentieth century exposed the remaining masonry and removed the lateral support of the walls provided by the soil fill. In addition, it introduced various mortar caps made of cement to protect the wall tops and to allow visitors to walk on the walls for viewing (Fewkes 1917). These impermeable caps readily shed water but inadvertently caused basal erosion from high water concentration and poor drainage and when cracked, allowed water to flow into the wall cores, leading to reduced strength, erosion, subsidence, and structural collapse. Although wall caps have been replaced over the years and cracks are repaired, the hard surfaces still shed large quantities of water to the base of the walls or into the core through breaches in the surface, especially at the cap/stone interface. The wall caps do not reduce the volume of water available to the walls through the top, bottom and sides; they merely divert it from the top to the wall faces and concentrate the unabsorbed volume into the soil at the wall base. Rather, one needs to look at the cyclic wetting and drying of the wall in its entirety to best address the deterioration of the wall by moisture. Soft caps partially address this problem by retaining water for re-evaporation without entry to the wall system.

Given the environment's severe annual, seasonal and daily temperature fluctuation

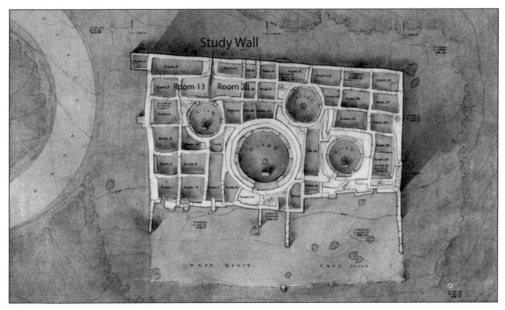

Figure 11.2   Site Plan of Far View House (Nusbaum 1934) MVRC.

Figure 11.3   Test wall in winter.

and water feed from standing snow in winter (and the likelihood that these conditions will become more extreme with climate change), it has been hypothesized that the past interventions have more than likely contributed to structural instability while at the same time have required high maintenance. Improvement in the preservation and display of the walls as well as reduction in long term maintenance cost and labor demands consideration of new intervention methods that would better manage water collection and disposal at the site. In order to measure the effectiveness of any new intervention, the performance must be quantitatively compared to that of the existing mortar capping system. Data from materials testing and environmental monitoring were analyzed to confirm hypotheses on the sources and mechanisms of deterioration and to evaluate the effectiveness of the proposed new intervention techniques.

## METHODOLOGY

In order to compare methods for the protection of the wall tops and the prevention of water penetration into the wall core, two types of caps were installed on top of the test wall (Figure 11.2, 11.3, and 11.4): a hard mortar and stone cap based on the Park's current methods of stabilization, and an experimental soft vegetative cap. Temperature and moisture probes were installed to monitor and evaluate the performance of each cap, the wall masonry and surrounding ground/fill.

Both caps were installed on top of the existing stabilized wall dating from the previous repair campaign in 1983 by K. Fiero. The wall was divided into two sections of roughly equal length, one for the hard cap on the north side and the other for a soft cap on the south side of the wall. The two caps were separated by a water impermeable geomembrane separator.

The hard cap was installed using the same method that has traditionally been used by the Park's stabilization crew at Far View House. One part Quickrete® type N masonry cement (ASTM C91) to three parts angular masonry sand was mixed with an added colorant (Colortech® # 52) at a ratio of one part colorant to 24 parts mortar mix for the stone setting mortar. The wall faces were overpointed with soil mortar amended (all parts by volume) with Rohm and Haas Rhoplex™ E330 (an acrylic emulsion admixture) at a ratio of one part E330 to two parts water. The soil mortar is a fine sandy loam blended from local red silt loam (two parts), yellow clay loam (one part), and medium to fine masonry sand (one part). Once the soil mortar foundation was prepared, a sandstone course was laid parallel along the edge of the wall top. Afterward, the gaps were pointed and the wall core was filled with the mortar. In the second course of the reconstructed wall top, an RH/Temperature probe was buried in the mortar and the remainder of the courses was prepared using the same method.

The soft cap foundation was prepared using the same soil mortar for overpointing the hard cap. The number of stone courses was the same as the hard cap but the fill materials were different. A 200 ml HDPE geonet layer by Poly-Flex, Inc. was first laid to provide reversibility of the system for any future intervention.

Next the first stone course was laid with an interior buffer zone filled with $0.6 - 1.2$ cm grade gravel of approximately $25 - 35\%$ void ratio to accommodate water drainage. An RH/Temp probe was placed in the gravel. Next, a drainage and separation layer was laid on top of the gravel and the sandstone course using a textured 60mil HDPE geomembrane from Poly-Flex, Inc.

Above this, a single sided 6 ounce per square foot geocomposite by the same supplier was installed consisting of a 200mil geonet facing down and geotextile facing up to prevent soil from accumulating within the net matrix. The layer also functioned as a

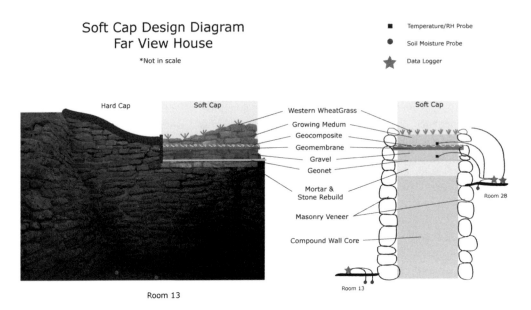

Figure 11.4  Test wall diagram.

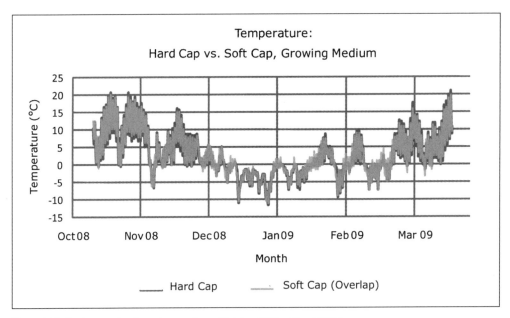

Figure 11.5  Hard cap and soft cap temperatures (October 2008 – March 2009).

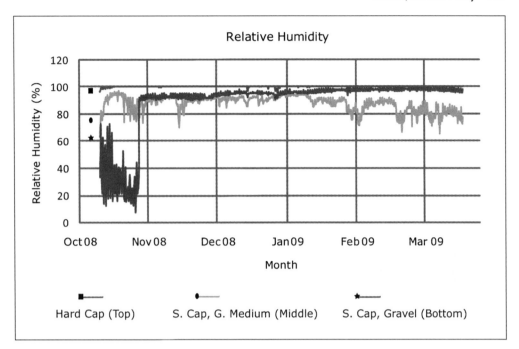

Figure 11.6  Hard cap and soft cap relative humidity (October 2008 – March 2009).

capillary break. The edges of each geofabric layer were pointed with soil mortar. Additionally two stone courses were laid in soil mortar on the geocomposite layer and the second RH/Temperature probe was installed on top of the geocomposite layer. Weep holes were added using two plastic straws on each side roughly at one third interval at the same level in the soft cap section. The wall core was then filled with turf consisting of growing medium and local western wheat grass (*Pascopyrum smithii* (Rydb.)) burying the probe. The choice of western wheat grass was based on the recommendation by the Park's natural resources staff due to its low maintenance, drought tolerance, weed competitiveness, and adaptability. The cap was then manually watered to ensure the survival of the grass in dry weather.

To record temperature and moisture level change, HOBO® Temperature/RH smart sensors (model number S-THA-M006) by Onset Computer were used. This device measures RH level between 0 to 100% and temperature between 0 and 50°C with an accuracy of plus or minus 3%, and plus or minus 4% in a condensing environment. It has a resolution of 0.5% RH at 25 °C. The reading may drift plus or minus 1% per year and an additional reversible drift of 3% is expected when the average relative humidity is above 70%. The probe responds to change within 5 minutes. It has an operating range between -40 to 75°C although its reading loses accuracy at temperatures below 0 and above 50°C. When the RH sensor becomes saturated, the sensor will read 100% and it is difficult to dry out the probe once wet. While short-term condensation/water exposure has no effect at or below 25°C, at 30°C or above it will lead to permanent positive drift. The effects of condensation/water exposure vary with the length of exposure time, the number and frequency of exposures, and the operating temperature. When the RH sensor is subjected to cyclical exposure to water and air, its terminals will likely corrode, reducing its sensitivity and reduction in RH reading. In an environment with high humidity between 0 and 25°C for over 1000

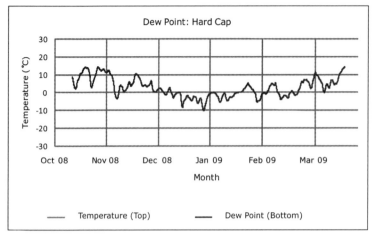

Figure 11.7  Hard cap dew point (October 2008 – March 2009).

Note: overlap bewteen two lines is such that they show up as one line.

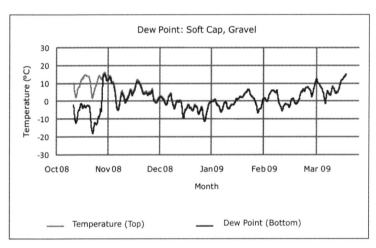

Figure 11.8  Soft cap gravel layer dew point (October 2008 – March 2009).

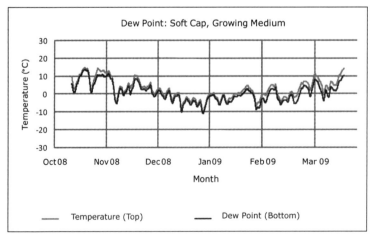

Figure 11.9  Soft cap growing medium dew point (October 2008 – March 2009).

hours, the drift of plus 3% RH will result. The THA sensor is also sensitive to insects, a possibility at least within the gravel matrix.

The data were then stored in HOBO® Micro Station Data logger model H21-002 encased in a Pelican™ Case for protection against weather. The data logger was programmed to record temperature and RH level change every hour. Wires were sheathed in plastic plumbing tubes to protect them from possible damage by animals. At the same time, each end of the tubes was tightened with a rubber cork and then sealed with adhesive in order to prevent moisture entry from the exterior.

## DATA

Based on the results of the testing program, the following observations can be made about the building materials from Far View House (Tables 11.1 and 11.2):

- The sandstone readily absorbs water and dries rapidly.
- The sandstone has high frost resistance when not restricted.
- The sandstone significantly loses strength when water saturated. This has significant implications for sandstone walls, since wetting under lateral pressure can lead to ruptures, compromising the stability of the wall structure.
- The sandstone that has undergone cyclical freeze-thaw loses flexural strength. However, its modulus of rupture is very similar to that of wet samples, indicating that cyclical freeze-thaw has not affected the sandstone structurally.
- The sandstone does not undergo severe reduction in compressive strength even after wetting. The sandstone performs relatively well under compression.

The clay content of the original soil mortar has high shear strength as demonstrated by its low to medium plasticity index. When combined with fibrous organic materials (plant ash), the soil mortar complements the sandstone's compressive strength by imparting good shear and tensile strength (Houben and Guillaud 1994: 82).

Environmental Monitoring Data

*Temperature.*

The temperatures recorded by all monitors show very similar patterns (Figure 11.5). The temperature data for the hard cap display a broad U shape pattern with a range from low -10°C to low 20°C. The soft cap growing medium follows a similar pattern and range as the hard cap, but takes on a slightly more stretched U shape with less daily temperature fluctuation, especially in the fall and spring. The soft cap gravel fill in general yielded the greatest daily and seasonal temperature range, most pronounced in the fall and spring, covering 0 to 20°C.

*Moisture.*

In order to understand the change in the moisture content of the wall using RH readings (Figure 11.6), dew point temperatures for each cap were calculated by the HOBOware™ software from Onset. An error in either temperature or RH reading from the sensors will result in an error in the calculated dew point. The data were then compared to the recorded temperatures to understand how the moisture content varied depending on the type of cap. In other words, the moisture content of an air space in the wall is expressed as dew point that denotes the transition temperature between liquid water and water vapor. Temperature readings higher than dew point indicate that the net phase transition of water favors the vapor form where the drying rate is faster than the condensing rate. The greater the gap between the temperature and the dew point is, the lower the moisture content. Conversely, the smaller the gap, the slower the drying process is than the condensing process. In this case, the moisture content increases.

*Hard Cap.*

The probe in the hard cap almost immediately reached 100% RH and remained at that level. There were few very minor drops in RH level not more than a degree, but their impact on overall RH level was insignificant. These drops only occurred when the temperature dropped below freezing.

The graph for dew point temperature very closely approximates the change in temperature level (Figure 11.7). The dew point graph slowly rose above the freezing point and reached 20°C by mid March. The difference remained close to zero almost all the time, indicating the saturated environment inside the hard cap.

*Soft Cap Gravel.*

RH inside the soft cap gravel varied markedly before and after October 28th, 2008 at 11am, jumping from 38% to 71%. By the end of the day, the level reached 90%. The reading hovered around 90% for the rest of the monitoring period.

The difference between the recorded temperature and dew point temperature also reflects this change in moisture level (Figure 11.8). Before October 28th, the difference ranged from -10 to 20°C. However, after this date, the difference drastically dropped to below 2°C, and by the time the data were retrieved, the level was further down at below 0.5°C.

For most of the winter, the dew point stayed below freezing, indicating the water vapor had deposited on surfaces as frost, not as dew. After mid February, the dew point went above the freezing point and gradually rose, reaching almost 15°C by mid March.

*Soft Cap Growing Medium.*

The relative humidity level was high for the probe located in the growing medium. It stayed between 80 and 100%. Unlike the constant readings from the hard cap, the reading from the soft capping showed some

variations. While the RH level showed a gradual drop since January, the level remained above 70% almost all the time.

While the dew point temperature closely tracks the temperature reading, there is a noticeable temperature difference (Figure 11.9). The gap is small during winter but wider in the fall and in the spring. In particular the gap in the spring continued to widen and corresponded to the lowering of RH level. It shows that the gravel had begun to dry with the temperature rise.

## DISCUSSION

The comparison between the soft and hard cap temperature regime demonstrates the soft cap's ability to dampen temperature fluctuation (Figure 11.5.) It is especially more pronounced during the fall and spring when there is a greater fluctuation in daily temperature. This result is in agreement with other experiments that show the heat buffering ability of vegetative covers (Liu 2005; Viles 2007) which would be advantageous in reducing thermal movement and cracking of wall tops.

Both the soft cap and hard cap experienced humid conditions in the bottom layer. The lack of ventilation and the evaporation of water would increase the hydrostatic pressure on the wall top and invite unnecessary biological growth. After one year, the newly installed hard cap has not shown any cracks nor has the soft cap shown any cataclysmic failure in holding back water from the top. In addition, it is yet to be determined whether the vegetation has successfully survived the weather and how its root system would affect the stone courses through possible root jacking. The most critical data suggesting water penetration would occur in spring, from April to June, when all the accumulated snow melts, exposing the soil surface to direct drying.

The moisture introduced during the installation of both caps affected and determined the RH level of each cap and

| Sandstone | Test results | Test |
|---|---|---|
| Color | Brownish yellow 10 yr 6/6 | Munsell Soil Color Chart |
| Texture | Fine - 320 grit | ACL Texture Calibration System |
| Hardness | 1.5-2.5 | Mohs Hardness Test |
| Density | 1.79 | ASTM C97-96 |
| Porosity (g/cm$^3$) | 21.98 | Normal 7/81 |
| Water absorptivity (g/(cm$^2$/sec$^{0.5}$)) | 0.0648 | Normal 11/85 |
| Drying (g/(cm$^3$ hr)) | 0.022 | Normal 29/88 |
| Frost resistance | High | RILEM Test No V.3 |
| Modulus of rupture, dry (psi) | 414 | ASTM Standard C99-87 |
| Modulus of rupture, wet (psi) | 181 | ASTM Standard C99-87 |
| Modulus of rupture, after frost | 187 | ASTM Standard C99-87 |
| Dry-wet strength ratio | 2.285 | |
| Compressive strength, dry (psi) | 2650 (//), 2220 ( ) | ASTM Standard C170-90 |
| Compressive strength, wet (psi) | 1950 (//), 1210 ( ) | ASTM Standard C170-90 |
| Dry-wet strength ratio | 1.4 (//), 1.8 ( ) | |
| Soluble salt | Sulfates | Ion Strip (Merckoquant®) |

| Original Soil Mortar | Test results | Test |
|---|---|---|
| Color | Brown, 7.5 yr 5/4 | Munsell Soil Color Chart |
| Hardness | 1.5-2.5 | Mohs Hardness Test |
| Plastic limit (%) | 20 | Atterberg Test |
| Liquid limit (%) | 34 | Atterberg Test |
| Plasticity index | 14 | Atterberg Test |
| Clay mineralogy | Most likely Kaolinite/Illite | Atterberg Test |
| Soluble salts | Nitrates | Ion Strip (Merckoquant®) |

Table 11.1  Far View House original building materials properties.

therefore compromised the effectiveness of the RH probes for evaluating the performance of the caps. The readings, however, provide important information for future design and implementation of the caps. Improvement in soft cap design will require an informed installation that blocks moisture retention and a more appropriate selection of moisture probes to effectively evaluate the performance of the capping intervention.

First, the penetration of moisture into the gravel fill showed that despite the installation of a horizontal geocomposite layer for drainage of water and for redirecting water horizontally, moisture still continues to penetrate from other areas – probably from the side wall joints through suction – into the fill level below the geocomposite, an area expected to be dry. It prompts an improvement in the design of the cap drainage system for more effective moisture control.

The repointing mortar also introduced excess moisture into the system during drying. The probes from both gravel fill and hard cap showed a huge spike in RH soon after the installation of the caps, quite possibly due to evaporating moisture from the mortar mixture. This suggests that when mortar is applied in the wall construction, extra time should be allowed for drying the mortar first, before installing any additional layer on top, in order to minimize the moisture entrapment.

Water can also enter into the gravel fill through an incomplete seal between the barrier and mortars. This also requires attention during the installation process, since the mere presence of a liquid water barrier does not ensure the complete seal-off of water.

At the same time, in order to better evaluate the moisture level inside the caps, a soil moisture reader that records the absolute amount of liquid water should replace the RH/temperature probe that measures only the relative amount of water vapor inside the wall. While an RH/temperature probe has an advantage of measuring two variables using one probe, it introduces many sources of error, particularly when it is installed in an enclosed system that does not allow ventilation. For an environment with ample sources of moisture, in this case a snow cap, in addition to watering of the vegetation layer during the initial phase of cap installation, the RH probe does not provide effective readings. A more robust RH/Temperature probe and a soil moisture probe would better resist against salt and insect damage. Furthermore, the soil core of the wall dries out slowly when the temperature is low and the surface area for evaporation is greatly reduced by the surrounding stone masonry.

Therefore, more effective measurement could be achieved using separate temperature and soil moisture sensors that could monitor environmental variables in absolute terms.

Finally, the above study also helps to understand the results from the research in England with greater implications for soft capping design. The moisture content graph for July 2005 at Byland Abbey shows that while a hard cap provides initial protection against precipitation by reducing the overall water absorbed, it dries slowly, resulting in greater overall moisture content for an extended period of time (Torraca 1998). A soft cap absorbs more water but dries faster due to exposure. Therefore, although there is greater water intake by the soft cap during rain events, it will give up moisture faster than a hard cap. As long as there is a long enough recovery period for a soft cap to dry, it will be a better choice than a hard cap in managing moisture and temperature fluctuation. This suggests that a soft cap should perform very well in dry climates with concentrated precipitation events. Considered in combination with water-susceptible materials such as soil based mortars, soft caps appear to be a far better solution to the problem of wall protection in the arid Southwest than hard caps.

The analysis also points to the need for designing a soil layer with appropriate particle size ratios customized for local climate as well as being an optimal growing medium. The use of soil mortar at Mesa Verde was effective, since not only does it provide shear and tensile strength to the masonry walls, it is also more compatible with a dry climate by giving up moisture easily. The fibrous organic additives in the mortar and cap act as wicks to accelerate drying just as straw functions in adobe. They also impart tensile strength as well as protecting soil mortar from erosion by rain. The cap design would need to address this either through the use of appropriate plant type or geosynthetics.

## CONCLUSION

The effects of hard and a soft capping on a masonry wall top were qualitatively and quantitatively evaluated from late fall to early spring. From October 2008 to March 2009, temperature, moisture and wall movement, as well as weather conditions were monitored. The data did not provide conclusive results for determining the preferred system. This will require more time, at least one full year, to make a meaningful analysis. However, as a prototypical installation, valuable information was gathered on environmental conditions, material properties, deterioration mechanisms, moisture balance and monitoring methodologies. The sources of moisture and thermal energy were evaluated in order to understand the enabling factors for deterioration mechanisms of the stone and earthen mortar wall. From the documentation of the past interventions and the observations made in the laboratory and in the field, possible deterioration mechanisms were analyzed and their effects ranked in order of significance. In addition, the research showed the difficulties in installing a monitoring system that predictably records what is desired. These findings will be incorporated into designing a new soft cap and monitoring system in the next phase of research.

One of the most significant findings from the study is the need to understand how water moves through the wall system, especially as it is influenced by grade and associated fill. Moisture uptake by the wall from the adjacent soil fill and ground occurs for an extended period of time, especially as snow drift melt, and ultimately may be more damaging to the overall stability of the wall than moisture entering from the top through occasional precipitation events. This has direct implication on selecting and integrating several preservation methods.

### Implement regular maintenance on site

Moisture continues to damage the site despite nearly a century-long effort to manage it. The installation of cappings on wall tops is only one component of maintenance. A regular inspection of walls is fundamental in order to understand the type and rate of deterioration at the site, to provide timely intervention and eventually to implement a preventive preservation program. Far View House represents one of the key mesa top ruins with links to other sites in the Southwest. The mesa top ruins do not receive protection against precipitation, solar radiation, and wind unlike alcove sites that are relatively well protected. Given the site's archaeological importance, this study argues for a systematic program of architectural documentation and conditions survey and assessment in preparation for renewed stabilization work.

### Select compatible intervention materials

The various cementitious hard caps have shown clear defects since their initial application in 1916. Although initially remedial against moisture penetration, they pose visual, physical, and chemical threats to the walls. The development of cracks, the ingress of liquid water and retardation of water evaporation as well as the supply of soluble salts into the vulnerable and still largely original building materials are only some of the inherent material characteristics of hard caps that activate or exacerbate deterioration mechanisms. These conditions have been observed and noted at various other archaeological sites resulting in severe consequences to the masonry structures over time. Updating information on the field performance of soil-cement and other mortar mixes could help develop intervention materials more compatible with the original construction systems.

| Name | Soil Mortar #1 | Soil Mortar #2 | Current Joint Repair | Cement Repointing |
|---|---|---|---|---|
| **Location** | 3" under Modern Repair | 3"-8" from Surface. SW lower corner. 6th Course from Grade | De-installed joint repair piece from wall top | Upper Core Bedding Mortar |
| **Surface appearance** | Fine fibrous organic material, burnt charcoal, fine white particles | Fine fibrous organic material, burnt charcoal, fine white particles | White angular, coarse grains, brown and black sub-round, coarse grains | White particles, other dark gray and brown particles |
| **Overall color** | Brown 7.5 yr 5/4 | Brown 7.5 yr 5/4 | Reddish yellow 7.5 yr 6/6 | White 2.5 yr 8/1 |
| **Texture** | Fine, 320 grit | Fine, 320 grit | Medium. 30-120 grit Avg. 80 grit | Medium. 80 grit, some 30 grit |
| **Hardness** | 1.5 -2.5. pale yellow finger nail scratch | 1.5-2.5. pale yellow finger nail scratch | 1.5 - 2.5. pale yellow finger nail scratch | 7. white streak by steel knife |

Table 11.2  Far View House repair mortars

**Control water movement both from wall tops and from grade and room soil**

Even if a properly installed cap system blocks water entry from wall tops, water can still enter the walls through other sources. The volume of the water available to the wall is not necessarily reduced by mere diversion of the water from the top of the wall. Water penetration from the wall faces needs to be controlled as well as moisture from grade and associated fill. The geosynthetic drainage system in Room 28 installed in 1983 needs an evaluation of its efficacy in controlling moisture content in this room as well as the entire site. Based on observations, the geosynthetics seem to have raised moisture levels underneath them, the opposite result of their intended use. This observation has a direct consequence on the moisture content of the masonry walls and their overall stability. The presence of the geosynthetic layer in the soft cap, therefore, needs critical assessment to determine whether it increases the uptake of moisture by the wall from grade and the lateral fill, depending on wall height and thickness. This is especially important for walls with associated differential fill, such as the case study wall, since moisture from the wet soil will travel into the wall and be forced to exit through the exposed wall faces, causing stone and mortar deterioration and undermining wall stability. These consequences can be observed where the wall displays clear evidence of these moisture related problems: mid wall bulge and cracking from lateral fill loading and constant wetting, and basal erosion of stones and mortar from rising damp, especially from snow bank melting.

## FUTURE RESEARCH

Future research should focus on four areas. The data show that the soft cap system requires further modification in its design and installation phase. This includes reducing moisture levels during the installation phase by completing all associated mortar work prior to placing the vegetative caps, planting the caps in the spring, taking full advantage of the establishment of the plant's evapo-transpirative mechanisms to reduce the need for relying on physical drainage systems, and maximizing the soil's ability to reduce moisture penetration by controlling soil texture and granulometry. Input from professionals on cap detailing would also reduce drip and water entry through the upper wall faces. The construction of wall mock-ups would also help assess the cap performance alone by reducing or controlling other variables.

In addition, the research needs to address effective monitoring methodology. For example, a non-contact monitoring system could offer helpful information. Available non destructive examination (NDE) methods need to be explored to map out void spaces and cracks as well as heat and moisture transfer in the wall within the confines of site access. Drawing on resources from industry would benefit this aspect. For a contact based monitoring system, the design should allow easy replacement of the probes in case of any monitor failure. Running a simulation model based on the data would also help visualize the dynamics of the system.

The research should also incorporate investigating the mechanisms of water entry into the wall from the associated soil fill and ground. Parallel efforts to reduce the amount of liquid water in the soil and to prevent water shedding from the wall tops could circumvent potential structural and material failures resulting from water absorption from below. This could be enhanced greatly installing time-lapse photographic monitoring on site to better understand the diurnal and seasonal patterns of moisture, especially in the form of snow deposition and melt.

Finally, better knowledge of Far View's masonry construction would benefit the long term maintenance of the site as well as generate a more accurate understanding of the site's original construction phases and the integrity of the masonry. Documenting the presence of intact ancient soil mortar walls (i.e., not rebuilt or stabilized) is critical in understanding the vulnerability of the walls to moisture and the effects of previous stabilization efforts. By investigating the original mortar technology through geo-physical and chemical analyses, it is possible to avoid introducing incompatible materials to the system. In addition, by observing how the materials on site behave under local environmental conditions, it is also possible to provide a sustainable and long term intervention program. Integrated research into both modern and original technology would bring about measurable benefits as a means to sustainable management of this important archaeological site.

## ACKNOWLEDGEMENTS

The authors would like to acknowledge the Mesa Verde National Park staff for their support for this project, especially Scott Travis, Julie Bell, Tim Hovezak, Preston Fisher, George San Miguel and Carolyn Landes for their support for this project. Edward Kavazanjian of Arizona State University assisted with the design and technical guideline. Alex Radin, John Hinchman, Denis Pierattini, and Victoria Pingarron Alvarez at the University of Pennsylvania offered valuable help for testing and equipment set-up. Our sincere gratitude goes to Poly-Flex, Inc. for generous material support. We would also like to give special thanks for their support to McMaster-Carr Supply Company, Onset Computers, and Pelican™ Products, Inc.

## CITED REFERENCES

Ashurst, J. A. 2007. Conservation of Ruins. London; Burlington, MA: Butterworth-Heinemann.

Demas, M. 2004. Site Unseen: The Case for Reburial of Archaeological Sites. IN Conservation and Management of Archaeological Sites 6 (3 & 4): 137 – 154.

Houben, H and H. Guillaud. 1994 Earth Comprehensive Guide. Villefontaine Cedex, France: CRATerre-EAG

Fewkes, J. W. 1917. A Prehistoric Mesa Verde Pueblo and its People. The Annual Report of the Smithsonian Smithsonian Institution. Washington, D.C.: Government Printing Office: 461-488.

Kavazanjian, E. Jr. 2004. The Use of Geosynthetics for Archaeological Site Reburial. IN Conservation and Management of Archaeological Sites 6 (3 & 4): 377-393.

2001. Design and Performance of Evapotranspirative Cover Systems for Arid Region Landfills. IN Proceedings: 36th Annual Western States Engineering Geology and Geotechnical Engineering Symposium 28-30 March 2001, University of Nevada, Las Vegas, Nevada.

Lechner, N. 2001. Heating, Cooling, and Lighting: Design Methods for Architects. New York: John Wiley and Sons.

Lee, Z., H. Viles, and C. Wood, editors 2009. Soft Capping Historic Walls: A Better Way of Conserving Ruins? English Heritage Research Project Report, p 9

Liu, K. and B. Baskaran. 2003. Thermal Performance of Green Roofs through Field Evaluation. IN Proceedings of Greening Rooftops for Sustainable Communities: Chicago 2003: May 29-30, 2003. Chicago: Ottawa (Canada) National Research Council Canada, Institute for Research in Construction.

Maekawa, S. 2004. Monitoring of Soil Moisture in Backfilled Soil for Conservation of an Ancestral Puebloan Great House in Chaco Canyon. IN Conservation and Management of Archaeological Sites 6 (3 & 4): 315-323.

Massari, G. 1993. Damp Buildings, Old and New. Rome: ICCROM.

Sass, O and H.A.Viles. 2006. How Wet Are These Walls? Testing a Novel Technique for Measuring Mositure in Ruined Walls. IN Journal of Cultural Heritage 7: 257-263.

*Stokely, S. 2007. Gordion 2007: Field Report for Terrace Building 2. Field report submitted to the University of Pennsylvania, Department of Historic Preservation.

Torraca, G. 1988. Porous Building Materials: Materials Science for Architectural Conservation. Rome: ICCROM. p.21

Viles, H. A. and C. Wood. 2007. Green Walls?: Integrated Laboratory and Field Testing of the Effectiveness of Soft Wall Capping in Conserving Ruins. IN Building Stone Decay: From Diagnosis to Conservation, by R. Prikryl and B.J. Smith, 309-322. London: Geological Society of London Special Publications.

*Matero, F, K. Wong and S. Stokely. 2008. Gordion 2008: Field Report for Terrace Building 2. Field Report submitted to the University of Pennsylvania, Department of Historic Preservation.

Wood, C. 2004. Rain Damage to Masonry: Wall Cappings and Towers. IN English Heritage Conservation Bulletin 45: 38.

Wood, C. 2005. Soft Capping: Justifying a Return to Picturesque. IN Context 90: 22-24.

# MOVING FROM LANDSCAPE CONNECTIVITY THEORY TO LAND USE PLANNING: URBAN PLANNING IN OAKVILLE, ONTARIO

*Paul F. J. Eagles, Elke Meyfarth O'Hara and Graham Whitelaw*

## ABSTRACT

The Provincial Policy Statement published under the Ontario Planning Act states that "The diversity and connectivity of natural features in an area, and the long-term ecological function and biodiversity of natural heritage systems, should be maintained, restored or, where possible, improved, recognizing linkages between and among natural heritage features and areas, surface water features and ground water features. Municipal planning must be consistent with this Policy. This Policy applies the concepts of protecting ecological cores and linkages within land use planning. The Town of Oakville proposed development in an area covering 3,000 hectares of farmland, woodland and wetlands to house 55,000 people. Two planning teams, one working for the Town and one for the area's land owners, developed competing subwatershed plans for the area, with ecological cores and linkages as a major component. Cores consisted of designated Environmentally Sensitive Areas, large woodlots, and significant wetlands. The two teams presented maps of cores and linkages to be protected from development, with similarities but some important differences in the linkage design. The paper outlines that the similarities and the differences between the two mapping schemes, are related to both landscape connectivity theory and to the conflict resolution processes involved in land use planning in an urbanizing area. Generally, the Town proposed much more land to be reserved as green space, compared to the landowners. All plans were prepared for defense in a hearing in front of the Ontario Municipal Board, an administrative tribunal dealing with land use matters under dispute. However, negotiation between the parties resulted in the Town's plan being accepted as the template for development before conflicting positions could be explored in the tribunal room. Our paper explores the limitations that the landscape connectivity literature has for land use planning within urbanizing areas.

## INTRODUCTION

Ontario, Canada operates under a policy-led planning system. Land use planning on private land is conducted under the authority of the Planning Act, the provincial legislation that sets the legal structure for municipal and private land use. The Provincial Policy Statement (PPS), created by the provincial government under the authority of the Planning Act, provides the broad policy framework that establishes direction on matters of provincial interest. The PPS provides policy direction in areas such as: land use density, housing, open spaces, infrastructure, energy, and resource use (Penfold 1998). While the province sets the overarching planning policies, planning decisions for private land are made at the municipal level. Municipalities are responsible for implementing the Planning Act via their Official Plans, zoning by-laws, and development application approval processes. The Ministry of Municipal Affairs and Housing (MMAH) is responsible for provincial-level plan input and review, policy development, and appeals. The Ontario Municipal Board (OMB) is an independent, adjudicative tribunal that is responsible for settling disputes over land use planning and other municipal issues. The OMB hears appeals and applications on land use planning under the Planning Act and other legislation. The OMB makes decisions, unless appealed to the cabinet or the courts. This quasi court is specifically designed to deal with the many technical issues that occur within land use planning. It is also designed to provide a forum for dispute resolution, and therefore to circumvent more costly and awkward court proceedings.

The purpose of this paper is to apply landscape planning theory for landscape connectivity amongst ecological core areas to a practical situation and determine if and how it works. This is done through a review of the application of the PPS policy for natural heritage protection, and specifically the concept of ecological linkages, during land development planning using a case study; the urban expansion of the Town of Oakville in southern Ontario. This is a useful case study because the plan was developed and debated over almost a decade (2000 to 2008), with two teams of ecologists and planners working separately from each other in designing the natural heritage system for a major urban development in a rural area. The case study analysis develops an understanding of how the science of landscape connectivity and linkages was applied within the context of the PPS. It is the authors' opinion that the approach of using two different planning teams insured that the assumptions and scientific bases were fully explored and evaluated.

This is a story of ecologists, municipal planners, and development planners working within current scientific theory and public policy to hammer out an urban development plan through conflict resolution processes. Before outlining the case study, it is necessary

to briefly explain the scientific theory, as well as the legal and policy structure in place at the time.

## A Brief History of Landscape Connectivity Planning In Ontario

Within this paper, the urbanizing landscape will be perceived as consisting as cores of environmentally sensitive areas (significant forests and wetlands), connected by linkages (river valleys and other corridors of natural habitat), each buffered by narrow strips of land. Landscape connectivity is a concept that refers to a landscape's structural and functional continuity, allowing for the flow of water, nutrients, energy, organisms, genes, and disturbances at many spatial and temporal scales. The loss of landscape connectivity leads to ecosystem fragmentation, which in turn contributes to a decline in biodiversity and threatens some species. The buffer is a narrow strip of land with compatible ecological features and functions that is placed between the cores and linkage areas and the adjacent developed lands. Over time, the theories of landscape connectivity between and amongst natural areas have increasingly been implemented within land use planning (Bennet et al. 2006; Meyfarth O'Hara 2009).

The concept of landscape connectivity as being important in ecosystem functioning was first recognized in Ontario's land use planning system in the 1970s and by the 1990s was present in a range of government reports, regional and provincial strategies. By 1995, the theory of landscape connectivity using linkages between important areas of core habitat had moved from theory to practice and was well established within Ontario's polices, programs and provincial legislation. For example, in 1995 the Lake Ontario Greenway Strategy called for the need to recognize the importance of habitat connections within the Oak Ridges Moraine (Meyfarth O'Hara 2009). By 2002, provincial policies and land use planning legislation had been established which recognized and protected the landscape connectivity value of the Oak Ridges Moraine, a massive landscape feature of approximately 200,000 hectares running east-west through southern Ontario north of Toronto (Whitelaw and Eagles 2007; Whitelaw et al. 2008).

There was considerable movement from theory to practice in planning for landscape connectivity in Ontario between 1970 and 2008, as scientific concepts were increasingly adopted by the planning profession and by government policy makers. A suite of stakeholders worked on this movement, including university professors and students, government policy-makers, and influential environmental groups. The movement of conservation biology principles developed by scientists into the larger society, including the concepts of landscape connectivity, was enhanced by a growing public awareness, which in turn contributed to rising pressure on the Ontario government to reform its land use planning policies and its land use planning system. The theory of landscape connectivity was increasingly included in key land use planning legislation and policies and is now an accepted part of planning for natural heritage in Ontario (Meyfarth O'Hara 2009). Landscape connectivity, as a concept, gained acceptance in Ontario to the point that it transcended political change. For example, the change in provincial government from the environmentally-responsible and socialist New Democratic Party, in power from 1990 to 1995, to the pro-business Progressive Conservative Party elected in 1995, led to major changes in policy and legislation but the concept of "natural connections" remained in the revised PPS in 1997. Following the election of the centrist Liberal Party in 2003, the concept of landscape connectivity was again retained, and then further strengthened in the revised PPS of 2005.

The following section will examine the Provincial Policy Statement and its policies on landscape connectivity.

## The Provincial Policy Statements of 1996, 1997 And 2005

Section 3 of the Ontario Planning Act gives the provincial government power to create a Provincial Policy Statement (PPS) through a process similar to that for the creation of a Regulation under an Act. Over the time of the planning of the case study area in Oakville Ontario, three different versions of the PPS were in force. For the first period of the case study up to 2005, Section 2 of the Ontario Planning Act required that all land owners, developers and municipalities were required to "have regard to" the policies of the PPS. In the second period after 2005, all plans must "be consistent with" the PPS. In the following section, the development of the natural heritage polices within the PPS over time are outlined.

Section 2.3 of the PPS 1996 (amended 1997) concerned Natural Heritage. Policy 2.3.1 stated that natural heritage features and areas will be protected from incompatible development. The PPS identified seven natural heritage features that were to be protected: 1) significant wetlands, 2) fish habitat, 3) significant woodlands south and east of the Canadian Shield, 4) significant valley lands south and east of the Canadian Shield, 5) significant portions of the habitat of endangered and threatened species, 6) significant wildlife habitat, and 7) significant areas of natural and scientific interest (Table 12.1). The PPS did not permit development and site alteration in significant wetlands south and east of the Canadian Shield or in significant portions of the habitat of endangered and threatened species. Development and site alteration was permitted in fish habitat, significant wetlands in the Canadian Shield, significant woodlands south and east of the Canadian Shield, significant valley lands south and east of the Canadian Shield, significant wildlife habitat, and significant areas of natural and scientific interest; if it was demonstrated that there would be no negative impacts on the natural features or the ecological functions for which the area was identified. Section 2.1 of the PPS 2005 concerns Natural Heritage. Policy 2.1.1 states that natural features and areas shall be protected for the long term. The PPS 2005 identified eight natural heritage features that are to be protected: 1) significant wetlands; 2) significant coastal wetlands; 3) fish habitat; 4) significant woodlands south and east of the Canadian Shield; 5) significant valley lands south and east of the Canadian Shield; 6) significant portions of the habitat of endangered and threatened species; 7) significant wildlife habitat; and, 8) significant areas of natural and scientific interest. A comparison of these two lists of features is found in Table 12.1. When identified and mapped these 8 natural heritage features are considered to be ecological cores, which are significant lands that will form the core protection areas for a future urban greenspace system. Municipal infrastructure, such as roads, water lines and sewers, was not included in the definition of development and therefore did not have to adhere to the PPS limitations, providing a major loop hole for the intrusion of development into the cores and linkages. Agricultural uses were permitted to continue in all areas designated into any of these 8 categories. The Town of Oakville is located south of the Canadian Shield, and thus subject to the relevant policies for this area.

The Conservative Government in power provincially from 1995 to 2003 changed the PPS so as to weaken the natural heritage policies. The policies were shortened and simplified in order to make development easier (Winfield, 2003).This was part of broader effort by this right wing government to reduce environmental regulation throughout Ontario. Although "significant natural corridors" were removed from the list of natural heritage features to be protected, the concept of landscape connectivity was retained in the revised PPS. The revised Section 2.3.3 in 1997 stated, "The diversity of natural features in an area, and the natural connections between them should be maintained and improved where possible." This policy statement of "natural connections between them" is important as it suggests that the concept of landscape connectivity was sufficiently well accepted in Ontario and thus was able to withstand a significant change in political power between parties with very different views about land use planning. However, the phrase ``natural connections`` was not defined in the policy, leaving the application of this important principle to environmental planners and urban planners working on individual land use plans both for governments and private developers.

In 2003, the right wing Conservative Government was replaced by the more centrist Liberal Government. This made Ontario unique in Canadian history where three different political parties formed successive majority governments. The importance of land use planning in Ontario politics was highlighted when this new government proposed amendments to the Planning Act very soon after the election. These changes were passed by the Ontario Parliament in 2004. In addition, a new PPS was created in 2005 coinciding with the effective date of the implementation of Section 2 of the amended Planning Act, which required that planning decisions on applications that are subject to the new PPS "shall be consistent with" the new policies (OMMAH, 2007). The new rule that all land use plans "shall be consistent with the PPS" was seen by many to be stronger wording than the older rule that the plans must "have regard to" the PPS. The new policies of the new PPS were intended to "fulfill the government's commitment to provide strong, clear policy direction on land use

Figure 12.1 Aerial photograph of the North Oakville case study area. The photo shows the entire case study area, consisting of the lands between the roads on 3 sides and the river valley at the bottom of this figure.

The site is composed mostly of farm fields with a few scattered forest areas (Google Maps, 2010).

Figure 12.3 North Oakville cores and linkages proposed by the Town of Oakville. The cores are the large blocks filled with cross lines. The linkages are linear features filled with hatched lines. The watercourses are coloured. The site boundary is a purple hatched line (Town of Oakville, 2008).

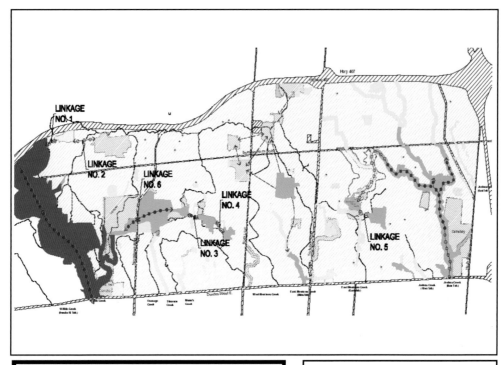

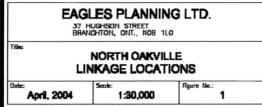

Figure 12.2 North Oakville cores and linkages as
proposed by the North Oakville Landowners. The
cores are the large blocks of light and dark green,
while the linkages are shown in red (NOMI, 2004).

planning to promote strong communities, a clean and healthy environment, and a strong economy (OMMAH 2007).

A noted above, the PPS of 2005 does not permit development and site alteration in 8 categories of natural lands unless it has been demonstrated that there will be no negative impacts on the natural features or their ecological functions. Development and site alteration is not permitted on adjacent lands to the natural heritage features and areas, unless the ecological function of the adjacent lands has been evaluated and it has been demonstrated that there will be no negative impacts on the natural features or on their ecological functions. As mentioned above, infrastructure is allowed and existing agricultural uses are permitted to continue. The identification of these 8 categories leads to the identification of areas for conservation, typically called core areas in planning documents.

The concept of landscape connectivity in the PPS 2005 was expanded so that Section 2.1.2 stated;

The diversity and connectivity of natural features in an area, and the long-term ecological function and biodiversity of natural heritage systems, should be maintained, restored or, where possible, improved, recognizing linkages between and among natural heritage features and areas, surface water features and ground water features.

Section 2.1.2 considerably strengthened the concept of linkages as the phrase: "recognizing linkages between and among natural heritage features and areas, surface water features and ground water features" is more expansive and directive than the previous phrase: "natural connections between them."

Section 2.1.2 requires that "the connectivity of natural features" and the "linkages between and among natural features and areas, surface water features and ground water features" be identified in planning

documents. Importantly, linkages should be "maintained, restored or ... improved" which provides direction to encourage linkage restoration within linkages. This concept of linkage restoration became important in the case study discussed below.

Therefore, for the purpose of urban development planning the natural heritage system consists of core areas, linkages between and amongst the cores, and buffers beside all cores and linkages.

Studies to identify cores and linkages are typically completed by field ecologists. These studies are then used by planners in the overall land use allocation process during urban development planning. However, the 2005 PPS, similar to the 1996 and 1997 PPS did not define connectivity or linkages. The Natural Heritage Reference Manual produced by the Ministry of Natural Resources defines an ecological linkage as a "pathway, connection or relationship between natural features and areas" (OMNR 1999, p. 49). This reference manual is only advisory as it does not have legal or policy standing under Ontario law. The lack of precision in the definitions and the scientific literature led to debate amongst ecologists and planners in regards to the functions, locations, and widths of linkages in many urban planning efforts.

Legally, the 1997 version of the PPS was in force at the time of the onset of planning for North Oakville. However, all parties to the planning effort addressed core, buffer, and linkage provisions meeting the intent of the 2005 PPS, even thought it was the developers' legal right to utilize the earlier, weaker version. In the following section we outline how the natural heritage policies were implemented in the case study area.

## Case Study: North Oakville East Secondary Planning Process

The North Oakville East Secondary Plan was part of one of the largest planning process underway in Ontario in the period from 2000 to 2008. This Secondary Plan, which is an outline of the proposed urban structure for the entire area, was developed for the northward urban expansion of the Town of Oakville. This area was designed to provide residences for 55,000 and workspace for 35,000 people upon completion (Town of Oakville 2008). The planning was the focus of long and intense negotiations between the municipality, many private landowners and developers, as well as community groups. The planning process included two separate subwatershed plans, with each team of planners presenting a different Natural Heritage System for the suburban site. The case provides an interesting example of the challenges of planning for landscape connectivity at the municipal level.

## Study Setting

The North Oakville Lands consist of 3,000 hectares in the northern part of the Town of Oakville. The focus of this case study is the North Oakville East Lands, which include the lands east of Sixteen Mile Creek (Figure 12.1). The North Oakville East Lands consist of a predominantly agricultural ecosystem, composed of open fields that have been in this state for almost 200 years. There are also scattered remnants of forest, wetlands and modified stream valleys (NOMI 2004). The remnant vegetation in the North Oakville East Lands displays a typical pattern for Southern Ontario and has significant implications for the identification, delineation and restoration potential of a local Natural Heritage System (Town of Oakville 2006). A few urban land uses are interspersed throughout the area, including retail, institutional (schools and a land fill site), public, and private open space uses (NOMI 2004).

Part of the North Oakville East Lands is located on the Trafalgar Moraine, which forms a defined ridge that separates the East Sixteen Mile Creek from other watersheds in the area (Town of Oakville 2006). Some groups opposing the urban expansion

attempted to equate development on this site with development on the Oak Ridges Moraine and thereby introduce the very strong land use restrictions found on that moraine (Whitelaw and Eagles 2007). Several scientific studies undertaken on the North Oakville portion of the Trafalgar Moraine concluded that it was not geologically unique or scientifically significant and its geology does not pose a constraint on urban land uses and development (Stantec et al. 2004), and this conclusion was accepted by the Town of Oakville and Province of Ontario. Therefore, the importance of the moraine component of the plan area faded and was given little subsequent emphasis.

## Planning History

On May 29, 2002, the elected Council of the Town of Oakville approved Official Plan Amendment 198 (OPA 198) to urbanize the 3,000 hectares of countryside in the northern portion of the municipality. Two important targets in OPA 198 for the North Oakville Lands included 55,000 residents and industrial space for 35,000 workers. The Town's plan for the North Oakville Lands proved to be highly controversial. On the same day that OPA 198 was approved, Member of Provincial Parliament Mike Colle introduced a private member's bill in the Ontario Parliament, The Trafalgar Moraine Protection Act, to protect and preserve the Trafalgar Moraine from development. Colle, who previously introduced a successful 1999 bill that called for the protection of the Oak Ridges Moraine, hoped to temporarily freeze the proposed development on the North Oakville Lands. However, as pointed out above, scientific evidence provided by the landowners resulted in a low emphasis being placed on this moraine and this legislation was never passed. Local environmental groups and a coalition of landowners appealed to the Ontario Municipal Board in June 2002 and asked for a hearing.

On August 6, 2003, the Town reached agreement with three Non-Government Organizations, Clear the Air Coalition, Oakville Green Conservation Association Inc. and the Residents Association North of Dundas (Sorensen Gravely Lowes 2004). Accordingly, OPA 198 was modified to reflect the settlement of those parties and the changes were approved by the OMB on September 12, 2003. OPA 198 established a general framework for the preparation of more detailed secondary plans and identified the need to prepare separate secondary plans for the areas west and east of Sixteen Mile Creek (Sorensen Gravely Lowes, 2004). OPA 198 also identified a number of requisite studies to be undertaken prior to approval of a secondary plan, including an analysis of the linkage component of the natural heritage and open space system.

The Town began working towards secondary plans with the initiation of the North Oakville Creeks Subwatershed Study in January 2002. The purpose of the Study was to "develop a subwatershed plan that allows sustainable development while ensuring maximum benefits to the natural and human environments on a watershed basis" (Town of Oakville 2006). The Town had two major goals for their secondary plans: 1) the plans had to ensure the preservation of a sustainable natural heritage system that could maintain a diversity of species and landscapes in an urban context; and, 2) the plans had to provide for a walkable, compact, and diverse community, developed in a New Urbanist form (Town of Oakville, 2006). This first goal when combined with the PPS requirements for a natural heritage system would lead to the identification of a system of core and linkages to be protected from development.

At the same time, a group of landowners and developers, known as the North Oakville Management Inc. (NOMI), worked on their own North Oakville East Subwatershed Study, starting in August 2000. This company assembled an interdisciplinary team of consultants to address the range of environmental issues in the study area subwatershed, including linkages for the natural heritage system. Although NOMI's subwatershed study was conducted in parallel with the Town's subwatershed study, NOMI participated throughout the Town's subwatershed process by sharing information, providing input on the Town's technical reports, attending public information sessions and participating in the Town's Technical Advisory Committee for their study (Stantec et al. 2004). By 2004 the NOMI planning team had their version of the subwatershed plan completed with attached maps showing the natural heritage system, including linkages (Figure 12.2).

In May 2003, the Town initiated the Inter Agency Review (IAR) to make recommendations regarding a sustainable natural heritage system for North Oakville. The IAR had input during the development of the Secondary Plan from the MMAH, the Region of Halton, the Ministry of Natural Resources, and Conservation Halton. In September 2003, the Town released the IAR report which presented guiding principles and a map of a conceptual natural heritage system for the Town to consider in the next step of developing the North Oakville Secondary Plan. The report recommended the creation of a natural heritage/open space system comprised of core natural areas, following the 8 categories of special lands identified in the PPS, connected by a system of linkages between these core natural areas, all of which was protected by buffer strips of land (Figure 12.3).

The number and extent of the proposed core areas in the IAR developed by the government agencies far exceeded the recommendations from an earlier report in 2000 previously undertaken by consultants retained by the Town to identify a natural heritage/open space system for North Oakville (Sorensen Gravely Lowes 2004).

Neither the landowners (NOMI) nor their consultants were part of the process that led to the completion of the IAR report. While NOMI agreed in principle with the protection and management of "environmental core areas versus isolated, scattered, small environmental pockets," it did not concur with the "extent and location of the proposed core areas and linkages or allowable uses in the core areas." NOMI found that:

The IAR Report lacks supporting scientific justification; presents inaccuracies in mapping; and is based to some degree on technical information/analysis with which the Landowners' Subwatershed Study team does not concur. Until scientific rationale for the core area delineations and conceptual linkages is presented and reviewed, the IAR cores and linkages are not supported as appropriate environmental lands for protection in the Subwatershed Study or the Secondary Plan (Stantec et al. 2004).

In addition to concern over the lack of scientific support for the IAR Report's linkage locations and width, NOMI expressed "serious concern over the lack of opportunity for Landowners' input to the IAR Report" (Stantec et al. 2004).

In September 2003, the Town invited a team of new urbanists to conduct a ten-day charrette in which four different new urbanist designs were sketched for North Oakville. NOMI presented its proposed development concept for North Oakville East and the concept formed the basis for one of the four plans carried through the charette and its numerous public sessions (Sorensen Gravely Lowes, 2004). These plans served as input for the Town's draft North Oakville Secondary Plan (Town of Oakville 2006). From here a secondary plan was developed by the Town using only those reports prepared by the Town's various planning efforts, which laid out in schematic form all major land uses for the area. The Town then completed a draft North Oakville East Secondary Plan to present at the OMB

as the Town's response to the Secondary Plan proposed by NOMI (Town of Oakville 2006). Pre-hearings for the OMB took place in 2005. This meant that at this time in 2005 there were two competing visions of the entire area, one prepared by the Town and another by the consultant team working for the land owners and developers.

It is worth noting that up to this point both teams of planners and consultants were working within the same science base and the same provincial laws and policies. Therefore, the many differences that occurred in the design of a natural heritage plan and strategy occurred within this coherent science and policy structure.

Therefore, by 2006 there were two separate Natural Heritage Plans produced, one by the Town, with support of some government agencies, and one by the landowners. In general the Town's plan outlined much larger cores and much larger linkages, than did NOMI. Exact measurements of all core areas are not available to the authors, but the Town's core areas appeared to be approximately 25% larger than the NOMI core areas. The Town's linkages were typically 233% larger than the NOMI linkages (100 meters wide compared to 30 meters wide). Some of the details are discussed below. At this point in 2006, it appeared that the differences would be explored and choices made through a long hearing, possibly lasting a year or more, in front of the Ontario Municipal Board. Such a hearing would cost several millions of dollars in legal and consulting fees for all parties. Therefore, a negotiated settlement was attempted.

Starting in May of 2006 the Town started a divide-and-conquer strategy amongst the development companies. The town negotiated with the developers individually, rather than through the umbrella NOMI organization. This approach was very successful from the Town's point of view and by August 14, 2007 a comprehensive settlement was reached between the Town and most of the landowners. It was agreed that the Town's Secondary Plan with the expansive natural heritage system would be put forward as the preferred solution at the OMB hearing. The landowners also agreed to donate to the Town, free of charge, all of their lands in the Town's proposed Natural Heritage System. These concessions by the landowners would significantly decrease both the time and expense of an OMB hearing and would get development started faster, but a considerable cost to the landowners. Further settlements were reached during the subsequent hearing on a few minor issues and the remaining issues were decided upon by the Board. The North Oakville East Secondary Plan was approved by the OMB in February 2008. Therefore, from the start of detailed planning in 2000, it took 9 years to design the layout of the urban expansion.

The Town was ultimately successful in the negotiation because it was broke apart the NOMI structure of many different development companies working together and negotiated with each individually. The Town was able to separately negotiate critical development issues such as the location of high density urban uses, the order of development, the location of infrastructure, the location of school sites and storm water management ponds, and all other urban form issues. Since these issues were critically important to the financial return of each development company, they became more important than the amount of land given to green space. No individual company could afford to not be part of the massive land allocation negotiation. This effort involved political and legal negotiation, the balancing of complex issues by lawyers and politicians with financial, density and timing benefits offered to developers to offset the loss of developable land through greenspace creation.

Therefore, the major differences between the two natural heritage visions were never

tested in front of the OMB. In order to gain an understanding of the differences in the two proposed natural area systems, a comparison is presented in the next section of the paper using an adapted Ahern's framework for landscape ecological planning, modified somewhat for landscape connectivity by other literature (Ahern 1995; Ahern 1999; Ahern 2005; Bennett 2003; Crooks and Sanjayan 2006; Hilty et al. 2006; Kleyer at al. 1996; Linehan and Gross 1998; Opdam et al. 2002). Each of the sections in headed by a question based on Ahern's framework.

## Comparison within Analytical Framework

In this section of the paper, comparisons of the planning processes for the Town's Secondary Plan and NOMI's Secondary Plan are presented separately. Documents reviewed include the Town's North Oakville Creeks Subwatershed Studies Draft Analysis Report (Town of Oakville 2003), the Town's North Oakville Creeks Subwatershed Studies (Town of Oakville 2004), the Town's North Oakville Creeks Subwatershed Study (Town of Oakville 2006), the Town's North Oakville East Secondary Plan (Town of Oakville 2008), NOMI's North Oakville East Secondary Plan (NOMI 2004), NOMI's Input to the North Oakville East Secondary Plan (Stantec et al. 2004), the Landowners North Oakville East Secondary Plan Planning Assessment Report (Sorensen Gravely Lowes 2004), and the North Oakville Natural Heritage Inventory and Analysis (LGL 2000).

## Was the planning process interdisciplinary and public?

Town of Oakville: The Town's study team included an interdisciplinary roster of consultants, with assistance from staff of the Town of Oakville, the Regional Municipality of Halton, the Halton Region Conservation Authority, and the Ontario Ministry of Natural Resources (Town of Oakville 2006). The Inter Agency Report

(IAR) report made recommendations for a natural heritage system in North Oakville. These data were then fed into The North Oakville Creeks Subwatershed Study which also included input from public participation for the purpose of identifying the key issues, developing a vision and objectives, discussing analysis findings for characterization, and development of a management and greenspace strategy (Town of Oakville 2006). The main process for input was from key stakeholders on the Technical Advisory Committee. Other methods for public participation included public meetings, a Steering Advisory Committee, Council meetings, and a design charette.

North Oakville Management Inc. (NOMI): NOMI's interdisciplinary study team of consultants consisted of experts in surface water, hydrology, hydraulics, natural heritage, natural heritage linkages, geology, hydrogeology, quaternary geology, and fluvial geomorphology (Stantec et al. 2004). This planning process was not public, although public comments raised during the design charette were used to modify the development concept (Sorenson Gravely Lowes 2004). NOMI also participated in the Town's Technical Advisory Committee.

### Were landscape connectivity goals and assessments defined?

Town of Oakville: The Town's Subwatershed Study provides the following assessment of ecological linkages on the North Oakville East Lands:

The North Oakville lands contain a variety of habitat types including agricultural fields, pasture, hedgerow, pioneer vegetation, mature woodlands, wetlands, and valleys, and have been described as a remnant agricultural landscape (Gore & Storrie and Ecoplans, 1996; LGL, 2000). The function of these lands has been influenced by urbanization to the immediate south, and by the local road network throughout the area. These roads have increased the amount of habitat fragmentation and have created barriers to ground travelling wildlife within the area and to areas adjacent to the subject lands. Connectivity between some northern and southern patches of habitat within the site appears to be maintained by the vegetated creek corridors, mainly Sixteen Mile Creek and Joshua's Creek (Town of Oakville 2004, 4E-75).

Thus, the Town's Subwatershed Study indicates that the ecological functions of the study area lands are impaired by urbanization to the south and the local roads have increased habitat fragmentation and block wildlife movement both within the study and from the study area to adjacent area. Existing regional connectivity is maintained by two north-south creek corridors, mainly Sixteen Mile Creek and Joshua's Creek.

The Subwatershed Study identifies, based on field mapping, five types of existing habitat connections:

1. Agricultural fields and open field habitats;
2. Hedgerows – Generally single rows of trees, sometimes double rows, often shrub-dominated or mixed;
3. Riparian habitats associated with watercourses that are primarily meadow and/or marsh habitats;
4. Stepping stones created by proximity of habitat types with little connecting habitat; and,
5. Connectivity created by contiguous woodland habitats (Town of Oakville 2004, p. 4E-76).

Thus, the Town identified a variety of existing linkages in the study area: open fields, hedgerows, riparian corridors, stepping stones, and woodlands.

Landscape connectivity is identified as an important component of the Town's North Oakville East Secondary Plan (NOESP). The Town's NOESP begins with a vision statement. Policy 7.2.2 Vision states:

North Oakville's development as an urban community shall reflect Oakville's distinct historical roots and small-town heritage and Trafalgar Township's village rural heritage, with nodal development, prestige industry, and green linkages continuing to define Oakville's unique landscape...

The character and pattern of the community will be significantly influenced by a planned natural heritage and open space system. This natural heritage and open space system is designed to protect the natural environment, provide a balance between active and passive recreation needs and contribute to the quality of life in North Oakville and the Town as a whole. A key component of the system will be the provision of an opportunity for residents and employees to use an extensive open space trail system.

The vision statement for the Town's NOESP identifies three goals for the natural heritage system: protecting the natural environment; providing recreational opportunities for the human community; and, contributing to overall quality of life in the area. These goals are supported by Policy 7.2.3 General Development Objectives, which are intended to guide future development of the planning area. The first three objectives feature landscape connectivity:

1. To establish as a first priority of the Town, a natural heritage and open space system, within the context of an urban setting, the majority of which is in public ownership.

2. To create a sustainable natural heritage and open space system which provides a balance between active and passive recreational needs and links to the existing open space system within the Town.

3. To identify, protect and preserve natural heritage features within the natural heritage component of the natural heritage and open space system and ensure that their use respects their functional role as natural areas within the ecosystem.

These objectives confirm that the first priority of the Town was the establishment of a natural heritage and open space system, and that this system was also designed to provide recreational needs, and that natural heritage features and functions within the system are to be protected.

Goals for landscape connectivity are found in Policy 7.3.5 Natural Heritage and Open Space System, which identifies Linkage and Optional Linkage Preserve Areas as areas designed to link Core Preserve Areas together to "maintain and enhance their environmental sustainability". These linkages "follow natural features whenever possible and are intended to be of sufficient size and character, including buffers, to ensure the functionality and sustainability of the Natural Heritage component of the System". However, there is no mention of the specific purpose of each linkage, the species it is intended for, or any justification for the proposed widths. "Environmental sustainability" is a vague goal for linkages. Linkages are species-specific, multi-scale and multi-functional, and they can function as conduit, habitat, filter, barrier, source and sink, often simultaneously, depending on the perspective of the target species. Planning for connectivity should therefore be based, at least partially, on the known behaviour of target species (Bowne et al. 2006).

A cultural goal for landscape connectivity in the NOESP is the provision of recreational opportunities for the human community via a trail system. This goal is supported by the General Development Objectives, intended to guide the future urban development of the Planning Area, which lists its second objective as:

To create a sustainable natural heritage and open space system which provides balance between active and passive recreational needs and links to the existing open space system within the Town.

The trail system is thus intended to be part of the natural heritage system. It not clear to the authors of this paper how the goals for landscape connectivity and recreation trail connectivity were used in the choice of lands identified as linkages. It appears that the ecological justification for linkages was used by the Town to identify land that could later be used for a trail system. If this is true, then part of the justification of the linkages was not ecological but recreational, even though this was not clearly identified in the selection of the linkages or in their width.

North Oakville Management Inc. (NOMI): NOMI's consultants, as part of their subwatershed study team, produced a report on natural heritage linkages. Appendix K of NOMI's Subwatersheds Study is a 93-page report on Linkages and Buffers on the North Oakville East Lands, which provides an assessment of regional and local connectivity of the study:

The North Oakville East ecosystem is now relatively isolated from other ecosystems by barriers that include urbanization to the south and the east as well as major transportation corridors to the south (Dundas Street, also known as Highway 5), east (403 Expressway Link) and north (407 Expressway).

There is some internal, local connectivity from the western portion of the North Oakville East lands to the 16 Mile Creek valley system to the west. Functional regional connectivity occurs only in and through the 16 Mile Creek system. There is no functional regional connectivity elsewhere on the North Oakville East lands. Some opportunities exist to maintain local, on-site connectivity amongst natural ecosystem elements (NOMI 2004 Appendix K, pp. 4-5)

NOMI's assessment is that the study area was too isolated by major roads and urbanization to provide regional connectivity across the North Oakville site to the larger area of the Halton Region, except for the one north-south valley corridor of 16 Mile Creek. NOMI felt that opportunities did exist to maintain local, on-site connectivity within the case study site.

The vision statement for NOMI's secondary plan features landscape connectivity as an important component of the long-term vision for the proposed communities in the study area. Policy 4 states:

The character and pattern of each Community will be highly influenced by a planned natural heritage / open space system which protects the natural flora of the area while providing extensive habitat for native animals and providing areas for passive and active recreational use. This natural heritage / open space system affords residents the opportunity to use an extensive open space trail system, which travels through mature woodlot blocks, around wetlands, through parks, along stream corridors and along safe and enjoyable streetscapes.

The above vision statement identifies three goals for the planned natural heritage/open space system: protecting natural flora; providing extensive habitat for native animals; and, providing areas for passive and active recreational use via an open space trail system. These goals are supported by the Environment and Open Space policies under the General Development Objectives in the NOMI plan. Among these objectives are:

1. To establish as a first priority, a natural heritage/open space system within the context of an urban setting that protects, preserves and, where appropriate, enhances significant natural heritage features, functions and linkages.

2. To create a sustainable natural heritage/open space system which provides for both active and passive recreational needs as well as pedestrian connections within the community and to the existing open space system south of Dundas Street.

3. To balance the natural ecological needs with housing and employment needs of the Town, the ability to create compact transit supportive communities and the social,

recreational and economic needs of Oakville residents.

4. To evaluate through the Subwatershed Study the significance of all natural heritage features and functions within the North Oakville East Secondary Plan area and to establish a policy framework for more detailed levels of evaluation at succeeding stages of the planning process.

5. To promote wooded urban squares as special focal points within the community.

6. To protect significant valleys and stream corridors while recognizing that many other stream corridors within the Secondary Plan Area are intermittent and have been modified by agricultural activities and may be further modified, realigned or consolidated.

These objectives confirm that the first priority of the NOESP is the establishment of a natural heritage and open space system that protects, preserves, and, where appropriate, enhances linkages. This system is intended to balance ecological needs with the needs of the human community, including passive and active recreational use. Landscape connectivity is thus identified as an important component of NOMI's NOESP. Like the Town's NOESP, biotic and cultural goals are presented but abiotic goals for landscape connectivity are not presented.

The Greenland Policies of the Land Use Plan for the NOESP include a section on linkages. Policy 1.1.1.2 e) vii Linkages states:

Linkages identify existing Natural Areas and potential Restoration Areas that currently provide a natural linkage function for wildlife species typical of the Secondary Plan Area. The primary function of linkages is to maintain connectivity for wildlife populations and/or habitats that are naturally continuous. Linkages should not be established between areas that previously did not exhibit functional connectivity.

The primary biotic goal of linkages is to maintain connectivity for wildlife populations and habitats that are naturally continuous.

A cultural goal for landscape connectivity in NOMI's NOESP is the provision of recreational opportunities for the human community via a trail system. Policy 1.11.4 entitled The Transportation and Transit Network includes a section on trails. The policy on cycling and pedestrian trails states:

The Urban Design and Open Space Guidelines establish a potential pedestrian and cycling trail system. This trail system provides connections within Greenland Area designations, along the boulevards of arterial roads, and along portions of collector roads that are critical to the continuity and connectivity of the trail system. Cycling trails are primarily located within open space lands and street boulevards and not located within roadways of high volume arterial roads. Cycling along local roads within neighborhoods will be facilitated by an interconnected street and open space system.

The proposed trail system will connect with Greenland and open space lands. Further in the Subwatersheds Study, the potential for the trail system is expanded upon:

Along with considering the need for wildlife connectivity, it is important at this stage in the design process to consider the need for human connectivity elements on the North Oakville East lands east of 16 Mile Creek. The careful planning that has gone into designing ecological linkages, buffers and natural areas for the site can be used as the groundwork for designing a detailed trails plan for the North Oakville East lands. A trail system can provide opportunities for recreation (e.g. walking, bicycling, roller blading), education (e.g. nature walks for school children) and natural history (e.g. bird-watching, plant identification, wildlife-viewing) (NOMI 2004 Appendix K, p. 80).

Two additional cultural goals for landscape connectivity are identified by NOMI: education and natural history. However, it is clear in the NOMI approach that the linkages were chosen for their ecological functioning, not for their potential for recreational linkage. The recreational trail design, when it would be designed in the future, would have to fit into the matrix of cores and linkages produced by ecological mapping.

### What approach to linkage identification and design was employed?

Both the Town and NOMI employed an intuitive, natural heritage system approach based on a system of core areas, linkages, and buffers. In general, the approaches were similar, but there were significant differences that are outlined below. Each approach is presented separately, and then compared.

Town of Oakville: Existing linkages were identified based on field mapping, aerial surveys, and wildlife observations. The Town's Policy 7.4.7 Natural Heritage Component of the Natural Heritage and Open Space System states that the length, width and general location of the linkages were defined based on factors established through the North Oakville Creeks Subwatershed Study, including:

• Composition of potential linkage feature;
• Character of the surrounding habitats;
• Presence and size of discontinuities; and,
• Required buffers.

It is important to note that these factors do not include a specific purpose of each linkage, target species, or species requirements, all of which are key factors affecting the likelihood of linkage success (Bennett, 2003). Whereas the Town's NOESP presents the process for determining width of stream corridors in detail (see Appendix 7.4 Stream Components), there is no similar rationale given for the widths assigned to the linkages.

The Town's Subwatershed Studies refers to several important linkage design considerations, but does not put them into practice. For example, it states, "Ecological linkages must be designed with an understanding of the species that will use

the connection" (Town of Oakville 2004 p.6-18). However, there is no mention of which species were used for the design of the Town's linkages. It states, "A diversity of linkage types and a measure of redundancy in the linkage network should be considered to provide a range of movement opportunities" (Town of Oakville 2004 p.6-18). It then states that all linkages should be 100 meters wide, except in one case where 70 meters is used. Despite recognizing the importance of having a variety of linkage types and despite previously acknowledging the existence of several types of linkages on site, including stepping stones, the Town predominantly used one type and one width of linkage. The 100 meter width may be excessive and may not be ecologically necessary given the existing low-level of linkage function in some locations on the site. Many of the proposed linkages contain roads, driveways, buildings and fences, all of which are serious barriers to connectivity that the Town does not address. Some of the proposed linkages lead to nowhere (there is no end habitat other than a major road) and two of the proposed linkages do not function in support of any linkage goal. In addition, the Town does not have ecological restoration plans for their proposed linkages, yet their intent is for the linkages to become forested and have linkage functions as per the PPS requirement to explore ecological restoration in linkages.

North Oakville Management Inc. (NOMI): The linkages were determined using aerial photographs, maps, and the habitat requirements of target species selected for the study area. Subsequent field observations "suggest that these linkage sites provide the highest probability of movement for species that require forested ecosystems" (NOMI 2004, Appendix K, p. 8). To deal with the varying degrees of connectivity in the study area, a flexible, three-level system of linkages was employed:

By using a variety of linkage types that includes both strips of forest, wetland and

| Provincial Policy Statement 1996 | Provincial Policy Statement 2005 |
|---|---|
| *Landscape Connectivity is defined as:* | |
| "The diversity of natural features in an area, and the natural connections between them should be maintained, and improved where possible." | "The diversity and connectivity of natural features in an area, and the long-term ecological function and biodiversity of natural heritage systems, should be maintained, restored or, where possible, improved, recognizing linkages between and among natural heritage features and areas, surface water features and ground water features." |
| *Natural Heritage Features and Areas protected from incompatible development:* | |
| Significant wetlands<br>Fish habitat<br>Significant woodlands south and east of the Canadian Shield<br>Significant valleylands south and east of the Canadian Shield<br>Significant portions of the habitat of endangered and threatened species<br>Significant wildlife habitat<br>Significant areas of natural and scientific interest | Significant wetlands<br>Significant coastal wetlands<br>Fish habitat<br>Significant woodlands south and east of the Canadian Shield<br>Significant valleylands south and east of the Canadian Shield<br>Significant portions of the habitat of endangered and threatened species<br>Significant wildlife habitat<br>Significant areas of natural and scientific interest |
| *Development and Site Alteration not permitted in:* | |
| Significant wetlands south and east of the Canadian Shield<br>Significant portions of the habitat of endangered and threatened species | Significant wetlands in Ecoregions 5E, 6E and 7E<br>Significant portions of the habitat of endangered and threatened species<br>Fish habitat<br>Significant coastal wetlands |
| *Development and Site Alteration Permitted if it has been demonstrated there will be no negative impacts on the natural features or their ecological functions:* | |
| Fish habitat<br>Significant wetlands in the Canadian Shield<br>Significant woodlands south and east of the Canadian Shield<br>Significant valleylands south and east of the Canadian Shield<br>Significant wildlife habitat<br>Significant areas of natural and scientific interest | Significant wetlands in the Canadian Shield in Ecoregions 5E, 6E and 7E<br>Significant woodlands south and east of the Canadian Shield<br>Significant valleylands south and east of the Canadian Shield<br>Significant wildlife habitat<br>Significant areas of natural and scientific interest |
| *Agrilcultural Uses permitted in:* | |
| Agricultural uses permitted to continue in all areas | Existing agricultural uses permitted to continue in all areas |
| Development and Site Alteration Permitted on Adjacent Lands if it has been demonstrated there will be no negative impacts on the natural features or their ecological functions | |
| Infrastructure not included in the definition of development | |

Table 12.1   Comparison of 1996 and 2005 Provincial Policy Statements

field habitat as well as stepping stones or patches of habitat that provide resources and assist animals in moving across a landscape, planners are better able to maintain connectivity at different spatial scales and take into account the mosaic of habitats now present at the site...Having a variety of linkage types provides options for a wide variety of species (NOMI 2004 Appendix K, pp. 17-18).

NOMI planning for connectivity uses a variety of species at a variety of scales using a three-level system of linkages consisting of:

• Level 1 Linkage is the highest functional connectivity, with existing habitats linked by similar ecosystem communities, for instance, two woodland areas connected via a forested hedgerow. Level 1 linkages are suitable for species intolerant of habitat disturbance and/ or with low to moderate dispersal capacities.

• Level 2 Linkage also provides continuous connectivity, but habitats are linked by somewhat different ecosystem communities, for instance, two woodland areas connected via a wetland or a drainage feature. Level 2 linkages are suitable for some species utilizing the protective woodland areas and those with moderate to high dispersal capacities.

• Level 3 Linkages are between patches or stepping stones of habitat that provide resources for some species to move through the landscape. Level 3 linkages are suitable for species tolerant of disturbance in linkage and ones that are mobile with high dispersal capacities, typically birds, squirrels, etc. This level of linkage does not require a defined terrestrial corridor between the stepping stones. This type of linkage will be enhanced as drainage features are naturalized, with the planting of parklands and stormwater ponds, and as the urban woodland canopy develops (Stantec et al. 2004 p.32).

The three-level system of linkages was designed to maintain functional connectivity for a variety of species at a variety of scales,

using two types of ground linkage and one type of aerial linkage, the stepping stone type.

NOMI's plan for linkages also included a 32-page section on planning for new roads, which discussed potential measures to mitigate the impact of new roads on ecological connectivity on the North Oakville East site, including wildlife crossing designs for the six target species of the North Oakville East lands. Therefore, the NOMI linkage plan attempted to deal with the issue of linkage blockage created by new road construction during development and current road operation.

### Comparison of the Town's and NOMI's Approach

The Town and NOMI have similar approaches, but with four distinct differences (Table 12.2). First, whereas NOMI assesses each linkage individually and employs a three-level system of linkages with varying widths and types, the Town plans for 100 meter wide linkages at virtually all locations. Second, NOMI's approach protects existing linkages, while the Town protects existing linkages, creates new linkages, and in a few locations, proposes new linkages that are non-functional. Third, NOMI's design is based on six target species, whereas the Town does not specify target species. Fourth, while NOMI's approach is strongly referenced to scientific literature, no clear justification is given for the Town's 100 meter wide linkages. In fact, when the authors cross checked the linkage references used in the Town's Subwatershed Studies, none of the references actually supported the Town's use of 100 meter wide linkages in the study area. For example, the Town states that Henry at al. (1999) "reported that corridors should not be less than 100m wide, as this will not create any 'core' habitat for interior or sensitive species." This is false. Henry et al. (1999, p. 647) actually report that, "Landowners and land managers often ask what the minimum corridor width should be for wildlife.

| NOMI | Town of Oakville |
|------|------------------|
| Flexible due to three-level system of linkages connecting habitat patches. The linkages were assessed individually. Category 1 and 2 linkages were a minimum of 15 meters. Category 3 was stepping stone linkages. Buffers of 7.5 meters on both sides of Category 1 and 2 linkages. | Inflexible due to 100 m-wide linkages connecting most types of habitat patches irrespective of the patches being connected. No additional buffers to linkages. One linkage 70 m wide. |
| Protect existing linkages between habitat patches. | Protect existing linkages between habitat patches. New linkages created between currently isolated habitat patches. |
| Linkage design based on requirements of six target species: white-tailed deer, red fox, deer mouse, Eastern garter snake, American toad and gray squirrel. | No target species specified. |
| Detailed road redesign proposed to facilitate linkage functions. | No detailed road redesign proposed where linkages abutted roads. |

Table 12.2   Comparison of the Approaches to Linkages on North Oakville East Lands.

Although this may seem like a reasonable question, in reality, there is no magic width, above which wildlife thrives and below which they are nonexistent". Therefore, this key reference was misapplied, and, perhaps, misunderstood.

Neither group did detailed field studies to evaluate the actual linkage functions across the entire site, in order to measure wildlife movement over the year.

The NOMI linkages varied in width, with a minimum of 15 meters, plus 7.5 meters of a buffer on both sides, leading to an effective width of 30 meters. The Town proposed 100 meter wide linkages, including the buffer, at virtually all locations. Given the significant reduction in the area available for development, the developers seriously questioned the Town's expansive linkage strategy. However, through negotiation the developers got the Town to agree to use up to half of the width of the linkage area for stormwater management facilities at certain locales. This meant that the developers had more developable land, thus somewhat reducing their concern about developable land lost to greenspace.

Table 12.2 compares the two approaches to linkages on the North Oakville East Lands. This comparison provides an example of an important issue: landscape connectivity planning faces constraints in terms of interpretation. In this case, planners for the Town and NOMI, using the same scientific literature, background data and maps, came up with two landscape connectivity plans for the study area that are largely similar but have significant differences. These differences would have led to a very interesting debate had both Secondary Plans been submitted to the OMB, which of course never happened.

What types of planning strategies were employed: offensive, defensive, protective, or opportunistic?

Town of Oakville: According to Ahern's typology, the planning strategies employed by the Town's NOESP constitute an offensive strategy. An offensive strategy is a vision or landscape configuration, which requires restoration or reconstruction to rebuild landscape elements in previously disturbed or fragmented landscapes (Ahern 2005). The Town's emphasis on creating new linkages represents a future landscape for the North Oakville East Lands that must be realized through restoration. The offensive strategy requires the displacement or replacement of intensive land uses (e.g. urbanization, agriculture) with extensive land uses that "put nature back" into the landscape (Ahern 2005). According to Ahern, this strategy is rarely practiced because it is expensive, uncertain, and, politically sensitive. However, it was successfully used here when a phalanx of government agencies placed their support behind the Town's strategy and the Town successfully negotiated with the landowners in a divide-and-conquer strategy.

According to the Town's Subwatersheds Studies Draft Analysis Report:

The current lack of forested connections and gaps between forested blocks indicates that the feasibility of creating forested connections would require considerable plantings. The existing discontinuities created by roadways are also an impediment to creation of a continuous forested connection throughout the Study Area (Town of Oakville 2003 p. 35).

Therefore, the Town's own estimation, their proposed strategy of creating linkages will be expensive and labour-intensive, with uncertain results. Furthermore, no restoration plan was developed, largely due to concern about the cost of the restoration activities.

North Oakville Management Inc. (NOMI): According to Ahern's typology, the planning strategies employed by NOMI's NOESP constituted a defensive strategy. A defensive strategy is employed when the existing landscape is already fragmented and core areas are already limited in area and isolated (Ahern 2005). The defensive strategy seeks to control and stop the negative processes of

fragmentation or urbanization (Ahern 2005). The defensive strategy is often appropriate as a last resort but can also be described as reactionary and ineffective, if the root causes of negative landscape change remain active (Ahern 1995).

### Were alternative scenarios for landscape connectivity evaluated?

Ultimately, only the Town's scenario, in the form of the Secondary Plan, was put before the OMB since the NOMI landowners accepted the Town's approach during negotiations. It appears that he NOMI landowners accepted the Town's interpretation in order to gain faster development approvals and higher development density on lands outside the Natural Heritage System. *Is there a landscape connectivity plan?*

Town of Oakville: Yes.

North Oakville Management Inc. (NOMI): Yes.

*Is there a policy of adaptive planning and management?*

Town of Oakville: The Town's Subwatersheds Studies includes an Implementation Report, which it refers to as a "living document" that can be refined using an Adaptive Environmental Management (AEM) approach:

AEM means making decisions as part of an on-going process. Monitoring the results of actions provides a flow of information that may indicate the need to change a course of action or change the document. The management strategy also includes recommended policies that should be incorporated into Official Planning documents such as the NOE-SP. Over time, government policies on relevant issues, such as terrestrial systems and SWM, will evolve. This strategy should always be applied with reference to the most recent applicable policies (Town of Oakville 2004, pp. 7-1 – 7-2).

However, the Town's NOESP does not mention a policy of Adaptive Environmental

Management or any form of adaptive planning and management.

The planning period for the NOESP is from 2006 to 2021 and it will be reviewed, at a minimum, every 5 years. The NOESP states that a program shall be established by the Town, in consultation with the Region of Halton and Conservation Halton, to monitor the development in the Planning Area on an annual basis, in accordance with directions established in the North Oakville Creeks Subwatershed Study. If it actually occurs, monitoring the proposed linkages will be difficult; however, as there are no specific goals given for the Town's linkages and it will therefore be difficult to establish monitoring protocols in the absence of any goals. Additionally, there was a disagreement amongst the Ministry of Natural Resources, Conservation Halton and the Town of Oakville about who would pay for the long-term monitoring of the linkage functions given its large expense. To our knowledge, no monitoring effort was agreed upon. The documents do not state who would be the manager of these natural heritage lands after development, but it is most likely to be the Parks and Recreation Department of the Town of Oakville. However, comment on how this Department would carry out such management was notably absent in the planning documents. At this point, it appears that the Town was in the process of creating a large greenspace system without stated policies on site management, ecological restoration, or ecological monitoring. Such plans would have to be developed in the future.

North Oakville Management Inc. (NOMI): The Implementation section of NOMI's NOESP included a policy for environmental monitoring;

The Town shall undertake regular monitoring of the health of the natural heritage/open space system within the North Oakville East Secondary Plan Area. The indicators to be monitored and the nature of the monitoring program (s) will be set out in the Subwatershed Study (NOMI 2004 p.39).

As noted by NOMI's Subwatersheds Study, the implementation of monitoring is initially often the responsibility of development proponents, while in the long term, the local municipality, conservation authority, or Ministry of Natural Resources is responsible for funding and carrying out the monitoring. Ultimately, the management of the Natural Heritage System will be the responsibility of the Town, not NOMI, and while the NOMI documents do make some recommendations for management, ultimately, NOMI indicated it is the Town's responsibility.

## SUMMARY

This case study reveals that the urban development process in Ontario contains high levels of emphasis placed on the conservation of cores, buffers, and linkages. Cores are relatively easy to define since they are outlined by identifiable ecological features, such as forests and wetlands. Buffers are placed adjacent to these cores, but there is much debate on the width of these buffers and the land uses that should occur in them after development. This case study shows significant differences amongst planners and ecologists in the interpretation of landscape connectivity theory as applied to ecological linkages within land development. This debate occurred because the scientific literature on linkage design does not provide sufficient information to allow for the precise definitions of linkage functions, locations, and widths that are needed in land use planning. In the absence of scientific precision, political and legal negotiation makes such decisions.

The case study shows that the decision on which approach would be used for linkage location and width largely took place during backroom negotiations between legal counsel, where multiple tradeoffs were made involving development density and timing, rather than in open discussion or in front of an administrative tribunal on the merit of the competing natural heritage visions. Therefore, it is not possible to understand in detail how and why many aspects of the Town's linkage plan were chosen. In this case, the final decision on the entire open space system, including cores, buffers, and linkages, was made by lawyers representing the various parties, not by ecologists. In fact, throughout the final processes leading to the agreement for the approved outline of the urban form, the ecologists on both planning teams were prohibited by the lawyers from talking with each other.

However, the result was a very substantial natural heritage system composed of cores, linkages, and buffers all owned by the Town and ultimately managed by the Town's Parks and Recreation Department. Significantly, there was little policy developed to guide the long term management of the natural heritage lands by the parks department after transfer from the current land owners to the Town, a common problem in natural heritage planning in Ontario (McWilliam, 2007). This lack of precision in the development of coherent policy for the management of the ecological features in the greenspace system leads to confusion when this management actually takes place, often leading to ecological degradation (McWilliam et al. 2010). This natural heritage system will create a very large open space and park system for the developing community but since the Official Plan goals for residences for 55,000 people and workspace for 35,000 were retained, the developed lands will have much higher urban density than would have occurred with a smaller park system in place. Therefore, the future citizens of this area will see much higher density of development in the areas where they live and work, but very low density in an expansive greenspace system.

The case study also reveals that approved linkage plans may not include all the necessary elements for long-term linkage success. For example, no ecological restoration plan was prepared for the extensive areas of farmland incorporated in the linkage areas. The lack of target species and specific goals for the linkages in North Oakville will mean that monitoring will be very difficult, as it will not be possible to create a monitoring plan that is based on the approved functions of these linkages. The only statement about ecological monitoring in the Town's Secondary Plan was that monitoring will be done the ensure that: "The health of the Natural Heritage component of the Natural Heritage and Open Space System is being maintained. (Town of Oakville, 2008, p.109)" Stepping stone linkages, which are recognized in the literature (Baum et al. 2004; Bennett 2003; Minor and Urban 2007; Hashimoto 2007; Rahel et al. 2008; Van Langevelde et al. 2002; Williams et al. 2004) were not accepted in this case study as only on the ground linkages were approved.

The case study reveals that in Ontario it is possible for linkages to be approved that go nowhere and thus are not truly linkages. For example, some linkages ended at multilane highways with no plans given for redesigning the highways to improve linkage function across those barriers.

The case study also reveals that the linkage concept can be used by municipalities to gain a considerable amount of open space land during the development process, probably in excess of that actually needed for linkage functions. In the North Oakville case study, some of the land identified by the Town of Oakville as linkages did not have linkage function according to the landowners' consultants. This suggests that the linkage concept under the current Ontario Provincial Policy Statement may be abused so as to gain open space and parkland without the municipality having to buy the land.

This case study is an example of the difficulty of using linkage theory within an adversarial decision-making process. Given the lack of clarity of the theory as applied to specific situations within urban planning, the final decisions rely on political and legal negotiation involving a wide range of financial, timing, and development options.

## LITERATURE CITED

Ahern, J. 1995. Greenways as a planning strategy. In Greenways: The beginning of an international movement, edited by J.G. Fabos and J. Ahern, pp. 131-155. Elsevier: Amsterdam.

Ahern, J. 1999. Spatial concepts, planning strategies, and future scenarios. In Landscape ecological analysis: Issues and applications, edited by J.M. Klopatek and R.H. Gardener, pp. 175- 201. Springer-Verlag: New York, NY.

Ahern, J. 2005. Theories, methods and strategies for sustainable landscape planning. In From landscape research to landscape planning: Aspects of integration, education and application, edited by B .Tress, G. Tress, G. Fry, and P. Opdam, pp. 119-131. Springer: Amsterdam.

Baum, K. A., K. J. Haynes, F. P. Dillemuth, and J. T. Cronin. 2004. The matrix enhances the effectiveness of corridors and stepping stones. Ecology 85: 2671-2676.

Bennett, A. F. 2003. Linkages in the landscape: The role of corridors and connectivity in wildlife conservation. IUCN: Gland, Switzerland and Cambridge, UK.

Bennett, A. F., K. R. Crooks, and M. Sanjayan. 2006. The future of connectivity conservation. In Connectivity conservation edited by K.R. Crooks and M. Sanjayan, pp. 676- 694. Cambridge University Press: Cambridge, UK.

Bowne, D. R, M. A. Bowers, and J. E. Hines. 2006. Connectivity in an agricultural landscape as reflected by interpond movements of a freshwater turtle. Conservation Biology 20: 780-791.

Crooks, K. R. and M. Sanjayan, 2006. Connectivity conservation: Maintaining connections for nature. In Connectivity conservation edited by K.R. Crooks and M.

Sanjayan, pp. 1-27. Cambridge University Press: Cambridge, UK.

Gore & Storrie and Ecoplans Ltd. 1996. Sixteen Mile Creek Subwatershed Plan. Report to the Regional Municipality of Halton. Burlington, ON.

Hashimoto, H. 2007. Connectivity analyses of avifauna in urban areas. In Landscape ecological applications in man-influenced areas: Linking man and nature systems, edited by S. K. Hong, N. Nakagoshi, B. Fu, and Y. Yorimoto, pp. 479- 488. Springer: Amsterdam.

Henry, A.C., D. A. Hosack, C. W Johnson, D. Rol, and G. Bentrup. 1999. Conservation corridors in the United States: Benefits and planning guidelines. Journal of Soil and Water Conservation 54: 645-650.

Hilty, J. A., W. Z. Zidicker, and A. M. Merenlender. 2006. Corridor ecology: The science and practice of linking landscapes for biodiversity conservation. Island Press: Washington, D.C.

Kleyer, M., G. Kaule, and J. Settele. 1996. Landscape fragmentation and landscape planning with a focus on Germany. In Species survival in fragmented landscapes, edited by J. Settele, C.R. Margules, P. Poschlod and K. Henle, pp. 138-151. Kluwer Academic Publishers: Dordrecht.

LGL Limited Environmental Research Associates (LGL). 2000. North Oakville Natural Heritage Inventory and Analysis Town of Oakville: Oakville, ON.

Linehan, J. R. & M. Gross, 1998. Back to the future, back to basics: The social ecology of landscapes and the future of landscape planning. Landscape and Urban Planning 42: 207-223.

McWilliam, W. J. 2007. Edge effects within municipal forests: Are municipal policies effective in limiting edge-resident encroachment in suburban woodlands. Ph. D. Thesis, University of Waterloo, Waterloo, Ontario. 271 pp.

McWilliam, W., P. F. J. Eagles, M. Seasons, and R. Brown. 2010. Assessing the degradation effects of local residents on urban forests in Ontario, Canada. Journal of Arboriculture and Urban Forestry 36(6): 253-260

Meyfarth O'Hara, E. 2009. Moving from landscape connectivity theory to land use planning practice: Ontario as a case study. PhD thesis, University of Waterloo, Waterloo, Ontario. 381 pp.

Minor, E. S., and D. L. Urban, 2007. Graph theory as a proxy for spatially explicit population models in conservation planning. Ecological Applications 17: 1771-1782.

North Oakville Management Inc. (NOMI). 2004. North Oakville East Subwatersheds Study. North Oakville Management Inc.: Oakville, ON.

Ontario Ministry of Municipal Affairs and Housing (OMMAH). 2007. Provincial Policy Statement. Ontario Ministry of Municipal Affairs and Housing: Toronto, ON.

Ontario Ministry of Natural Resources (OMNR). 1999. Natural Heritage Reference Manual for Policy 2.3 of the Provincial Policy Statement. Ontario Ministry of Natural Resources: Toronto, ON.

Opdam, P., R. Foppen, and C. Vos. 2002. Bridging the gap between ecology and spatial planning in landscape ecology. Landscape Ecology 16: 767-779.

Penfold, G. 1998. Planning Act Reforms and Initiatives in Ontario, Canada. In The cornerstone of development: Integrating environmental, social, and economic policies, edited by J. Schnurr and S. Holtz, pp 149-176. International Development Research Centre: Ottawa, ON.

Rahel, F. J., B. Bierwagen, and Y. Taniguchi. 2008. Managing aquatic species of conservation concern in the face of climate change and invasive species. Conservation Biology 22: 551-561.

Sorensen Gravely Lowes. 2004. The Landowners' North Oakville East Secondary Plan Planning Assessment Report. North Oakville Management Inc.: Oakville, ON.

Stantec Consulting Ltd., Bird and Hale Ltd., Beatty and Associates, P. F. J. Eagles Planning Ltd., K.W.F. Howard, Lorant Consulting, C. Eyles, N. Eyles, Menzies Consulting, Aquafor Beech Ltd., and Ecoplans Ltd. 2004. Input to the North Oakville East Secondary Plan. North Oakville Management Inc.: Oakville, ON.

Town of Oakville. 2003. North Oakville Subwatersheds Studies. Draft Analysis Report. Town of Oakville: Oakville, ON

Town of Oakville. 2004. North Oakville Creeks Subwatershed Studies. Town of Oakville: Oakville, ON.

Town of Oakville. 2006. North Oakville Creeks Subwatershed Study. Town of Oakville: Oakville, ON.

Town of Oakville. 2008. Official Plan Amendment Number 272; North Oakville East Secondary Plan. Town of Oakville: Oakville, ON.

Van Langevelde, F., G. D. H. Claasen and A. G. M Schotman. 2002. Two strategies for conservation planning in human-dominated landscapes. Landscape and Urban Planning 58: 281-295.

Whitelaw, G., and P. F. J. Eagles. 2007. Planning for linkages and corridors on private land: The Oak Ridges moraine experience. Conservation Biology: 675-683.

Whitelaw, G. S., P. F. J. Eagles, R. B. Gibson, and M. L. Seasons. 2008. Roles of environmental movement organizations in land use planning: Case studies of the Niagara Escarpment and Oak Ridges Moraine, Ontario, Canada. Environmental Planning and Management: 801-816.

Williams, P., L. Hannah, S. Andelman, G. Midgley, M. Araujo, G. Hughes, L. Manne, E. Martinez-Meyer, and R. Pearson. 2004. Planning for climate change: Identifying minimum-dispersal corridors for the Cape Proteaceae. Conservation Biology: 1063-1074.

Winfield, M.S. (2003). Smart Growth in Ontario: The Promise vs. Provincial Performance. The Pembina Institute.

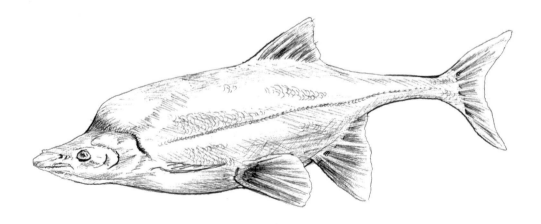

# RECOVERING ENDANGERED FISH IN THE SAN JUAN RIVER USING ADAPTIVE MANAGEMENT

*Mark C. McKinstry*

## ABSTRACT

The San Juan River Basin is the second largest of the three sub-basins that comprise the Upper Colorado River Basin. From its origins in Colorado, the San Juan River flows approximately 560 km to Lake Powell, intercepted along the way by Navajo Reservoir where the water is stored for use in the San Juan and Rio Grande basins and for delivery to the lower Colorado River Basin. The San Juan River Basin Recovery Implementation Program (SJRIP) was begun in 1992 and authorized by Congress in 2000 to protect and recover endangered fishes (Colorado pikeminnow [*Ptychocheilus lucius*] and razorback sucker (*Xyrauchen texanus*]) while allowing water development to proceed. The SJRIP is a nationally recognized effort which has served as a model to address other Endangered Species Act issues throughout the country. Aggressive efforts are being implemented through Program Elements and comprehensive plans to: construct fish passages, fish screens, and propagation facilities; restore and enhance aquatic habitat; improve water use efficiency; stock native fish and control non-native fish species; and conduct hydrologic evaluations that allow the adoption of natural-flow mimicry. Specific tasks to assist in recovering the two fish species are listed in the Program's Long-Range Plan. All activities conducted through the SJRIP are evaluated by two technical committees (Hydrology and Biology) comprised of experts in endangered fish management, fish ecology and biology, geomorphology, hydrology, and habitat management. A Coordination Committee oversees the activities of the two technical committees and has responsibility over recovery actions. The Bureau of Reclamation and U.S. Fish and Wildlife Service work cooperatively to carry out the management activities approved by the Program participants. To ensure a scientifically defensible approach to recovery actions, the SJRIP uses standing peer reviewers to evaluate progress of the individual projects and the Program as a whole. Management activities are formulated in an adaptive management framework whereby models are created to help understand the system and identify potential actions, hypotheses are developed to test the actions, and management actions are then conducted with subsequent monitoring of the fish and habitat used to inform, and modify, future activities. The adaptive management approach has allowed the SJRIP to make progress towards recovery in the face of uncertainty about various aspects of the biology and ecology of these endangered species. The Department of the Interior recognized the SJRIP and its sister program, the Upper Colorado River Recovery Implementation Program, with a Cooperative Conservation Award in 2008, citing the programs' excellence in conservation through collaboration and partnerships.

## INTRODUCTION

From its origins in the San Juan Mountains of Colorado, the San Juan River flows ~ 560 km to Lake Powell (Figure 13.1). The San Juan River sub-basin is the second largest of the three sub-basins that comprise the Upper Colorado River Basin. It drains about 98,400 square km of southwestern Colorado, northeastern Arizona, northwestern New Mexico, and southeastern Utah. In its upper reaches, the river traverses rugged terrain and has a relatively high gradient. The river emerges from canyon-bound reaches shortly after entering New Mexico and flows through a broad floodplain for much of its course in New Mexico and Utah. About 113 km upstream of Lake Powell, the river again enters canyon reaches for the remainder of its course. Depending on the elevation of Lake Powell, a ~ 10 m tall waterfall just upstream of where the river enters Lake Powell can dramatically separate the river from the lake habitat. This waterfall precludes movement of fish from the lake to the river, but not vice versa. The USFWS has designated approximately 289 km of the lower river near Farmington, NM to Lake Powell as critical habitat for endangered fish species, the Colorado pikeminnow (*Ptychocheilus lucius*), and razorback sucker (*Xyrauchen texanus*) (U. S. Fish and Wildlife Service 1994).

In 1922, the seven basin states of Utah, Colorado, Wyoming, New Mexico, Arizona, Nevada, and California signed a compact dividing the Colorado River between the Upper and the Lower Colorado River basins. In 1948, the Upper Basin states (Wyoming, Colorado, Utah, and New Mexico), together with Arizona, signed an agreement apportioning the Upper Basin share between the states. Each of the States and the Bureau of Reclamation, under the authority of the Colorado River Storage Project Act (CRSPA; P.L. 84-485), initiated the development of the waters of the Upper Colorado River Basin. The passage of CRSPA in 1956 allowed for the construction of many large mainstem impoundments on the Colorado River and various tributaries including Flaming Gorge

Dam on the Green River, Aspinall Unit on the Gunnison River, Glen Canyon Dam on the Colorado, and Navajo Dam on the San Juan River.

While the construction of these impoundments was essential for the development of water storage and flood control, and allowed the Upper Basin States to develop their water resources, their construction and operation altered natural river ecosystems and the native floral and faunal communities (Poff and Zimmerman 2010; Mueller and Marsh 2002; Collier et al. 1996). As a result, natural riverine habitats were altered, migration routes were blocked, and selective chemical treatments were applied to eradicate native species in favor of nonnative sport fish species. The alteration of the rivers in the Colorado River system ultimately led to drastic declines in several fish endemic to this system. Eventually, various programs were developed to work towards the recovery of the native fish, especially those listed as endangered under the Endangered Species Act of 1973 (P.L. 93-205). These programs (from downstream to upstream) include the Lower Colorado River Multi-Species Conservation Program (http://www.lcrmscp.gov/), the Glen Canyon Dam Adaptive Management Program (http://www.usbr.gov/uc/rm/amp/), the San Juan River Basin Recovery Implementation Program (SJRIP; http://www.fws.gov/southwest/sjrip/), and the Upper Colorado River Recovery Implementation Program (http://www.fws.gov/mountain-prairie/crrip/index.htm).

The San Juan River historically provided habitat for Colorado pikeminnow and razorback sucker; however, fishery surveys indicated that both species were essentially extirpated by 1992 (see Holden 2000). Colorado pikeminnow were listed as endangered in 1967 (U.S. Fish and Wildlife Service 2002a) and razorback sucker were listed in 1991 (U. S. Fish and Wildlife Service 2002b). In addition, in 2009 New

Mexico listed the roundtail chub (*Gila robusta*) as endangered; Colorado classified the flannelmouth sucker, (*Catostomus latipinnis*), bluehead sucker (*Catostomus discobolus*), and roundtail chub as species of special concern; and Utah listed the roundtail chub and Colorado River cutthroat trout (*Oncorhynchus clarkipleuriticus*) as sensitive species.

### San Juan River Basin Recovery Implementation Program

In 1992 a Cooperative Agreement to establish a long-term program to recover the endangered species in the San Juan River was signed by the states of Colorado and New Mexico, along with the USFWS, Reclamation, Bureau of Indian Affairs, Bureau of Land Management, and the Navajo Nation, Jicarilla Apache Nation, Southern Ute Indian Tribe, and Ute Mountain Ute Indian Tribe. The passage of P.L. 106-392 in 2001 provided long-term funding support from Power Revenues, Congressional appropriations, and the states of Colorado and New Mexico to support the recovery of endangered Colorado pikeminnow and razorback sucker under the formation of the SJRIP, although it was anticipated that actions taken under this Program would provide benefits to other native fishes in the Basin and prevent them from becoming endangered in the future.

The specific goals of the SJRIP are:
• To conserve populations of the Colorado pikeminnow and razorback sucker in the Basin consistent with recovery goals established under the Endangered Species Act,
• To proceed with water development (Table 13.1) in the Basin while maintaining compliance with Federal and State laws, interstate compacts, Supreme Court decrees, and Federal Trust responsibilities to the Southern Utes, Ute Mountain Utes, Jicarillas, and the Navajos.

Downlisting and delisting of Colorado pikeminnow and razorback sucker can occur throughout the Colorado Basin if several demographic parameters are met. A population of Colorado pikeminnow in the San Juan River can contribute to delisting if over a seven-year period beyond downlisting the population is self-sustaining and exceeds 800 adults such that the trend in adult point-count estimates does not decline significantly and the mean estimated recruitment of age-6 naturally produced fish equals or exceeds mean annual adult morality. The population of razorback sucker in the San Juan River can contribute toward downlisting the species if the point estimate for adults exceeds 5,800 adults, the trend in adult point estimates does not decline significantly, and the mean estimated recruitment of age-3 naturally produced fish equals or exceeds mean annual adult mortality. Certain site-specific management tasks for each species must also be identified, developed, and implemented to minimize or remove threats to each species.

### Long-Range Plan

To guide recovery efforts in the San Juan River Basin a Long-Range Plan (LRP; http://www.fws.gov/southwest/sjrip/pdf/ DOC_Long_Range_Plan_2009.pdf) has been developed. The LRP is the Program's research, monitoring and implementation document. Using the research information provided from past studies and Program evaluation reports, the LRP outlines a multi-year approach for guiding the research and monitoring programs and recovery actions necessary to achieve the Program's goals. The LRP details the progression and priority of implementing recovery actions within the San Juan River Basin that are expected to result in recovery of the San Juan River populations of Colorado pikeminnow and razorback sucker and contribute to recovery and delisting of both species throughout the Colorado River Basin. As these actions are completed, they

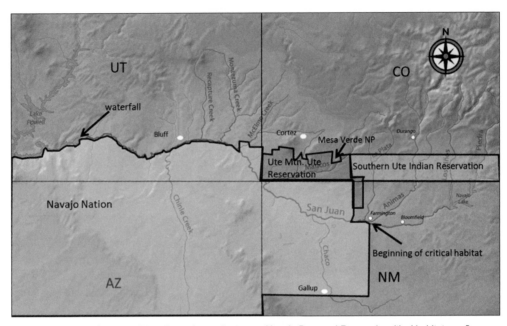

Figure 13.1  Map of San Juan River illustrating major towns, Navajo Dam and Reservoir, critical habitat, confluence of Animas River, and waterfall near Lake Powell.

| State | Number of Consultations | Acre-feet/Year |
|---|---|---|
| New Mexico | 18 | 617,216 |
| Colorado | 94 | 217,456 |
| Utah | 12 | 9,140 |
| **Total** | **124** | **843,812** |

Table 13.1  Number of consultations and acre-feet/year of water currently covered by the SJRIP in accord with the goal of continued water development while recovering endangered fish species.

constitute milestones marking progress of the Program toward achieving the goal of recovery of the endangered fish species. The LRP is used as a basis for scheduling, budgeting and implementing program research, monitoring, capital projects, and other recovery action activities. Activities within the LRP are segregated into elements: protection, management and augmentation of habitat (Bliesner et al. 2008); water quality protection and enhancement (Abell 1994); managing interactions between native and non-native fish species (Davis and Furr 2008; Elverud 2008); monitoring and data management (Ryden 2008a); protection of genetic integrity and management and augmentation of populations (Ryden 2008b, c); information and education; and Program management. All documents related to the SJRIP can be found at http://www.fws.gov/southwest/sjrip/index.cfm.

## Adaptive Management

The SJRIP uses adaptive management (Walters 1986; Williams et al. 2007) to guide research, monitoring (Nichols and Williams 2006), and management actions. Adaptive management involves the recognition of uncertainty in regards to ecological processes and uses conceptual models, development of management actions that are hypothesized to affect fish populations, developing integrated research and monitoring projects to detect effects of management activities, and finally, the use of results, conclusions and recommendations from research and monitoring activities to update conceptual models, modify management actions, and propose new experimental actions. The development of a population model in the SJRIP was critical in the development of stocking goals to guide rebuilding populations of the endangered fish (Holden 2000). Currently, the annual goals for the SJRIP are to stock 11,400 razorback sucker $\geq$ 300 mm total length (TL), and >300,000 age-0 (50mm) and 3,000 age-2 (300mm

TL) Colorado pikeminnow. The stocking program for the endangered fish is a good example of where adaptive management is being used to increase knowledge and improve management for the fish. Recent information on survival, movements, and growth of the endangered species is now being integrated with the population model to make changes in fish stocking. For example, analyses by Bestgen et al. (2009) and Zelasko et al. (2010) have shown that razorback suckers had higher survival if they were stocked in the river at a larger size (> 300mm) and if they were stocked in fall or spring and not winter, although some of the results were confounded by weak sample sizes and unusually high flows during one particularly large stocking event that may have led to those fish being flushed downstream to Lake Powell. Using the findings and recommendations from these studies, as well as additional information from river-wide monitoring, the SJRIP conducted a series of workshops in 2009 to develop an experimental stocking plan for larger (> 400 mm) razorback sucker that would be stocked during two different seasons (fall and winter) and at two different locations in the river. This experimental stocking is taking place during 2010-2011.

Monitoring efforts also have shown that razorback sucker are surviving and reproducing in the river (Ryden 2008a), but recruitment of wild-spawned larval fish has not yet been documented (Brandenburg and Farrington 2008), suggesting that a large number of fish may be disappearing past the waterfall and using the river and inflow area in Lake Powell. To address this finding a new project to look for razorback sucker in the lower river below the waterfall and at the inflow area to Lake Powell is scheduled to begin in spring 2011. A similar project has been ongoing in Lake Mead at several inflow areas, including the Colorado, Virgin, and Muddy Rivers, and has led to the discovery of several large groups of razorback suckers that are spawning and recruiting in Lake Mead (Albrecht et al. 2011).

Colorado pikeminnow are also surviving in the river but few adults have been captured and survival of younger age classes is now thought to be lower than originally hypothesized in the conceptual model. Durst (2009) conducted a cost: benefit analysis to examine return rates of the two sizes of Colorado pikeminnow that the Program stocks. The original conceptual model predicted that larger (>200 mm) Colorado pikeminnow would have higher survival and be more cost effective for the Program to stock, even though rearing costs for the larger fish were substantially greater. Durst's (2009) analysis showed that stocking the smaller fish was far more cost effective since the larger number of small fish (~100 mm) that were reared for the same costs of the smaller number of large fish led to a greater number of fish captured through monitoring efforts in the river. The Program is now in the process of updating the model and adjusting the stocking plan by eliminating the stocking of larger Colorado pikeminnow to incorporate these new findings.

One factor that has both assisted and hampered the recovery of the two fish species is the presence of a10m waterfall at the bottom of the river where it meets Lake Powell. The waterfall was formed when the water elevation in Lake Powell declined due to drought and the river moved into a new channel consisting of reservoir sediments and cliffs. The river now runs over a cliff before dropping into a channel and flowing to the lake. While the waterfall forms a barrier to upstream movement of nonnative fish, it also serves to truncate habitat and forms a sink for native (endangered and non-endangered) larval fish and prevents the movement of adult endangered fish back into the river that have passed over the waterfall.

An integrated fish monitoring and nonnative fish-removal program currently provides the majority of information for both the native and nonnative fish community (Davis and Furr 2008; Elverud 2008; Ryden 2008a). Using information collected during monitoring and nonnative fish-removal efforts the Program adjusts stocking plans, nonnative removal efforts, and the construction and operation of capital facilities. A working hypothesis that the Program is currently testing is that reduction of nonnative catfish and common carp to levels ~ 20% of pre-reduction levels is hypothesized to reduce predation and competition on the native fish and provide conditions favorable for recovery (Haines and Moode 2007; Davis and Furr 2008; Elverud 2008). Common carp have been reduced to ~ 20% of pre-removal levels based on catch/effort statistics during the last four years and there are indications that adult channel catfish are also declining (Davis and Furr 2008; Elverud 2008). However, while spawning and reproduction of the two endangered fish have been documented for several years, to date the monitoring program has detected few fish of either species recruiting to the population (Ryden 2008). Only in recent years have stocking goals been met, but the lack of recruitment is cause for concern and questions are now being asked in regards to whether the San Juan River provides enough habitat above the waterfall to recover these two fish.

Hypotheses have also been developed related to specific habitat types that are believed to be important to early life-stages of Colorado pikeminnow and razorback sucker. Due to reduced peak spring flows and the introduction of non-native vegetation (i.e., salt cedar [*Tamarix spp.*] and Russian olive [*Elaeagnus angustifolia*])that acts to stabilize the river channel, the San Juan River has had a reduction in complex habitat types such as backwaters, embayments, secondary channels, and low-velocity habitats (Bliesner and Lamarra 2006). A pilot project is now planned for winter 2011-2012 to experimentally renovate several backwaters and secondary channels to improve retention of early life stages of the fish (Stamp et al. 2006). Habitat creation and alteration is expensive and a test of this hypothesis is necessary before a large-scale habitat project is undertaken.

## Peer Review

A critical component of the SJRIP is the use of a standing peer-review panel comprised of experts in fish and habitat ecology. The peer-reviewers assist the participants in making scientifically defensible decisions to guide management and research actions. The standing peer-review panel is often supplemented using other reviewers with specific expertise in certain topic areas. The peer-reviewers help resolve questions about methodology and interpretation, and they assist the Program managers in making decisions about where to direct critical resources. The peer-reviewers develop an annual report to the Program that outlines their concerns and recommendations for future actions. Another beneficial activity within the SJRIP has been the programming of workshops to examine in finer detail the methodology and results of specific projects and, subsequently, what those projects are providing in terms of information and progress toward recovery of the two fish. Peer reviewers play a critical role in these workshops by assisting Program participants with evaluating ongoing and potential projects for their contribution toward the Program's goals and objectives.

## CONCLUSIONS

The SJRIP is a nationally recognized program coordinated by the U.S. Fish and Wildlife Service that uses adaptive management to assist in the recovery of two endangered fish species. The Program is guided by an overall Program Document that defines the goals and objectives of the Program, and the responsibilities of the participants. The LRP details the actions considered necessary to reach those goals

and objectives and is used to guide the development of annual work plans. The Program is overseen by a Coordination Committee that makes final decisions based on recommendations from the Biology and Hydrology Committees. These decisions are then implemented by the many partners in the Program. Active adaptive management has led to some important questions being answered about the potential for recovery of Colorado pikeminnow and razorback sucker in the San Juan River Basin.

## LITERATURE CITED

Abell, R. 1994. San Juan River Basin water quality and contaminants review: volumes I and II. University of New Mexico, Museum of Southwestern Biology, Albuquerque, NM. http://www.fws.gov/southwest/sjrip/pdf/DOC_waterqualityreviewvol1.pdf

Albrecht, B. A., P. B. Holden, R. B. Kegerries, and M. E. Golden. 2011. Razorback sucker recruitment in Lake Mead, Nevada-Arizona, why here? Lake and Reservoir Management 26:336-344.

Bestgen, K. R., K. A. Zelasko, and G. C. White. 2009. Survival of hatchery-reared razorback suckers (Xyrauchen Texanus) stocked in the San Juan River Basin, New Mexico, Colorado, and Utah. Final report prepared by Larval Fish Laboratory, Department of Fish, Wildlife, and Conservation Biology, Colorado State University, Fort Collins, Colorado 80523

Bliesner, R., E. De La Doz, and V. Lamarra. 2008. Hydrology, geomorphology, and habitat studies : 2007 annual report to San Juan River Basin Recovery Implementation Program. http://www.fws.gov/southwest/sjrip/pdf/DOC_Hydrology_geomorphology_habitat_studies_2007.pdf

Bliesner, R. and V. Lamarra. 2006. Hydrology, geomorphology and habitat studies, 2005 Final Report. Prepared by Keller-Bliesner Engineering, LLC, for the San Juan River Basin Recovery Implementation Program, U.S. Fish and Wildlife Service, Albuquerque, NM.

Brandenburg, H. and M. Farrington. 2008. Colorado pikeminnow and razorback sucker larval fish survey in the San Juan River during 2007. American Southwest Ichthyological Researchers L.L.C., Albuquerque, NM. http://www.fws.gov/southwest/sjrip/pdf/DOC_Colorado_pikeminnow_razorback_sucker_larval_fish_survey_San_Juan_River_2007.pdf

Collier, M., R. H. Webb, and J. C. Schmidt. 1996. Dams and rivers: A primer on the downstream effects of dams. U.S. Geological Survey, Circular 1126, Tucson, AZ. 94 pp.

Davis, J. E., and D. W. Furr. 2008. Non-native species monitoring and control in the upper San Juan River, New Mexico, 2007. U.S. Fish and Wildlife Service, Albuquerque, NM. http://www.fws.gov/southwest/sjrip/pdf/DOC_Non-native_species_monitoring_control_upper_San_Juan_River_2007.pdf

Durst, S.L. 2009. Evaluation of age-0 versus age-1+ Colorado pikeminnow stocking. White paper report to the San Juan River Basin Recovery Implementation Program Biology Committee. San Juan River Basin Recovery Implementation Program, USFWS, Albuquerque, NM.

Elverud, D. 2008. Nonnative control in the lower San Juan River 2007. Utah Division of Wildlife Resources, Moab, UT. http://www.fws.gov/southwest/sjrip/pdf/DOC_Nonnative_control_lower_San_Juan_River_2007.pdf

Haines, G. B., and T. Moode. 2007. A review of smallmouth bass removal in Yampa Canyon, with notes on the simulated effort needed to reduce smallmouth bass in the Green River subbasin. Report prepared for Upper Colorado River Recovery Implementation Program. U.S. Fish and Wildlife Service, Vernal, UT. http://www.coloradoriverrecovery.org/documents-publications/technical-reports/nna/HainesModdeModelingReport.pdf

Holden, P.B. 2000. Program evaluation report for the 7-year research period (1991-1997). San JuanRiver Basin Recovery Implementation Program, USFWS, Albuquerque, NM. http://www.fws.gov/southwest/sjrip/pdf/DOC_ProgramEvaluationReport.pdf

Mueller, G. A., and P. C. Marsh. 2002. Lost, a desert river and its native fishes: a historical perspective of the lower Colorado River. U.S. Geological Survey, Information and Technology Report USGS/BRD/ITR—2002—0010. Denver, CO. 69 pp.

Nichols, J.D., and B.K. Williams. 2006. Monitoring for conservation. Trends in Ecology and Evolution 21:668-673.

Poff, N. L., and J. K. H. Zimmerman. 2010. Ecological responses to altered flow regimes: a literature review to inform the science and management of environmental flows. Freshwater Biology 55:194-205.

Ryden, D. 2008a. Long term monitoring of subadult and adult large-bodied fishes in the San Juan River: 2007. U.S. Fish and Wildlife Service, Grand Junction, CO. http://www.fws.gov/southwest/sjrip/pdf/DOC_Long_term_monitoring_sub-adult_adult_large-bodied_fishes_San_Juan_River_2007.pdf

Ryden, D. 2008b. Augmentation of the San Juan River razorback sucker population: 2007. U.S. Fish and Wildlife Service, Grand Junction,CO. http://www.fws.gov/southwest/sjrip/pdf/DOC_Augmentation_San_Juan_River_razorback_sucker_population_2007.pdf

Ryden, D. 2008c. Augmentation of the San Juan River Colorado pikeminnow population: 2007. U.S. Fish and Wildlife Service, Grand Junction, CO. http://www.fws.gov/southwest/sjrip/pdf/DOC_Augmentation_Colorado_pikeminnow_San_Juan_River_2007.pdf

Stamp, M., J. Grams, M. Golden, D. Olsen, and T. Allred. 2006. Feasibility evaluation of restoration options to improve habitat for young Colorado pikeminnow on the San Juan River. Grant Agreement No. 04-FG-40-2158 with the Bureau of

Reclamation, Prepared by BIO WEST, Inc. for the San Juan Recovery Implementation Program, U.S. Fish and Wildlife Service, Albuquerque, NM.

U.S. Fish and Wildlife Service. 1994. Determination of critical habitat for the Colorado River endangered fishes: razorback sucker, Colorado squawfish, humpback chub, and bonytail chub. Dept. of the Interior, U. S. Fish and Wildlife Service, Federal Register, 21 March 1994, 59:13374-13400.

U.S. Fish and Wildlife Service. 2002a. Colorado pikeminnow (*Ptychocheilus lucius*) Recovery Goals: amendment and supplement to the Colorado Squawfish Recovery Plan. U.S. Fish and Wildlife Service, Mountain-Prairie Region (6), Denver, CO. 71 pp.

U.S. Fish and Wildlife Service. 2002b. Razorback Sucker (*Xyrauchen texanus*) Recovery Goals: amendment and supplement to the Razorback Sucker Recovery Plan. U. S. Fish and Wildlife Service, Denver, CO. 78 pp.

Walters, C.J. 1986. Adaptive Management of Renewable Resources. Blackburn Press, Caldwell, NJ.

Williams, B. K., R. C. Szaro, and C. D. Shapiro. 2007. Adaptive management: The U.S. Department of the Interior technical guide. US Department of the Interior, Washington, DC.http://www.doi. gov/initiatives/AdaptiveManagement/ TechGuide.pdf

Zelasko, K.A., K.R. Bestgen, and G.C. White. 2010. Survival rates and movement of hatchery-reared razorback suckers in the Upper Colorado River Basin, Utah and Colorado. Transactions of the American Fisheries Society 139: 1478-1499.

# TOOLS FOR CONSERVATION AND COLLABORATIVE DECISION MAKING

# SPATIO-TEMPORAL MULTI-OBJECTIVE DECISION MAKING IN FOREST MANAGEMENT

*Boris Poff, Aregai Tecle, Daniel G. Neary and Brian W. Geils*

## ABSTRACT

Forest ecosystem management is the art and science of making decisions that involve numerous cultural, social, economic and environmental components interacting with one another. With many stakeholders involved, it is important that such an inherently multi-objective problem be evaluated using scientifically-based methods that are transparent, adaptable and inclusive for all interested parties. The Multi-Objective Decision Making (MODM) process is one such method that was developed in the 1940s but only has been applied to natural resource management since the late 1990s. We demonstrate the appropriateness and effectiveness of MODM for forest management by creating and comparing future scenarios incorporating values and priorities from an interdisciplinary management team managing for wildfire, silviculture, wildlife and invasive species. One benefit of MODM is the ability to model forest change with time and space, which is critical since forest ecosystem management analyses are needed at varying spatial and temporal scales. To do this, a conceptual model intended for forest managers and other decision-makers is constructed using one MODM technique, numerous mathematical response functions and two modeling programs – one spatial and one dynamic. Compromise Programming (CP) is the MODM technique utilized in this ecosystems modeling effort. Twenty-two mathematical response functions that represent major forest management objectives are expressed in terms of one decision variable, forest stand density. An individual-tree growth model (FVS), used by land resources managers to develop silvicultural and land management plans, is used as the dynamic modeling component, while a Geographic Information System (ArcGIS) facilitates the spatial aspect of the modeling effort. With the forest planning software available, it is becoming easy to solve a complex forest management problem in a spatio-temporal MODM framework. In this paper the authors demonstrate how forest managers can identify numerous feasible forest management alternatives in terms of overstory vegetation density management, as well as to what extent the identified management actions achieve the selected objectives over time and space.

## INTRODUCTION

The southwestern ponderosa pine (*Pinus ponderosa var. scopulorum*) forest ecosystem had been relatively free of large-scale disturbances for about hundred years until the end of the last century (Swetnam 1990; Swetnam and Baisan 1996). However, recent extremes in weather combined with forest conditions related to a century of fire exclusion can potentially lead to regime shifts that might be irreversible (Folke et al. 2004). A forest ecosystem is a unique combination of biological and physical structures with numerous cultural, social, economic and environmental components that interact with one another, and managing these forest resources is complex, even without attempting to capture the influence of extreme events like wildfire. The management of such a system involves many individuals or groups of stakeholders with different, often conflicting objectives (Bare and Mendoza 1988). Furthermore, management objectives related to forest resources do change over time and space and can be expressed either quantitatively or qualitatively (Kennedy and Koch 2004; Thomas 2006). Because management objectives apply to forest ecosystems that change with time and across landscapes (Buongiorna and Gilles 2003), the most appropriate management is adaptive, dynamic, spatially varied and multi-objective in nature. Land managers and decision-makers therefore require suitable models that can capture this complexity at the landscape scale and in the face of major uncertainties on how ecosystems respond to increasing human use (Steffen et al. 2004) and other disturbances (Paine et al. 1998; Jackson et al. 2001).

In addition to scientifically sound models that can capture complexity, land managers need models that are transparent and legally defensible. Public distrust in the bureaucratic decision-making processes of the Federal government has lead to the development of a number of laws, including the 1970 Clean Water Act, the 1973 Endangered Species Act, the 1976 Federal Land Policy and Management Act, and the Government Performance and Results Act of 1993 (Dowdle 2006). These legislative acts mandate public input and are aimed at limiting the decision-making discretion of federal land management agencies, such as the USDA Forest Service (USFS). In these laws individuals, businesses and environmental NGOs have legal recourse in any land-management process. When management decisions lack transparency and are based exclusively on expert opinion, they can be difficult to defend in court. Because the USFS manages public lands, its management decisions are often subject to

lawsuits (Keele et al. 2006). Such lawsuits have prevented the implementation of fuel reduction treatments in forests, which as a consequence have experienced catastrophic wildfires, especially in Arizona (Ostergren et al. 2006).

With the number of catastrophic wildfires increasing in the western US, however, the public is looking for ways that are less litigious and technical to engage the USFS in their decision-making processes. According to a 2003 survey conducted in north-central Arizona, Ostergen et al. (2006) found that the public, in general, has the desire for the USFS to continue to manage national forests for values other than commercial timber, especially as it restores the forest to its original more open landscape. Many are expecting the USFS to play a more active role in communicating with and involving the public. The MODM modeling effort presented in this paper can help the USFS to make clear, inclusive, objective, defensible and transparent forest management decisions.

Models are abstract representations of the real world and any model that describes the variation of one or more phenomena over the Earth's surface is a spatial model (Goodchild 2005). If modeling goals include describing past behavior or predicting future outcomes of management policies or actions on entire landscapes, then spatial modeling becomes essential (Risser et al. 1984; Costanza et al. 1990; Sklar and Constanza 1991; Constanza and Voinov 2003; Maxwell and Voinov 2005). Modeling ecosystems at a landscape scale is a complex process, which can be simplified using state-of-the-art modeling software. Advances in computer hardware and software technology over the past decade have allowed the integration of comprehensive land resources management and spatial analysis modeling on a single platform to visually interpret joint resource management outcomes. However, many current models are often informal, intuitive and supported by the experience of people running them. What is needed in today's forest ecosystem management is a transparent method that addresses multiple objectives and alternative management actions that vary across spatial and temporal scales, because decision makers and stakeholders are unlikely to trust a model that they don't understand (Maxwell and Voinov 2005).

It is important to include a time component in the analysis of environmental problems, where pattern changes respond to external forces, such as land cover changes. It is also essential to choose the correct spatial resolution, to ensure that variations, which occur within the dimensions of a cell or polygon, will be registered by either the data or the process. A well-known example of a landscape-scale forest ecosystem model that addressed issues of spatial scale is the Forest Ecosystem Management Assessment Team's (FEMAT) Management Plan for Old-Growth Ecosystems within the range of the Northern Spotted Owl in the Pacific Northwest of the United States (Vogt et al. 1997). FEMAT (1993) suggested four spatial scales, namely regional, physiographic province, watershed and specific site, other authors disagree. Körner (1993), Levin (1993), and Reynolds et al. (1993) have suggested that the primary consideration of an ecosystem management should be the identification of scale at which manageable ecological processes occur. Because units of forest ecosystem scales do not have functional boundaries, physical boundaries can be delineated based on management objectives (Vogt et al. 1997). Temporal changes in attributes of cell values can be computed for single cells or for neighboring cells of varying size and shape. Once a model is set up and run, the changes can be visualized as a film, providing extra understanding of the processes being modeled (Burrough et al. 2005). Here a spatio-temporal MODM model becomes a useful tool that can synthesize the challenges land managers and decision-makers face and

to allow them to address these challenges in a dynamic and adaptive framework at the manageable landscape scale.

## The Southwestern Ponderosa Pine Forest Ecosystem

The southwestern ponderosa pine forest ecosystem is one of the largest continuous stands of pine forest in the U.S., stretching from southern Utah through northern and central Arizona to south central New Mexico. This area constitutes about half of the 6.5 million ha of ponderosa pine forest ecosystem in the Rocky Mountains (USFS 1989). Over 65% of this forest ecosystem is under National Forest ownership (Conner et al. 1990), while the rest is in private, tribal, state and National Park ownership. The distribution of ponderosa pine is mostly affected by climatic factors, such as precipitation and temperature gradients, as well as fire regimes. However, anthropogenic influences such as grazing and fire-suppression have also some influence on tree distribution and density (Covington et al. 1997). While some wildlife species require high density to provide nesting, bedding, hiding and escape cover (Wagner et al. 2000) this same dense vegetation presents a management problem for reducing the risk of wildfires, and the spread of tree diseases and pathogens. Under these conflicting situations managers are often asked to achieve specific societal objectives while being required to meet ecological and economic constraints.

## The Upper Beaver Creek Watershed Fuel Reduction Project

An example of a USFS project in north-central Arizona that could benefit from more public involvement is the Upper Beaver Creek (UBC) Watershed Fuel Reduction Project.

The primary purpose of the UBC project is to reduce the risk of stand-replacing wildfire that would threaten people, private property and natural resource values. The

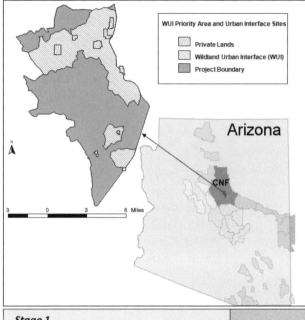

## Upper Beaver Creek Watershed Fuel Reduction Project

Figure 14.1  Location of the Upper Beaver Creek Watershed Fuel Reduction Project area in the Coconino National Forest (CNF) in north-central Arizona.

Private lands and WUI are indicated within the project area. Other National Forests in Arizona are also shown.

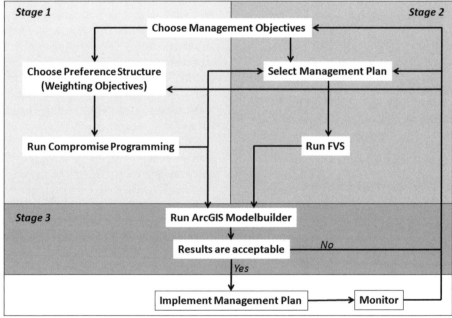

Figure 14.2    This flow chart illustrates the three stages of the modeling effort. In the first stage the MODM technique is used to determine a preferred management plan on the basis of management objectives and the sets of preference structures. In the second stage FVS is run based on the management objective values and the selected management plan. In the final stage ArcGIS Modelbuilder is run using the FVS and CP outputs as inputs. The results can then be evaluated, and if acceptable, be implemented and monitored. If the solution is not acceptable by the DM(s), then the management objectives, preference structure and/ or the selected management plan may be modified after either of these steps as part of the greater scheme of adaptive management.

USFS management team responsible for achieving these goals believes that the most effective way to do so is to begin restoring this fire prone ecosystem to pre-European settlement conditions (USFS 2006).

The intent of the UBC project is to show changes from the existing undesirable conditions to some desired future conditions and demonstrate improvement in the overall forest fire regime and overall forest conditions (USFS 2006). Past research has shown that forest thinning to reduce fuel loads is less likely to result in stand-replacing crown fires (Fulé et al. 2001; Omi and Martinson 2002). Hence, the proposed actions of the USFS consist of a variety of vegetation management, mechanical fuel reduction, and prescribed burning activities over the next 20 years to reduce the potential for stand-replacement wildfire and to begin restoring the forest to a fire-adapted ecosystem, in which low-intensity ground fires may occur in 2-10 year intervals. A detailed description of the proposed actions is given in the "Upper Beaver Creek Watershed Fuel Reduction Project: Proposed Action Report" (USFS 2006). The proposed actions are designed to comply with Forest Plan standards and guidelines, as amended (USFS 1987).

The desired condition for the UBC Watershed Fuel Reduction Project area is defined by the Coconino National Forest Plan (USFS 1987) as a landscape where fire conditions are moving toward achieving the desired fire regime that is appropriate for the vegetation type within the ecosystem. This is a relatively vague description and leaves the definition of "the desired condition for the ponderosa pine type" up to USFS interdisciplinary (ID) team interpretation. While the professional experience of the ID team, which consists of experts from various disciplines, can be extensive, their value judgments might lack objectivity and transparency, potentially leaving other groups and stakeholders in the project area to question the specifics of the USFS ID team's decisions. While most members of Wildland Urban Interface (WUI) communities in Arizona understand the benefits of fuel reduction treatments, there is a high influx of people from other communities who are not as familiar with the forest ecosystem in north-central Arizona (Ostergren et al. 2006). This group of people would benefit from a more transparent decision-making process by the USFS. Even long-time residents, who previously distrusted USFS decisions, would be more likely to accept USFS decision if the decision making process was more transparent (Ostergren et al. 2006). A modeling effort, as the one suggested in this paper, is objective, transparent and therefore defensible in legal and social settings. The MODM technique in itself is objective, because it converts the management objectives, both quantitative and qualitative, into empirically and scientifically developed mathematical response functions. Disclosing the decision makers' preference structure in the form of weights assigned to the various management objectives in a given project area achieves transparency and allows members of the affected communities to see how the experts arrive at their decisions.

## Terminology

The following section defines several important terms and expressions as used in this paper. An Achievement Level (AL) is an indicator of how well a forest stand meets the most preferred solution given the management objectives, scaled from 0 to 100 percent. The Decision Variable in this paper is tree density level expressed in basal area (BA), with various discrete levels of the decision variable forming management alternatives. Compromise Programming (CP) is a distance-based technique used for evaluating a multiple objective management problem to arrive at a compromise and fair solution (Tecle et al. 1988). A Management Objective is the desired forest management

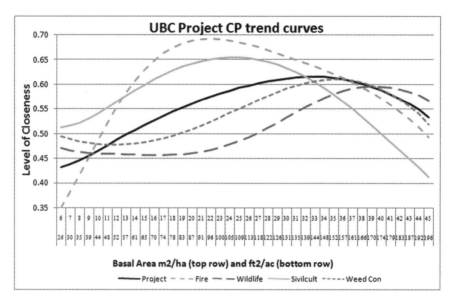

Figure 14.3 The sensitivity of the CP algorithm to changes in weights along the range of the decision variable values. For example, at TBA values of 21 m²/ha (90 ft²/ac) decision makers managing fire consider the conditions most favorable, while those managing wildlife find the conditions least favorable.

direction, such as slowing the spread of Dwarf Mistletoe (DMT) or maintaining or increasing habitat for Mexican Spotted Owl (MSO). Multi-Objective Decision-Making (MODM), is a technique for arriving at a management decision that satisfies multiple objectives, or that satisfies the desires of multiple decision makers (Tecle et al. 1998). Objective Weights are a set of weights given to all management objectives to represent Decision Makers' (DM's) preference structure among the objectives. A Response Function is a mathematical formulation of a management objective or forest ecosystem component response to certain management action(s). In this paper a Simulation represents the outcome of forest vegetation growth as predicted by the Forest Vegetation Simulator (FVS) after thinning and prescribed burn treatments have been applied.

## METHODS

The modeling approach in this study consists of developing objective response functions, using two sets of objective weights assigned by an USFS ID team in a workshop setting, and utilizing three different models: one MODM technique, a vegetation growth model, and GIS modeling software. These approaches were combined to evaluate the objective functions in an integrated manner through time and space. The objective response functions link the objectives and preferences of decision makers to the density of forest stands expressed in tree basal area. The MODM technique used is Compromise Programming (CP), and the vegetation growth model and the GIS software are the Forest Vegetation Simulator (FVS) and ArcGIS, respectively. FVS is an individual-tree growth model used by forest managers to develop land management plans. ArcGIS is a Geographic Information System, used by land managers, other decision makers and analysts for creating, storing, analyzing and managing spatial data and associated attributes. To visualize differences in expected outcomes for different management scenarios, achievement levels for each stand within the project area were tied to GIS and mapped spatially so users can view the projected outcomes over time. The outcome of the modeling effort is expected to assist forest resource managers to arrive at a generally acceptable forest ecosystem management scheme.

The modeling effort demonstrated in this paper provides results at three stages: (1) the MODM stage; (2) the temporal projection stage; and (3) the spatial (GIS) stage (see Figure 14.2).

Even though the user of this model might be ultimately interested in the output at the last stage, this interactive process allows the modelers to make modifications at either of the first two stages: the MODM stage and the temporal projection stage.

## Compromise Programming

Compromise Programming (Poff et al. 2010; Tecle et al. 1988; Tecle and Duckstein 1993; Zeleny 1973, 1982) is based on the concept of distance to arrive at a most satisfying solution to a particular problem. Here, distance is used as a proxy to measure preference, where the preferred solution is the closest non-dominated solution to an infeasible ideal point (Zeleny 1973). A non-dominated solution in a MODM problem is one that does not show any improvement in any one of the objective solutions without making at least one other solution worse, in terms of the given achievement level (Tecle et al. 1988; Tecle and Duckstein 1993). The ideal point represents the joint location of the maximum solutions of all individual objectives optimized separately. Therefore, arriving at a compromise solution can be viewed as minimizing a DM's regret for not obtaining the ideal solution. Weights are assigned by DMs and/or stakeholders

to management objectives to show their ranked, relative importance.

The general formulation of a CP technique is expressed as follows;

$$\min\left\{l_p=\left[\sum_{i=1}^{I} W_i^p\left(Z_i^*-Z_{ij}\right)^p\right]^{1/p}\right\}, j=1,...,J$$

where $l_p$ is the distance metric, for any $p$ in which $0 < p < \infty$. The equation is the measure of a solution's closeness to the ideal point $Z^*$, the set of all the maximum values of all objective functions. $Z_{ij}$ is the value of objective $i$ under a specific discrete value of decision variable $j$. $I$ is the number of objectives within categories and may range from one to six. $J$ is the number of discrete decision variable values. $Z_i^*$ is the maximum value for objective $i$. The weight $W_i$ in equation [1] signifies the importance of objective $i$ relative to the other objectives. The $p$ is the metric parameter. Different values of $p$ represent different aspects of a compromise programming algorithm. For $p = 1$, all deviations from $Z_i^*$ are directly proportional to their magnitude. For $1 < p <$ infinity, the largest deviation has the greatest influence. Varying $p$ from 1 to infinity allows to move from having a perfect compensation among the objectives to having no compensation among the objectives in the decision making process, respectively. The greater the conflict between different DMs is, the smaller the possible compensation (Zeleny 1974, 1982; Goicoechea et al. 1982; Szidarovszky et al. 1986).

To avoid scale effects and to make all objective function values commensurable, the objective functions are normalized by dividing the right hand side by the expression $Z_i^* - Z_i^{**}$, where $Z_i^{**}$ is the worst value of objective $i$. The normalized objective functions are expressed in the following manner:

$$Z_{ij}=(Z_{ij}-Z_i^*)(\underline{Z}_i^*-Z_i^{**}) i=1,...,I \text{ and } j=1,...,J$$

where the $Z_{ij}$ on the left hand side of the equation represents the normalized elements of the original payoff matrix $Z_{ij}$ on the right hand side of the equation. This normalization process guarantees the $Z_{ij}$ on the left hand side of the equation to have values between 0 and 1.

## Objective Response Functions

An objective response function represents the quantitative expression of a desired management direction for a forest ecosystem component, for example the expression for the cost of management to be minimized or any societal requirement of the forest system that needs to be optimized. The specific objective functions considered in this study are described in Table 14.1.

They are all expressed in terms of one decision variable, tree stand density in tree basal area (TBA in m²/ha and ft²/ac). The process by which individual response functions are selected and created is mostly dependent on the availability of forest management response data on the decision variable and can be a part of the MODM process. The response functions used here were created a priori as parts of the various management objectives and, therefore, are considered external to the MODM process.

Forest stands in north-central Arizona are defined by past management actions, mostly consisting of early logging. The logging practices typical for the area, fire suppression and the episodic nature of obtaining successful regeneration have contributed to the heterogeneous age distribution and varying class sizes of trees in these forests. The decision variable, tree stand density has TBA values that range from 6 to 45 m²/ha (26-196 ft²/ac). A TBA of 6 m²/ha is the minimum acceptable density level for the forest in the study area to qualify as "a forest" according to the United Nations Food and Agriculture Organization (FAO) guidelines (FAO 2006). The average upper limit of the majority of the data available for which response functions have been created is 45 m²/ha (Poff 2002). Using TBA as the decision variable has several advantages: (1)

availability of data on the variable related to many management objectives or ecosystem functions for constructing response functions (Brown et al. 1974); (2) many forest land managing agencies use TBA to quantify management activities (Edminster et al. 1991); (3) TBA is one of the major decision variables used in FVS (Dixon 2002), which is also utilized in this modeling effort. The disadvantage is that TBA as an expression of forest density does not make any distinction between the conditions of some forest characteristics such as tree size, clumpiness or age distribution. However, other forest density characteristics, such as trees per hectare or percent of canopy closure can easily be converted to TBA in the southwestern ponderosa pine forest (Severson and Medina 1983; Rupp 1995).

The objective response function values are normalized to avoid scale effects and to make all objective function values commensurable. To determine the most preferred forest management alternative in terms of achieving the desired objectives, the twenty-two objective response functions are grouped into nine objective categories on the basis of their similarity to minimize the number of related objectives having undue influence on the decision-making process. For example, the objectives expressing risk of flooding, water yield and water quality are combined into a desired hydrologic condition category (see Table 14.1). Compromise programming is adapted here to perform a sequential two-level evaluation of vegetation management actions based on TBA. The first level involves trade-off analysis within each of the nine objective categories consisting of more than one management objective. Then, a second-stage evaluation is performed to produce a trade-off analysis among the nine categories. A feature of CP is that it allows the analyst to select a distance parameter p as described above. The p-values of 2 and 1 were used in the first and second level of the CP, respectively. A more detailed

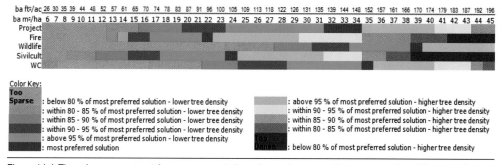

Figure 14.4 The columns represent the management alternatives, whereas each row represents a different preference structure. The most preferred solutions - in terms of management alternatives - for a given weighting scheme are colored in dark green.

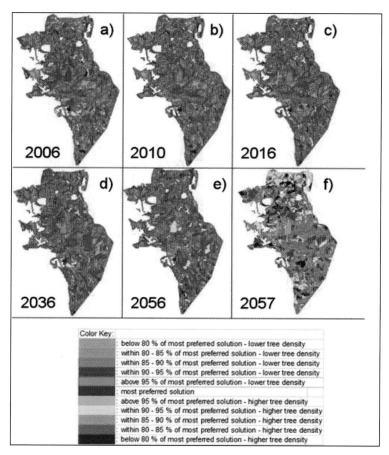

Figure 14.5    Shows how well the individual forest stands in the project area are meeting all management objectives simultaneously given the project's preference structure. a) Indicates the stands performance in 2006: The year in which the first treatment was simulated. b) – e) Indicate simulated performance at different time steps. Each figure shows how well the individual stands are meeting all management objectives simultaneously during a particular time step/project year. f) Represents the "no action alternative" and shows how the forest stands would meet all objectives if no treatment had been performed. White areas are either indicative of non-Forest Service lands or non-ponderosa pine dominated forest stands.

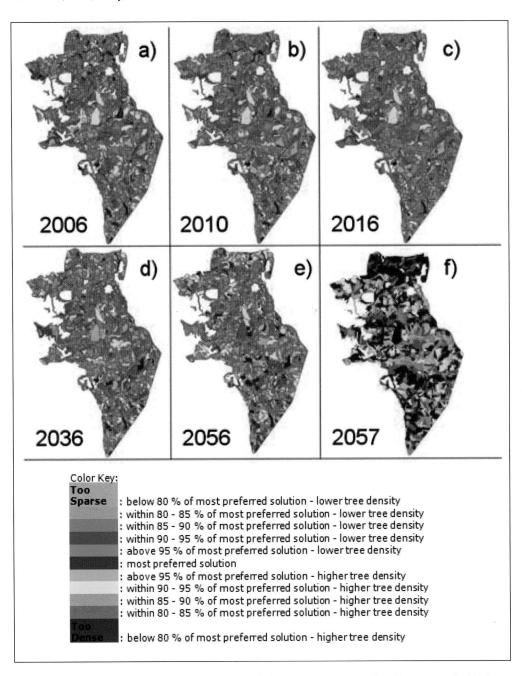

Figure 14.6  Shows how well the individual forest stands in the project area are meeting all management objectives simultaneously given the fire and fuels reduction preference structure. a) Indicates the stands performance in 2006: The year in which the first treatment was simulated. b) – e) Indicate simulated performance at different time steps. Each figure shows how well the individual stands are meeting all management objectives simultaneously during a particular time step/project year. f) Represents the "no action alternative" and shows how the forest stands would meet all objectives if no treatment had been performed. White areas are either indicative of non-Forest Service lands or non-ponderosa pine dominated forest stands.

explanation and description of this CP process and corresponding sensitivity analysis is given in Poff et al. (2010).

## Articulation of the Preference Structure

During a team meeting, the members of the Mogollon Rim Ranger District Interdisciplinary (ID) Team assigned weights to the different management objectives and their criteria as specified in Table 14.1. The ID Team consisted of a Team Leader and one person each from fire management, invasive/weeds control/understory vegetation, silviculture and two wildlife habitat management experts. The members of the ID team were instructed to assign two sets of weights by filling out a survey form listing the management objectives and the corresponding categories. The first set consists of within-category weights (WCW) (see Table 14.1 columns P-WL on the right) for use in the first level of the trade off evaluation process to find a compromise solution for each objective category. If an objective category only had one objective, this step is omitted for that particular category. The second set of weights is assigned to each one of the objective categories (see Table 14.1 columns P-WL on the left). These weights are used in the second level of the trade off analysis procedure to arrive at the most preferred solution, in terms of TBA.

The weights shown in Table 14.1 represent the preference structure of the ID Team. The members of the ID Team were asked to assign weights on the basis of their relative preferences on the given objectives. For example, based on her judgment, the wildlife experts used the weights to indicate the relative importance of each management objectives if the project area was primarily managed for wildlife. The CP analysis was then run for the four remaining preference structures in the same manner as for the first. This allows comparison of the CP results under each management scenario where the project area was primarily managed for fire prevention, wildlife, silviculture or weed control/understory vegetation.

## Forest Vegetation Simulator

In this study the USFS Forest Vegetation Simulator (FVS) was used as the vegetation growth model to evaluate forest conditions with time is for predicting forest stand dynamics. The model summarizes current stand conditions, and simulates future stand conditions under various alternative management scenarios (Dixon 2002). The resulting output is then used as input into forest planning models and other analysis tools such as GIS (McMahan et al. 2002). While FVS serves as the temporal projection tool in this study, other software with similar functions, such as the Tree and Stand Simulator (TASS) (Mitchell 1975) could be used instead. However, FVS was chosen because it is used by the forest land management agencies to simulate growth of southwestern ponderosa pine stands (Dixon 2002). In its most basic function, FVS computes future forest stand density based on a measured stand density and set of variables that influence stand growth projections, including physical variables such as slope and aspect, biological variables such as tree mortality, and management variables such as thinning or prescribed burning treatments. In FVS, stand conditions such as mortality are also influenced by conditions in neighboring stands. The ID Team identified burning and thinning prescriptions in the project area, which were simulated in FVS for the first twenty years. The overall time scale for the simulation is 50 years with two-year time steps for the first ten years and ten-year time steps for the remaining forty years. Further details on the treatment descriptions is found in the "Upper Beaver Creek Watershed Fuel Reduction Project: Proposed Action Report" (USFS 2006).

| Weight | | | | | Objective | Objectives | Criteria | WCW | | | | |
|---|---|---|---|---|---|---|---|---|---|---|---|---|
| **P** | **F** | **S** | **WL** | **WC** | Categories | | | **P** | **F** | **S** | **WL** | **WC** |
| 3 | 3 | 6 | 1 | 2 | Maximize Social Benefits | Asthetic Quality | Visual | 5 | 5 | 5 | 1 | 5 |
| | | | | | | Cultural Resources | Tree Density | 4 | 5 | 10 | 1 | 5 |
| | | | | | | Recreational Use | Willingness to pay | 1 | 1 | 5 | 1 | 1 |
| 9 | 1 | 9 | 5 | 9 | Minimize Insects & Diseases | DMT Infection | DMR | 5 | 1 | 9 | 3 | 6 |
| | | | | | | Pine Beetle | Attacked Trees | 8 | 1 | 9 | 3 | 6 |
| | | | | | | Bark Beetle | Hazard Rating | 8 | 1 | 9 | 3 | 6 |
| 5 | 5 | 8 | 5 | 8 | Min Exotics | Exotic Plants | Plants/ha | N/A | | | | |
| 5 | 1 | 7 | 7 | 7 | Max Forage | Herbage Production | Amount of Herbage | N/A | | | | |
| 7 | 1 | 4 | 1 | 3 | Max Timber | Timber Growth | Timber Yield | N/A | | | | |
| 4 | 5 | 2 | 1 | 4 | Min Costs | Costs | Cost of Thinning | N/A | | | | |
| 10 | 10 | 10 | 5 | 10 | Minimize Fire Hazard & Effects | Forest Fire | Crown Fuel | 9 | 10 | 10 | 5 | 8 |
| | | | | | | | Crown Fire | 10 | 10 | 10 | 5 | 8 |
| | | | | | | | Heat Intensity | 5 | 10 | 10 | 5 | 10 |
| 6 | 3 | 3 | 8 | 5 | Desirable Hydrological Condition | ↓ Flood Hazard | Peak Flow | 3 | 3 | 5 | 10 | 1 |
| | | | | | | ↑ Water Yield | Streamflow | 2 | 1 | 5 | 10 | 4 |
| | | | | | | ↑ Water Quality | Sedimentation | 6 | 5 | 5 | 10 | 5 |
| 8 | 5 | 5 | 10 | 10 | Optimize Wildlife Habitat | Non-Game Species | Abert Squirrel | 7 | 5 | 7 | 8 | 8 |
| | | | | | | Game Species | Deer (spp.) | 6 | 10 | 7 | 8 | 2 |
| | | | | | | | Elk | 1 | 5 | 7 | 1 | 1 |
| | | | | | | Predator Species | Mex. Gray Wolf | 1 | 1 | 7 | 10 | 1 |
| | | | | | | Threat./Endang. | MSO | 8 | 5 | 7 | 10 | 8 |
| | | | | | | USFS Sensitive | N. Goshawk | 7 | 5 | 7 | 10 | 8 |

Table 14.1   Weights range from 1 to 10; where 1 indicates the objective is least important compared to the other objectives and 10 represents the most important objective. The values given in this table represent the preference structure of the UBC ID Team. P = Project, F = Fire, S = Silviculture, WL= Wildlife, WC = Weed Control, WCW = Within-Category Weights.

## Geographic Information System (GIS)

ESRI's ArcGIS was the spatial analysis and display part of this MODM modeling endeavor. The ModelBuilder extension of ArcGIS allows the user to build a model using a diagram that resembles a flowchart (Krivoruchko and Gotway-Crawford 2005; Maidment et al. 2005; Miller et al. 2005). In this feature of ArcGIS, the model consists of a set of spatial processes that convert input data into an output layer. ModelBuilder itself is not dynamic in version 9.1, which was used here. However, temporal components can be simulated by running the model multiple times, using a regular time step (Miller et al. 2005) as used in this study. Using ArcGIS, levels of closeness were assigned to polygons based on their TBA values. This measure of closeness of a CP solution to the ideal point serves as a surrogate for the achievement levels of the management problem. Each layer represents a group of forest stands with a given Achievement Level based on the CP results. Modelbuilder then simulates associations of the achievement level with the decision variable TBA over time. Using the query function in Modelbuilder, eleven layers were created per weighting scheme per time step output in the FVS table. The eleven layers were grouped into one layer, leaving one layer per time-step, which displays the 11 sub-layers with each representing a color coded achievement level of the given weighting scheme on a stand by stand basis. Values that were outside the achievement level range displayed in Figure 14.3 were extrapolated, based on the trend curve shapes in Figure 14.2, to create achievement level layers where necessary.

There are a total of eleven time steps in this simulation process: one every two years from 2006 to 2016; one every ten years from 2016 to 2056; plus the stand conditions in 2057. The 2057 achievement level layer represents the "no-action" alternative, simulating forest growth as if no management action had been taken. The eleven sub-layers, for each time step, are colored coded as shown in Figures 14.4, 14.5 and 14.6.

## RESULTS

Presentation of the results of this modeling effort are two-fold: (1) The MODM technique arrives at a preferred tree basal area (TBA) based on the DMs preference structure and the objective function values; and (2) The FVS forest growth and treatment simulation part provides the TBA on a stand level basis, which is then matched with the CP results and displayed in ArcGIS. This provides a visual representation of how well all management objectives are achieved simultaneously through time for the entire landscape of the project area.

| Weighting Scheme | Preferred Basal Area in m²/ha | Preferred Basal Area in ft²/ac |
|---|---|---|
| Project | 32-34 | 139-148 |
| Fire | 22 | 96 |
| Wildlife | 39 | 170 |
| Silviculture | 24 | 105 |
| Weed/Understory | 35 | 152 |

Table 14.2   Ranges of preferred basal areas with respect to all 22 management objectives under each one of the five weighting schemes.

## Compromise Programming Results

The management alternatives (in the form of stand density expressed in $m^2$/ha and $ft^2$/ac TBA) are the discretized version of the decision variable displayed on the x-axis in Figure 14.3. The y-axis is a standardized level of closeness, which ranges from zero to one where 1 represents an infeasible ideal point. The trend curves in Figure 14.3 display the results of CP analysis for the five different ID Team preference structures. Each preference structure or weighting scheme of an ID team member is represented by one trend curve. The management alternative value under the highest point in each curve shows the most preferred forest stand density level with respect to all management objectives. The results, shown in Table 14.2, indicate the sensitivity of the CP analysis results to weights. Besides giving us the most preferred basal area for each weighting scheme, Figure 14.3 shows that the "Fire" and "Silviculture" preference weighting schemes have higher levels of closeness than the weighting schemes that favor the other ID Team members. They also have more defined curvatures to them indicating a narrower range of acceptable TBA.

The trend curve with the "Wildlife" favoring weights has the lowest level of closeness of all preference structures, mostly in the lower TBA range. This is probably due to the fact that the wildlife species considered by the response functions are forest species which consequently do better in denser forest stands (Patton 1984, 1991; Severson and Medina 1983). However, towards the end of the lower TBA range, the wildlife trend curve begins to come up again indicating the need of some species to have forage at the lower forest density. It is also interesting that the best TBA for the wildlife biologist is considered below 80% of the level of closeness by the silviculturalist and below 85% of the level of closeness by the fire manager. The trend curve with "weeds control/understory" favoring weights has

also some sinuosity to it, indicating the ID team's favor of controlling weeds (or invasive plant species) (high TBA) and desire for improved understory vegetation (low TBA). The trend curve with the more balanced project weights appears to favor stand densities that are good for wildlife habitat. The relatively flat curvature of the project trend curve is indicative of addressing all management objectives with similar importance, while the individual trend curves for the remaining preference structures are more to the point and have a narrow range of the best possible basal area. These results illustrate the CP algorithm's sensitivity to the assigned weights, as was expected.

Figure 14.4 displays the same results in a different manner, allowing for further interpretation. The columns represent the management alternatives (in the form of stand density expressed in basal area) that range from 6 to 45 $m^2$/ha (or 26 to 196 $ft^2$/ac), whereas each row represents the different preference structures articulated by the ID team members. The most preferred vegetation management alternatives – in terms of the decision variable values – for a given weighting scheme are colored in dark green. As the colors shift from dark green to yellow and orange, the stand densities become increasingly too dense to satisfy all management objectives simultaneously. On the other side, as the colors shift from dark green to green, blues and purple, the stand densities decrease steadily and fail to satisfy all management objectives. Each change in color away from the dark green signifies a five percent decrease in the CP achievement level. Red and pink signify achievement levels of less than 80%, either on the increasing or decreasing trends in tree basal area, respectively. The same color coding is used later in the ArcGIS analysis. Figure 14.4 provides data describing the most preferred basal area for each weighting scheme plus some additional

| Criteria | Criterion Scale | 144 ft2/ac 33 m2/ha | 96 ft2/ac 22 m2/ha | Best Value | Worst Value |
|---|---|---|---|---|---|
| Scenic Beauty Index | Ordinal (1-3) | 2.19 | 2.35 | 2.39 | 1.34 |
| Willingness-to-pay[1] | US$/ha | 53.94 | 65.96 | 67.45 | 0.00 |
| Willingness-to-pay[2] | US$/ha | 14.46 | 17.36 | 17.87 | 0.00 |
| Beetle Attacked Trees | % of TBA Killed | 1.86 | 1.41 | 0.72 | 2.19 |
| Bark Beetle Hazard Rating | Ordinal (1-12) | 8 | 6 | 2.31 | 10.81 |
| Dwarf Mistletoe Rating | 10-yr Infection Rate | 1.13 | 0.98 | 0.77 | 1.28 |
| Individual exotic plants | Plants/ha | 7.43 | 10.90 | 3.96 | 25.79 |
| Amount of herbage | kg/ha | 92 | 217 | 1298.64 | 59.65 |
| Timber Yield | m³/ha | 123 | 93 | 143.40 | 29.58 |
| Cost of Tree Removal | US$/ha | 396.36 | 759.96 | 1.35 | 1288.35 |
| Crown Fuel Load | t/ha | 10.85 | 7.11 | 1.67 | 14.93 |
| Heat Intensity | kJ/m² | 137 | 82 | 40.40 | 309.11 |
| Crown Fire | % Crown Burned | 57.91 | 30.52 | 8.57 | 98.38 |
| Sedimentation (H₂O Quality) | t/ha/yr | 3.60 | 7.34 | 0.00 | 12.78 |
| Streamflow | m³/sec | 7.74 | 10.15 | 17.44 | 6.76 |
| Peak Flow | m³/km² | 41.95 | 53.39 | 29.47 | 70.03 |
| Abert Squirrel Habitat | Ordinal (1-5) | 2.08 | 1.6 | 2.69 | 1.03 |
| MSO Habitat | Ordinal (1-5) | 1.84 | 0.81 | 2.79 | 0.00 |
| Mule Deer Habitat | Ordinal (1-5) | 3.17 | 4.1 | 4.32 | 2.85 |
| Northern Goshawk Habitat | Ordinal (1-5) | 3.81 | 1.92 | 3.87 | 0.00 |
| Rocky Mountain Elk Habitat | Ordinal (1-5) | 3.43 | 3.61 | 4.44 | 2.63 |
| Mexican Gray Wolf Habitat | Ordinal (1-5) | 3.09 | 3.29 | 4.37 | 2.74 |

[1]Willingness-to-pay for forest conditions based on forests as a cultural resource
[2]Willingness-to-pay for forest conditions based on forests as a recreational resource

Table 14.3  Values of the 22 forest management objectives for two different tree basal area levels (middle columns) determined using CP as well as the objectives' best and worst values (columns on left). Units for criteria vary by row and are given in the criterion scale column.

information. The achievement levels of the various TBA levels can be compared to the most preferred TBA for a given weighting scheme. All preference structures have a narrow classification for the "most preferred basal area," which is defined here as the TBA level falling within 10% below and above the most preferred basal area. However, the "most preferred basal area" varies noticeably for each preference structure or weighting scheme. The "Fire" and "Silviculture" weighting schemes have a narrower range of acceptable TBA, whereas in the "Wildlife" weighting scheme, the 10% below and above the most preferred basal area range stretches over 10 m$^2$/ha (or 50 ft$^2$/ac).

Given the management objective response functions and preference structure of the project manager, the CP result suggests a preferred basal area that is the best compromise between the needs of all the managers. This preferred tree basal area is 32 – 33 m$^2$/ha (139 – 144 ft$^2$/acre) for the project area, as shown in Figure 14.3 and Table 14.2. On the other hand, Table 14.3 lists the values of the 22 management objective response functions for a TBA level of 33 m$^2$/ha (144 ft$^2$/acre), and the "best" and the "worst" values for that management objective response function within a TBA that ranges from 6 to 45 m$^2$/ha (26 to 196 ft$^2$/ac).

This allows the DMs to see the value of each individual management objective that correspond to the most preferred TBA compared to all the possible values in the given range. Because the UBC data used in this study originated from a "fuel reduction" project, Table 14.3 also lists the values of the 22 management objective response functions that correspond to 22 m$^2$/ha (96 ft$^2$/acre) tree basal area, which is the preferred TBA for the "fire" preference structure, shown in Table 14.2.

## Spatio-Temporal Results

Sample results of the spatial-temporal analysis of the CP results are displayed in Figures 14.5 and 14.6. The figures display the CP analysis results for each individual stand in the project area through time and under the preference structure assigned by the UBC project manager (Figure 14.5) and the UBC Fire manager (Figure 14.6), respectively. Forest vegetation treatments were simulated in FVS from year 2006 (Figures 14.5a and 14.6a) to 2026. Because trees grow, stand densities generally increase after treatment, and as time progresses, the CP results for individual stands change with their increasing density. Some stands are cut to levels below the most preferred tree basal area and grow into that level and possibly beyond. Figures 14.5f and 14.6f illustrate what the CP achievement level of the stands in the project area would be if no treatment had taken place. These figures are based on FVS simulation without any treatment.

## DISCUSSION

When comparing Figures 14.5 and 14.6, one can see the similarities and differences in results between those favoring the overall project and those giving more weights to fire prevention activities. The FVS simulation that projects forest tree growth over the next 50 years for both weighting schemes is the same. However, because the most preferred basal area for the project weighting scheme is higher than that for the fire prevention (33 vs 22 m$^2$/ha), the treatments for the fire weighting schemes tend to reduce the stand density more than required by the project weighting scheme at first but the achievement levels for the latter are met in the future. When comparing Figure 14.5e with 14.6c, for example, it seems that the Project achievement levels of 95% of the preferred solution are met by 2056, whereas similar achievement levels are met in the Fire weighting scheme by 2016. Given the dynamics of tree growth in a forest, it is debatable whether or not reducing a stand's density to a level that falls below the "most preferred" TBA is desirable or not. With time, such a stand grows into its most preferred TBA level and remains there for a relatively long period, as illustrated in Figures 14.5 and 14.6. It appears to be more detrimental to not achieve the most preferred level by having stands that are too dense, because they will never "grow into" that level, unless a fire or insect outbreak reduces the live basal area within a stand. By comparing the maps showing the projected TBA scenarios, it is possible to achieve a working consensus between the DMs. For example, comparison of Figure 14.5 with Figure 14.6 might convince the Fire DM to thin forests more than they would consider ideal at the present time, as shown in Figure 14.5a, knowing that they can expect to approximate their ideal conditions sometime between 5 and 25 years (Figure 14.5 c and d). The no action alternative illustrates that the selected treatments have a desired influence in terms of achieving all management objectives for either preference structure.

Additional information can be obtained when comparing the values of the management objective in Table 14.3 to the weights assigned to them in Table 14.1. Those objectives generally weighted higher, compared to the objectives weighted lower have values that are closer to the most preferred value. However, there are some management objectives that received low weights from the ID team, but came close to their most preferred value, as in the case of creating/maintaining Mexican Grey wolf habitat. This indicates that some objectives benefit from the selected most preferred TBA even though their TBA requirements were given little consideration during the CP analysis process.

Some of the 22 response functions in this study represent ecosystem components that serve as a surrogate for another management objective. For example, the Abert Squirrel habitat response function may be representative of the squirrel's predator species and other bird species that also do better in denser forest stands. Even though 22 response functions are presented here, the DMs or land managers can choose to give varying importance to the different objectives. A fire manager who is concerned with reducing the fuel accumulation in the wildland urban interface may give higher weights to those response functions and objectives that reduce fire hazards at the landscape level as the fire manager of the USFS UBC ID team did. His desired achievement levels are very different from those of the wildlife team members.

As illustrated in Figures 14.5 and 14.6, the user can see how well a selected management prescription satisfies all management objectives simultaneously on a stand-by-stand basis, for a selected weighting scheme. The user can also identify the stands having too high TBA and those stands having too little and by how much. In other words, the eleven layers, one layer for each of the 11 time steps, provide visual input for multi-objective decision making in a spatio-temporal framework.

The ID team working on this project found the insights gained from this study useful and interesting. The most challenging part of their decision-making group effort was determining and assigning the weights to the individual and categories of management objectives. While the importance of the various management objectives and the implications of assigning high or low weights were discussed in a group setting, individual team members assigned weights independently based on their respective field of expertise and the resources they represent. The results illustrate the sensitivity and usefulness of this modeling approach to a real-world problem, where these types of decision are made by a group of often only four or five decision-makers.

However, because of the modeling effort used for this project, the authors hope that decision-making process provided enough transparency for the treatments chosen by the ID team to be accepted more easily by those not involved in this procedure.

## CONCLUSION

This paper presents an integrated modeling scheme that combines spatial and dynamic computer programs with a MODM model for evaluating multiple forest management objectives to select the most appropriate forest ecosystem management option at a landscape level. Various management objectives are addressed in a realistic and widely applicable setting. The decision variable for this model was tree basal area (in $m^2$/ha or $ft^2$/ac) of a ponderosa pine forest in northern Arizona. This stand density is currently above its historical level (USFS 2006). The modeling scheme can provide insight to the "best approach" to manage the forest to meet various ecosystem management objectives simultaneously over space and time. The tree basal area used as the decision variable links management objectives and ecosystem functions to management action alternatives expressed in the form of reduction of forest overstory trees. The outcome of CP indicates a most preferred tree basal area level of 33 $m^2$/ha (144 $ft^2$/acre) when all 22 management objectives are considered simultaneously. On the other hand, the "fuel reduction" weighting scheme indicates a TBA of 22 $m^2$/ha (96 $ft^2$/acre) tree basal area. When these different weighting schemes are evaluated in terms of their level of closeness to the most preferred alternative, the similarities and differences between the preference structures becomes even more apparent.

The modeling effort presented in this paper is designed to help land managers, stakeholders and other decision makers concerned with southwestern ponderosa pine forest ecosystem management during their planning and operational phases. It is interactive, allowing users to evaluate different preference structures at various stages in the planning process. Different weighting schemes of the response functions can be used to reflect various preference structures of one or more of DMs. The advantage of this model compared to the growth and yield models is its ability to visually display how well numerous management objectives are met simultaneously at a forest stand level and across an entire project area or landscape. It allows DMs and land managers to determine how far into the future some stands require additional treatment while other stands do not. By running multiple simulations, DMs have an additional, visual tool that can be used in team or public meeting environments to show how various forest treatments will affect a variety of management objectives and ecosystem components.

Future research and modeling efforts should attempt to link response functions to other forest management variables such as stand density index (SDI) or forest structural classes. It would also be advantageous to improve the dynamic interactiveness of the spatial aspect of this modeling effort. Once this has been accomplished, it would facilitate modeling the impacts of large "disturbances" such as climate change, which modelers and decision makers will have to deal with in the near future.

## LITERATURE CITED

Bare, B.B., Mendoza, G. 1988. Multiple objective forest land management planning: An illustration. European Journal of Operational Research 24, 44-55.

Brown, H.E., M.B. Baker, Jr., J.J. Rogers, W.P. Clary, J.L. Kovner, F.R. Larson, C.C. Avery, R.E. Campbell. 1974. Opportunities for Increasing Water Yields and Other Multiple Use Values on Ponderosa Pine Forest Lands, USDA Forest Service Paper RM-129. 36p.

Buongiorna J., Gilles, J.K., 2003. Decision Methods for Forest Resource Management. Academic Press, San Diego, CA.

Burrough, P.A., Karssenberg, D., van Deursen, W., 2005. Environmental modeling with PCRaster. In: Maguire, D.J., Batty, M., Goodchild M. F. (Eds.), GIS, Spatial Analysis and Modeling. ESRI Press, Redlands, CA. pp.333-356.

Conner, R.C., Born, J.D., Green, A.W., O'Brien, R.A., 1990. Forest Resources of Arizona. USDA Forest Service Research Bulletin INT-69. Intermountain Forest and Range Experiment Station, Ogden, UT.

Costanza, R., Sklar, F.H. White, M.L. 1990. Modeling coastal landscape dynamics. BioScience 40, 91-107.

Costanza, R., Voinov A. (Eds.), 2003. Spatial explicit landscape simulation modeling. Springer Verlag, New York, NY.

Covington, W.W., Fulé, P.Z., Moore, M.M., Hart, S.C., Kolb, T.E., Mast, J.N., Sackett, S.S., Wagner, M.R. 1997. Restoring ecosystem health in ponderosa pine forest of the Southwest. Journal of Forestry 95(4), 23-29.

Dixon, G.E. (Compl.), 2002. Essential FVS: A User's Guide to the Forest Vegetation Simulator. Internal Report, Fort Collins, CO: USDA Forest Service, Forest Management Service Center. (Last Revised: October 2004)

Dowdle, M.W. 2006. Public Accountability: Design, Dilemmas and Experiences. New York: Cambridge University Press.

Edminster, C.B., H.T. Mowrer, R.L. Mathiasen, T.M. Schuler, W.K. Olsen, F.G. Hawksworth. 1991. GENGYM: A Variable Density Stand Table Program System Calibrated for Mixed Conifer and Ponderosa Pine Stands in the Southwest. USDA Forest Service Paper RM-297. 32 p.

FAO (Food and Agriculture Organization of the United Nations), 2006. Global Forest Resources Assessment 2006: Progress towards sustainable forest management. FAO Forestry Paper 147; Food and Agriculture Organization of the United Nations; Rome, Italy.

FEMAT (Forest Ecosystem Management Assessment Team), 1993. Forest Ecosystem Management: An ecological, economic and social assessment. Washington D.C. Government Printing Office, No. 1993-793-071.

Folke, C., Carpenter, S., Walker, B., Scheffer, M., Elmqvist, T., Gunderson, L., Holling, C.S.  2004. Regime shifts, resilience, and biodiversity in ecosystem management. Annual Review of Ecology Evolution 35, 557-581.

Fulé, P.Z., McHugh, C., Heinlein, T.A., Covington, W.W. 2001. Potential fire behavior is reduced following forest restoration treatments. In: Vance, R.K., Covington, W.W., Edminster, D.C. (Compliers), Ponderosa Pine Ecosystems Restoration and Conservation: Steps Toward Stewardship. Proc. RMRS-P-22. Odgen, UT: USDA Forest Service, Rocky Mountain Research Station. pp. 28-35

Goicoechea, A.M., Hansen, D.R., Duckstein, L., 1982. Multiobjective Decision Analysis with Engineering and Business Applications. New York: John Wiley & Sons.

Goodchild, M.F., 2005. GIS and modeling overview. In: Maguire, D.J., Batty, M., Goodchild M. F. (Eds.), GIS, Spatial Analysis and Modeling. ESRI Press, Redlands, CA. pp.1-17.

Jackson, J.B.C., Kirb, M.X., Berher, W.H., Bjorndal, K.A. , Botsford, L.W., et al., 2001. Historical overfishing and the recent collapse of coastal ecosystems. Science 293, 629-638.

Keele, D.M., Malmsheimer, R.W., Floyd, D.W., Perez, J.E. 2006. Forest Service land management litigation 1989-2002. Journal of Forestry 104, 196-202.

Kennedy J.J., Koch, N.E., 2004. Viewing and managing natural resources as human-ecosystem relationships. Forest Policy and Economics 6, 497-504.

Körner, C., 1993. Scaling from species to vegetation: the useful of functional groups. In: Schulze, E., Mooney, H.A. (Eds.), Biodiversity and ecosystem function. Springer-Verlag. Berlin. pp. 117-140.

Krivoruchko, K., Gotway-Crawford, C.A., 2005. Assessing the uncertainty resulting from geoprocessing operations. In: Maguire, D.J., Batty, M., Goodchild M. F. (Eds.), GIS, Spatial Analysis and Modeling ESRI Press, Redlands, CA. pp.67-92.

Levin, S.A., 1993. Concepts of scale at the local level. In: Ehleringer, J.R, Field, C.B. (Eds.), Scaling Physiological Processes: Leaf to Globe. Academic Press, San Diego. pp.7-19.

Maidment, D.R., Robayo, O., Merwade V., 2005. Hydrologic modeling. In: Maguire, D.J., Batty, M., Goodchild M. F. (Eds.), GIS, Spatial Analysis and Modeling ESRI Press, Redlands, CA. pp.319-332.

Maxwell, T., Voinov, A., 2005. Dynamic, geospatial landscape modeling and simulation. In: Maguire, D.J., Batty, M., Goodchild M. F. (Eds.), GIS, Spatial Analysis and Modeling ESRI Press, Redlands, CA. pp.131-149.

McMahan, A.J., Courter, A.W., Smith, E.L., 2002. FVS-EMAP: A simple tool for displaying FVS output in ArcView GIS. In: Crookston, N., Havis R.N. (Compls.) Second Forest Vegetation Simulator Conference; 2002 February 12-14 Fort Collins, CO. Proc. RMRS-P-25. Ogden , UT: USDA Forest Service Rocky Mountain Research Station. pp. 57-61

Miller, I., Knopf, S., Kossik, R., 2005. Linking general-purpose dynamic simulation models with GIS. In: Maguire, D.J., Batty, M., Goodchild M. F. (Eds.), Spatial Analysis and Modeling ESRI Press, Redlands, CA. pp.113-129.

Mitchell, J.K. 1975. Dynamics and simulated yield of Douglas fir. Forest Science Monograph 17, 39.

Omi, P.N., Martinson, E.J., 2002. Effect of fuels treatment on wildfire severity. Joint Fire Sciences Program Report. Available online at www.cnr.colostate.edu/frws/ research/westfire/FinalReport.pdf; last accessed August 12, 2008.

Ostergren, D.M., Lowe, K.A., Abrams, J.B., Ruther, E.J. 2006. Public perception of forest management in North Central Arizona: The paradox of demanding more involvement but allowing limits to legal action. Journal of Forestry 104, 375-382.

Paine, R.T., Tegner, M.J., Johnson, E.A. 1998. Compounded perturbations yield ecological surprises. Ecosystems 1, 535-545.

Patton D.R. 1984. A model to evaluated Abert squirrel habitat in uneven-aged ponderosa pine. Wildlife Society Bulletin, 12, 408-414.

Patton D.R. 1991. The ponderosa pine forest as wildlife habitat, Chapter 8 in A. Tecle and W.W. Covington (eds.) Multiresource Management of Southwestern Ponderosa Pine Forests: The Status of Knowledge. Albuquerque, NM: USDA Forest Service, Southwestern Region, pp.361-410.

Poff, B., 2002. Modeling southwestern ponderosa pine forest ecosystem management in a multi-objective decision-making framework. MS Thesis, Northern Arizona University, Flagstaff, AZ.

Poff, B., A. Tecle, D.G. Neary, B.W. Geils. 2010. Compromise programming in forest management. Journal of the Arizona-Nevada Academy of Science 42(1), 44-60.

Reynolds, J.F., Hilbert, D.W., Kemp, R.R., 1993. Scaling ecophysiology from the plant to the ecosystem: a conceptual framework. In: Ehleringer J.R., Field C.B. (Eds.), Scaling Physiological Processes: Leaf to Globe. Academic Press, San Diego. CA. pp. 127-140.

Risser, P.G., Karr, J.R., Forman, R.R.T., 1984. Landscape ecology: Directions and approaches. Illinois Natural History Survey, Champaign, IL.

Rupp, D.E., 1995. Stochastic, Event Based and Spatial Modeling of Upland Watershed

Precipitation-Runoff Relationships. MS. Thesis. Northern Arizona University, Flagstaff, AZ.

Severson, K.E., Medina, A.L. 1983. Deer and elk management in the Southwest. Journal of Range Management Monograph No. 2. 64p.

Sklar, F.H., Costanza, R., 1991. The development of dynamic spatial models for landscape ecology. In: Turner, M.G., Gardner R. (Eds.), Quantitative methods in landscape ecology. Springer Verlag, New York, NY.

Steffen, W., Sanderson, A., Jäger, A., Tyson, P.D., Moore III, E., et al., 2004. Global Change and the Earth System: A Planet Under Pressure. Heidelberg: Springer-Verlag.

Swetnam, T. W., 1990. Fire history and climate in the southwestern United States. In: Krammes, J.S. (Tech Coor.), Effects of Fire Management of Southwestern Natural Resources. USDA Forest Service GTR-RM-191. pp. 6-17

Swetnam, T. W., Baisan, H.C., 1996. Historical fire regime patterns in the southwestern United States since AD 1700. In: Allen, C. (Tech Coor.), Fire effects in Southwestern Forest. USDA Forest Service GTR-RM-286. pp. 11-32

Szidarovszky, F., Gershon, M.E., Duckstein, L., 1986. Techniques for Multiobjective Decision Making in System Management. Elsevier Science Publishers, Amsterdam.

Tecle, A., Fogel M., Duckstein, L. 1988. Multicriterion analysis of forest watershed management alternatives. Water Resources Bulletin 24, 1169-1178.

Tecle, A., Duckstein, L., 1993. Concepts of multi-criterion decision making. In: Nachtnebel, H.P. (Ed.), Decision Support System in Water Resource Management. UNESCO Press, Paris.

Tecle, A., S. Bijaya, and L. Duckstein. 1998. A multiobjective decision support system for multiresource forest management. Nature & Resources 31(3):8-17.

Thomas, J.W., 2006. Adaptive Management: What is it all about? Water Resources – Impact 8, 5-7.

USFS (United States Forest Service), 1987. Coconino National Forest Plan, as amended (2006). USDA Forest Service, Southwestern Region, Albuquerque, NM.

USFS, 1989. An Analysis of the Lands Base Situation in the United Sates: 1989-2040. USDA Forest Service General Technical Report RM-181. Rocky Mountain Forest and Range Experiment Station, Fort Collins, CO.

USFS, 2006. Upper Beaver Creek Watershed Fuel Reduction Project; Proposed Action Report. USFS Coconino National Forest, Mogollon Rim Ranger District. Flagstaff, AZ.

Vogt, K.A., Gordon, J.C., Wargo, J.P., Vogt, D.J., Asbjornsen, H., Palmitto, P.A., Clark, H.J., O'Hara, J.L., Keeton, W.S., Patel-Weynand, T., Witten, E., 1997. Ecosystem: Balancing Science with Management. Springer, New York, NY.

Wagner, M.R., Block, W.M. Geils, B.W., Wenger, K.F. 2000. Restoration ecology: A new forest management paradigm, or another merit badge for foresters? Journal of Forestry 98, 22-27.

Zeleny, M., 1973. Compromise programming. In: Cochrane J.L., Zeleny, M. (Eds.), Multiple Criteria Decision-Making. University of South Carolina Press, Columbia, SC. pp. 263-301.

Zeleny, M. 1974. A Concept of Compromise Solutions and the Method of the Displaced Ideal. Computers and Operations Research 1, 479-496.

Zeleny, M., 1982. Multiple Criteria Decision Making. McGraw-Hill Book Company, New York.

# INNOVATIVE PLANTING METHODS FOR THE ARID AND SEMI-ARID SOUTHWEST

*David R. Dreesen, Gregory A. Fenchel, Danny G. Goodson, and Keith L. White*

## ABSTRACT

The Los Lunas Plant Materials Center has developed several unconventional but effective methods for establishing plants in arid and semi-arid regions of the southwestern US. These methods, which are relevant to riparian and upland restoration efforts in the Southwest, are summarized in this paper. For riparian restoration, several deep-planting techniques have proven successful for establishing woody riparian plants by rapidly connecting the root systems of these phreatophytes to the capillary fringe above a shallow water table. These techniques include deep-planting several stock types including long-stem rooted stock, dormant unrooted whip cuttings, and dormant unrooted pole cuttings. On upland xeric sites, woody plants have been established with minimal irrigation by employing long root systems along with embedded watering tubes used to apply water that is slowly released from hydrated hydrophilic starch-based polymers to replenish deep soil moisture. Finally, the key factors influencing the success of native grass seedings are discussed. New methods to enhance native grass establishment are being tested to prolong the availability of shallow soil moisture by using long-lasting organic mulches or hydrophilic starch-based polymers.

## INTRODUCTION

Restoration efforts in disturbed riparian and upland areas of the southwestern US are particularly challenging because of soil moisture limitations. One of the primary objectives of technology demonstrations performed by the Los Lunas Plant Materials Center (LLPMC) is to address more effective means of establishing native plants on arid and semi-arid disturbed lands. The development of plant resources and new planting techniques are part of the LLPMC's mission to advance conservation technology. These technologies are intended for application in conservation practices promoted by the USDA Natural Resources Conservation Service (NRCS) for environmental quality improvement and wildlife habitat enhancement. The service region of the LLPMC covers much of New Mexico and Arizona, as well as parts of Utah, Texas, and Colorado. The LLPMC is located near the convergence of the Colorado Plateau, the Chihuahuan Desert, and the High Plains, and in conjunction with other PMCs in Arizona, Colorado and Nevada develops new conservation methodologies using plants adapted to desert, grassland, woodland, and montane ecoregions of the Southwest.

Not only does the Southwest experience limited precipitation, but it is also subject to very high evaporation rates (Seiler 1997) due to high temperatures, low humidity, and high winds resulting in a very rapid loss of surface soil moisture. Soil moisture plays a key role in the success of revegetation efforts in arid regions. The general lack of precipitation at lower elevations is compounded by factors that limit the infiltration of precipitation including bare, fine-textured, or compacted soils as well as natural and fire-caused water repellency. If water does enter the soil profile, it can be lost to evaporation or deep infiltration before root uptake and plant transpiration can occur. In addition, root systems can redistribute moisture within the soil profile (Ryel et al 2004).

The primary objectives of the LLPMC planting strategies described in this paper are to maximize the use of existing soil moisture reservoirs and minimize or eliminate the need for irrigation to successfully establish plantings. Specifically these strategies include:

• Planting woody riparian species in areas with shallow groundwater (less than 2.5 m (8 ft)) using deep-planting techniques to immediately or rapidly connect nursery stock root systems with the capillary fringe (i.e., the zone of soil located immediately above the water table and wetted by capillarity) or the base of dormant cuttings with ground water.
• Planting upland xeric woody species using long-rooted containerized stock to access existing deep soil moisture. Watering tubes are embedded alongside the root ball of these woody plants to allow replenishment of deep soil moisture.
• Establishing native grass species from seed in arid regions is problematic; thus, we are testing methods that attempt to maximize infiltration, minimize evaporation, and prolong the availability of water in the surface soil long enough to allow germination and establishment.

## Riparian Planting Methods

Before addressing the revegetation of riparian plant communities, it is necessary to emphasize the importance of sediment erosion, transport, and deposition as well as over-bank flooding to the natural regeneration and functioning of these plant communities (Scott et al. 1997). It is possible to successfully establish riparian vegetation without over-bank flooding, but these revegetated areas will not be functionally self-sustaining without the reintroduction of over-bank flooding and channel change processes (Scott et al. 1997; LLPMC 2009). On flow regulated rivers, there are instances of natural regeneration on sandbars, but generally these new communities are prone to be scoured away by flood flows when they occur. For many riparian species to naturally regenerate, the roots of seedlings must extend below the receding water table and related soil drying front following over-bank flooding events (Mahoney and Rood 1991). In those situations where natural regeneration is not occurring, planting riparian species can establish some structural elements of these communities in the interim until over-bank flooding can be reinstated.

To better understand the deep-planting rationale, it is important to remember that phreatophytes are dependent on moisture supplied by shallow groundwater and their roots will proliferate in the capillary fringe. The depth and annual fluctuation of the water table will determine the feasibility of using deep-planting techniques as well as indicate which species might be best adapted to particular ground water conditions.

Because phreatophyte root systems exploit the capillary fringe, it is important to have some idea about the thickness of the fringe above the water table (Table 15.1). Very coarse alluvium may have a fringe thickness of 15 cm (6 in) or less. However, fine sands can have a capillary rise of several feet (VICAIRE 2006). Although the capillary rise in fine textured silts and clays may be many feet, these poorly aerated sediments may be a favorable substrate for only certain guilds of riparian species (Merritt et al. 2010)

## Planting Dormant Unrooted Pole Cuttings

Deep-planting techniques are aimed at minimizing or eliminating the need to irrigate riparian plantings. This can be accomplished by placing an unrooted pole cutting below the water table which

| Sediment | Grain diameter (cm) | Pore radius (cm) | Capillary rise (cm) | Capillary rise (in) |
|----------|---------------------|------------------|---------------------|---------------------|
| Fine silt | 0.0008 | 0.0002 | 750 | 300 |
| Coarse silt | 0.0025 | 0.0005 | 300 | 120 |
| Very fine sand | 0.0075 | 0.0015 | 100 | 40 |
| Fine sand | 0.0150 | 0.003 | 50 | 20 |
| Medium sand | 0.03 | 0.006 | 25 | 10 |
| Coarse sand | 0.05 | 0.010 | 15 | 6 |
| Very coarse sand | 0.20 | 0.20 | 4 | 1.6 |
| Fine gravel | 0.50 | 0.100 | 1.5 | 0.6 |

Table 15.1 The height of capillary rise in sediments of various textures (VICAIRE 2006).

provides hydration until adventitious roots can develop in the capillary fringe. Pole cuttings of cottonwoods (*Populus deltoides Bartram ex Marsh. ssp. monilifera* (Aiton) Eckenwalder, plains cottonwood; *Populus deltoides Bartram ex Marsh. ssp. wislizeni* (S. Watson) Eckenwalder, Rio Grande cottonwood; *Populus fremontii S. Watson*, Fremont cottonwood) and tree willows, such as Gooddings willow (*Salix gooddingii C.R. Ball*), are planted as dormant pole cuttings up to 5 m (16 ft) long (LLPMC 2005). Holes are dug below the water table with a front-end-loader-mounted auger, and the base of the dormant pole is inserted into the hole in contact with the ground water (see Figure 15.1). Tree guards constructed of 1.5 m (5 ft) tall poultry wire are installed to protect the pole cutting from beaver and other rodent damage; the base of the tree guard is buried when backfilling for anchorage. We have found that the pole cuttings will root in the capillary fringe as well as form adventitious roots near the soil surface (i.e., at depths of 25 to 50 cm). Cottonwoods and Gooddings willow do not tolerate high salt levels in the alluvium or groundwater. There are other riparian species (see Table 15.2) more tolerant of salts (Taylor and McDaniel 1998), but these cannot be recommended for planting as pole cuttings because they do not readily form adventitious roots.

Over the past 20 years, the LLPMC has installed well over 10,000 pole cuttings. If the plantings are located on a site where the auger can easily reach the depth of the groundwater, where the alluvium is well aerated, and where there is no excessive salinity (i.e., <2 to 3 dS/m), a 90% survival rate is expected. However, at sites where some or all of these factors are not optimal, then the survival rate can drop significantly.

Planting Dormant Unrooted Whip Cuttings

Nursery stock with stems generally less than 2 m (6.5 ft) in length, slender stems, and few branches are termed whips. A dormant, unrooted, whip cutting planting technique is used with shrub willows and redosier dogwood (*Cornus sericea L.*) which do not produce pole-sized cuttings (LLPMC 2007a). These species readily produce whip cuttings with lengths of 1.5 to 2.5 m (5 to 8 ft) and stem diameters less than 2.5 cm (1 in).

Figure 15.1  Planting a dormant pole cutting into a hole dug with a front-end-loader- mounted auger.  Note a poultry wire tree guard suspended on the top of the pole cutting, the bottom of which will be lowered  into the hole and secured with top 15-30 cm (6-12 in) of backfill.

Figure 15.2  Planting dormant whip cuttings of coyote (sandbar) willow with a rotary hammer equipped with a 0.9 m (3 ft) long and 2.5 cm (1 in) diameter bit.

| Salinity tolerance (< maximum EC tolerated) | Common name | Scientific name |
|---|---|---|
| Most tolerant (<14 dS/m) | Fourwing saltbush | *Atriplex canescens (Pursh) Nutt.* |
| Moderately tolerant (<8 dS/m) | Torrey wolfberry | *Lycium torreyi A. Gray* |
| | Screwbean mesquite | *Prosopis pubescens Benth.* |
| | Willow baccharis | *Baccharis salicina Torr. & A. Gray* |
| Somewhat tolerant (<2.5 dS/m) | Goodding's willow | *Salix gooddingii C.R. Ball* |
| Not tolerant (<2 dS/m) | Rio Grande cottonwood | *Populus deltoides Bartram ex Marsh. ssp. wislizeni (S. Watson) Eckenwalder* |

Table 15.2. Salinity tolerance of common floodplain species within the Middle Rio Grande Valley adapted from Taylor and McDaniel (1998).

We have frequently used this technique to plant coyote or sandbar willow (*Salix exigua Nutt.*) with good results. It is advisable to plant at different elevations above the shore line to buffer the planting from extreme changes in the ground and surface water levels. The preferred planting method for dormant whip cuttings is to use a rotary hammer drill with a 0.9 m (3 ft) long and 2.5 cm (1 in) diameter bit (see Figure 15.2). In finer-grained alluvium without appreciable cobble or rock, a team of two people can plant more than 800 whips in a day. These drills are also effective in penetrating frozen surface soils which are often encountered during the late-winter and early-spring planting window.

### Planting Rooted Stock with Long Root Balls or Long Stems

Deep-planting methods can be used to minimize or eliminate the need to irrigate rooted transplants by placing the existing root system of a containerized plant in contact with the capillary fringe (LLPMC 2007b). The technique for planting long-stem, rooted riparian stock includes several variants depending on the height of the growing container and the length of the stem (see Figure 15.3). In general, two container sizes are used for growing the rooted stock and are removed before planting; HDPE (high density polyethylene) or PVC tall pot containers are 76 cm (30 in) tall with a 10 cm x 10 cm (4 in x 4 in) top cross section and HDPE tree pot containers are 35 cm (14 in) tall with a10 cm x 10 cm (4 in x 4 in) top cross section (see Figure 15.4). Stem lengths greater than approximately 1.5 m (5 ft) are considered long stems. In many situations, 2.5 cm (1 in) diameter PVC watering tubes are placed alongside the root ball to apply water if the water table declines before the roots have extended sufficiently into the capillary fringe.

These tubes also serve as effective markers of plant placement to record survival and growth. The tall pot root ball, about 70 cm (28 in) long, provides a very effective nursery stock type for planting at deeper ground water sites. However, the production of tall pot stock is much more difficult and expensive than growing long-stem tree pot stock with stems 1.5 to 2.5 m long and root balls 30 cm (12 in) long. Tall

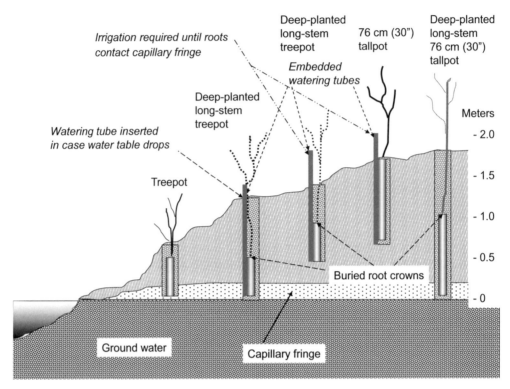

Figure 15.3   Several approaches to planting rooted nursery stock are used to connect root systems to the capillary fringe. Long-stem stock is grown in HDPE one-gallon treepots (35 cm rootballs) or in PVC or HDPE tallpots (76 cm rootballs). Rootballs are removed from growing containers before planting; root crowns can be buried up to 1.8 m (6 ft) deep. PVC watering tubes perforated along the lower third are installed to allow deep soil moisture to be replenished if the water table declines before roots have extended into the capillary fringe.

pot and long-stem stock as well as dormant pole cuttings are typically planted using a 2.4 m (8 ft) long, 23 cm (9 in) diameter, front-end-loader-mounted auger to dig holes down to the water table. The root crowns of both tall pot and tree pot stock are buried up to 1.8 (6 ft) below the soil surface.

We have found that all of the riparian species that we have tested not only tolerate, but thrive when the root crown is buried and the roots can immediately access capillary moisture. Riparian species are adapted to burial by sediments because this is a common occurrence in naturally functioning riparian systems (Merritt et al. 2010). A year or two after planting and upon excavation of long-stem stock with buried

root crowns, we have observed the rapid development of a secondary root system at shallow depths of 10 to 30 cm (4 to 12 in) (see Figure 15.5). This has been true for all the riparian species we have excavated so far including New Mexico olive *(Forestiera pubescens Nutt. var. pubescens)*, skunkbush sumac *(Rhus trilobata Nutt.)*, Emory's baccharis *(Baccharis emoryi A. Gray)*, netleaf hackberry *(Celtis laevigata Willd. var. reticulata (Torr.) L.D. Benson)*, boxelder *(Acer negundo L.)*, screwbean mesquite *(Prosopis pubescens Benth.)*, and desert false indigo *(Amorpha fruticosa L.)*.

The LLPMC has been using deep-planting methods for long-stem containerized stock for about five years, and survival rates

Figure 15.4  Long-stem plants of New Mexico olive *(Forestiera pubescens Nutt. var. pubescens)* in a HDPE treepot (left) 35 cm (14 in) tall with a 10 cm x 10 cm (4 in x 4 in) top cross section and a PVC tallpot (right) 76 cm (30 in) tall and 10 cm (4 in) in diameter. Note meter stick for scale.

Figure 15.5    Adventitious roots developing on a deep planted desert false indigo *(Amorpha fruticosa L.)*.

Note embedded 2.5 cm (1 in) diameter PVC watering tube.

generally have been between 70 and 90%. The main cause of mortality has been long-term inundation due to over-bank flooding of six weeks or more. If there is a severe decline in the water table, neglecting to water plants using the embedded watering tubes can result in significant mortality.

Figure 15.6 shows a riparian planting that utilized deep-planted stock to establish a diverse riparian community including Rio Grande cottonwood *(Populus deltoides Bartram ex Marsh. ssp. wislizeni (S. Watson))* and Gooddings willow pole cuttings *(Salix gooddingii C.R. Ball)*, coyote or sandbar willow *(Salix exigua Nutt.)* whip

cuttings, and long-stem tree pot and tall pot New Mexico olive *(Forestiera pubescens Nutt. var. pubescens)* and Emory's baccharis *(Baccharis emoryi A. Gray)*.

## Upland Planting Methods For Woody Species On Xeric Sites

Planting on upland xeric sites has been accomplished using nursery stock grown in 76 cm (30 in) tall pots.  The objective is to use the existing deep soil moisture reservoir to establish native species adapted to the site. To assure there is a sufficient amount of deep soil moisture to establish the transplants, a watering tube is placed in the planting hole alongside the root ball. The PVC watering tube has perforations along the bottom third of the tube to provide moisture to the lower portion of the root ball. Water can be pumped into 2.5 cm (1 in) diameter PVC tubes or starch-based hydrogel (i.e., hydrophilic polymer) slurries can be dispensed into 7.6 cm (3 in) diameter PVC tubes. This very viscous slurry has to be pumped through a minimum of 5.1 cm (2 in) diameter hoses (see Figure 15.7). The principal benefit of the hydrogel is the slow release of water over many weeks thereby reducing the number of times water has to be applied to the transplants. The starch-based hydrogel decomposes leaving no residual material that could be rehydrated. A synthetic polymer hydrogel should be able to be rehydrated, but we have not yet tested these products. The cost and difficulty in pumping polymers needs to be weighed against the benefit of fewer water applications. Whatever the source of water, the main objective is to create or replenish the deep-soil moisture reservoir. The material used for the watering tubes is not important as long as it conveys water to the bottom portion of the root ball. It has been suggested that cardboard mailing tubes might be effective and biodegradable.

Our first major demonstration of this planting methodology was carried out near Santa Fe, New Mexico starting in the fall of 2000. The New Mexico Department of Transportation (NMDOT) had reconstructed a highway, and had applied grass and forb seed as well as erosion control blankets to the surrounding disturbed areas. The LLPMC was asked to establish native shrubs without using a conventional drip irrigation system. Because of the problems involved with traffic control when irrigating from a mobile water tank, it was paramount to minimize the number of water applications. Therefore, the use of hydrated hydrogels rather than plain water was deemed necessary (LLPMC 2002).

In November and December 2000, the LLPMC planted 1,700 woody plants and inserted PVC watering tubes [7.6 cm (3 in) diameter] alongside each plant's root ball (see Figure 15.8). In this particular situation, the backfill consisted of cloddy and frozen soils and required an application of water into the planting hole to reduce soil voids during the planting process. Because of the very dry fall of 2001 and winter of 2002, hydrated starch-based hydrogel was applied through the watering tubes at budbreak in early May 2001 and again in the spring of 2002. This area experienced a severe drought during the period of 2001 through 2003 and received only 30% to 50% of the average precipitation of 13.5 inches. Six of the eight years between 2001 and 2008 received below average amounts of annual precipitation. After the first growing season, there was 96% survival rate. After eight years, the overall survival rate was 80%.

### Native Grass Seeding Methods

Although a prolonged reservoir of shallow soil moisture is the key to seed germination and seedling establishment, there are number of other important factors influencing the success of an unirrigated grass seeding in arid and semi-arid environments (Dreesen 2008) including season of growth, seeding depth, seed dormancy, soil compaction, soil texture, weed competition, and mulch application.

The time of year for seeding is determined by differences in optimum growth temperatures between cool-season and warm-season grasses and the seasonality of significant precipitation events.

Cool-season grasses have optimum growth temperatures around 21° C (70° F) (Jones 1997). They can be seeded before or during summer rains. Dormant seeding (i.e., soils too cold for germination) in the fall can be successful if ample soil moisture is available in the spring. Cool-season grasses in the West are typically found in mountain ecoregions or in cold desert ecoregions that receive winter moisture (such as the

Figure 15.6    A four year old planting along the Rio Grande near Bernalillo, NM established using Rio Grande cottonwood *(Populus deltoides Bartram ex Marsh. ssp. wislizeni (S. Watson))* and Gooddings willow pole cuttings *(Salix gooddingii C.R. Ball)*, coyote or sandbar willow *(Salix exigua Nutt.)* whip cuttings, and long-stem, deep planted New Mexico olive *(Forestiera pubescens Nutt. var. pubescens)* and Emory's baccharis *(Baccharis emoryi A. Gray)*.

Figure 15.7    Using a hydro-mulcher to mix and pump a slurry of hydrophilic starch-based polymer into 7.5 cm (3 in) diameter PVC watering tubes placed next to the rootballs of xeric upland shrubs grown in tallpots.

Figure 15.8 Fremont's mahonia (left) *(Mahonia fremontii (Torr) Fedde)* and Gambel oak (right) *(Quercus gambelii)* established using tallpot stock irrigated with hydrophilic starch-based polymer dispensed into embedded watering tubes adjacent to the rootball.

Great Basin). The grasses in the Stipeae tribe (including the genera Hesperostipa, Achnatherum, Nassella, and Piptochaetium) have a cool-season physiology but lack certain storage carbohydrates (Jones 1997). The upshot is these grasses typically do better than other cool-season grasses in warmer climates. Most of the species in this tribe are commonly called needlegrasses or ricegrasses.

Warm-season grasses have optimum growth temperatures near 32° C (90° F) with little growth occurring below 15° C (60° F) (Jones 1997). They are best seeded in anticipation of or during summer monsoons because they need warm temperatures as well as moisture to germinate and become established. They typically are found in western ecoregions with summer precipitation patterns, and at lower elevations

as in the Chihuahuan desert, the Sonoran desert, the southern high plains, as well as the Colorado Plateau.

Depth of seed placement is a crucial factor; there is a trade-off between shallow seeding depths to allow high rates of emergence versus the better moisture conditions with increasing depth.

For most native grasses, the recommended seeding depth is one-quarter to one-half inch. However, some large-seeded native species such as wheatgrasses can be seeded somewhat deeper – between one-half and one-inch deep. Some species such as Indian ricegrass (*Achnatherum hymenoides* (Roem. & Schult.) Barkworth) are adapted to deep burial. Because very small seed can be easily sown too deep, they are often surface-broadcast and incorporated into the soil by raindrop impact or mulching activities (Winkel et al. 1991).

Presence of dormant seed can be advantageous because it provides a reserve of viable seed if less than optimum moisture conditions occur during initial rainfall events.

Shallow compaction zones can limit the ability of seedling roots to follow the downward drying front resulting in seedling desiccation.

The establishment of grass seedlings is dependent on the drying rate of the surface soil. A study conducted in a Sonoran grassland of southeast Arizona showed that slow drying rates correspond to a high probability of grass seedling survival (Roundy et al. 1997). In contrast to soils undergoing fast drying, slow drying soils had uniformly high water content in the top 15 cm (6 in) of soil, up to three times the total soil water content, and it took four times as long for the drying front to exceed seedling rooting depth. High seedling survival rates are correlated with the high cumulative precipitation for the prior four days, low air temperatures, and low vapor pressure deficits.

Large soil voids can prevent adequate seed-to-soil contact and reduce capillary movement of soil moisture which can retard germination and growth.

The influence of soil texture on water infiltration and water storage capacity can be a key factor in seeding success. The term "inverse texture effect" has been proposed to indicate that in arid regions, coarse-textured soils have more useable soil moisture than fine-textured soils. The reasoning behind this inverse texture effect has to do with the depth of wetting in coarse soils which is related to the water storage capacity of the soil (Noy-Meir, 1973). The key is that much of the moisture in coarse-textured soil is deep enough to avoid being lost rapidly to evaporation. Sands have penetration depths two to four times those of fine-textured soils assuming an equivalent amount of infiltrated moisture. Another effect of soil texture is on water infiltration rates. Coarse textured soils can have infiltration rates two to six times those of fine-textured soils accentuating the inverse texture effect and the availability of soil water.

The importance of controlling invasive annual weeds prior to seeding is paramount. Often, the proliferation of annual weeds should preclude any seeding until the weed seed bank can be drastically reduced. A key is to prevent the weeds from going to seed in their first year of growth following disturbance. Herbicide application and mowing of weeds prior to seed set can be effective control methods. It is usually recommended to forgo any nitrogen fertilization with native grass seeding because nitrogen favors weed growth over grass growth (Schwinning et al. 2005).

To take maximum advantage of limited precipitation, the application of mulch can be an essential step in arid climates. Applying mulch provides one of the few opportunities to conserve soil moisture and increase infiltration. Native grass hay is one of the better mulch materials. The mulch should be applied dense enough to shade the soil and prevent wind desiccation, but not so dense as to retard grass emergence. In addition, the grass hay or grain straw should be crimped into the soil to limit transport by wind or water. Straw of annual grains is often used as mulch, but it degrades rapidly. Hydromulch and erosion control blankets are expensive, but they may be the only feasible alternatives on steep slopes. A product such as Woodstraw® (registered trademark of forestconcepts), which are wood strands shaped like grain straw, is much heavier and degrades much more slowly as well as being much less libel to being moved by wind. Wood chips can be an effective mulch if applied in a thin layer after seeding. Gravel or rock mulch can also be effective if not applied too deeply (Winkel et al 1991).

The LLPMC has recently conducted some seeding trials to determine if the purported attributes of wood strand longevity and resistance to erosion of Woodstraw® make it a more effective mulch than conventional crimped straw or hay. Early results indicate that this mulch does not readily move or rapidly degrade. We also have installed a trial to determine if the addition of starch-based hydrophilic polymer in the seed furrow enhances grass seed germination and establishment by increasing soil moisture storage adjacent to the seed. Initial seedling counts do not indicate a benefit from polymer application.

The development of species for restoration seeding should consider several attributes that might help to increase seeding success. The following traits would generally be considered disadvantageous agronomic characteristics, but might increase the long-term success of arid-land seeding.

• Delaying germination after the initial precipitation events, which are followed by long dry periods, may allow a viable seed bank to be retained. This strategy is used by some invasive grass species to outcompete native species by preserving seed viability through initial wetting and drying cycles (Abbott and Roundy 2003).

• Increasing the time from seed imbibition to the loss of desiccation tolerance (i.e., when an imbibed seed can dry out but still be viable) should prolong seed viability if alternating wetting and drying cycles occur.

• By purposefully selecting for dormancy, it might be possible to retain substantial numbers of viable seed through several growing seasons. A mixture of low and high dormancy seed could be planted to provide a seedbank able to germinate and establish over a range of wetting and drying cycles.

The past development of native grass releases may have selected against these disadvantageous agronomic characteristics. An additional characteristic which could benefit arid land seeding is the ability to emerge from deeper soil depths. As an example, the blue grama (*Bouteloua gracilis (Willd. ex Kunth) Lag. ex Griffiths*) cultivar 'Alma' was selected for greater emergence from deeper soil depths (Wynia 2007).

Large-scale seeding in arid environments is a costly endeavor with a poor chance of success. Many species, especially some shrubs and forbs, are difficult to establish even under optimum conditions; this often results in poor species diversity following a seeding. An alternate strategy is to establish seed source islands to provide long-term natural seed dissemination (Reever Morghan et al. 2005). These islands could be established using intensive cultural practices which would be impractical and extraordinarily expensive on a large scale. Such practices could include planting seedlings, irrigation for establishment, fencing to limit herbivory by large and small animals, and weed control.

## CONCLUSIONS

Effective planting methodologies in arid and semi-arid regions are dependent upon taking advantage of existing soil moisture reservoirs, conserving existing soil moisture, or efficiently providing sufficient soil moisture to enable establishment. Shallow ground water and its associated capillary fringe provide an easily exploited moisture source for riparian restoration efforts if appropriate planting techniques are employed and suitable nursery stock is available. The success of upland plantings of woody species in arid situations can be enhanced if methods of exploiting existing deep soil moisture are used and if techniques of replenishing these reservoirs are employed. The chances of successfully seeding native grasses on arid lands can be improved by understanding the seed and soil factors which affect establishment and by manipulating these factors to make the most effective use of seedbed soil moisture.

## LITERATURE CITED

Abbott, L.B. and B.A. Roundy. 2003. Available water influences field germination and recruitment of seeded grasses. Journal of Range Management 56(1): 56-64.

Dreesen, D.R. 2008. Basic guidelines for seeding native grasses in arid and semi-arid ecoregions. U. S. Department of Agriculture, Natural Resources Conservation Service, Los Lunas Plant Materials Center, Los Lunas, New Mexico. 4p.http:// www. nm.nrcs.usda.gov/programs /pmc/brochures/ BasicsOfSeedingNativeGrasses.pdf

Jones, T.A. 1997. The expanding potential for native grass seed production in the western United States. p. 89-92. In Landis, T.D. and J.R. Thompson (Tech. Coords.). National Proceedings, Forest and Conservation Nursery Associations. U.S. Department of Agriculture, Forest Service Gen. Tech. Rep. PNW-GTR-419. Pacific Northwest Research Station, Portland, OR.

LLPMC. 2002. Tall pot transplants established with hydrogel. A revegetation technique without intensive irrigation for the arid Southwest. U. S. Department of Agriculture, Natural Resources Conservation

Service, Los Lunas Plant Materials Center, Los Lunas, New Mexico. 22p. http://www.nm.nrcs.usda.gov/programs/pmc/reports/projects/tallpot.pdf

LLPMC. 2005. The pole cutting solution. Guidelines for planting dormant pole cuttings in riparian areas of the Southwest. U. S. Department of Agriculture, Natural Resources Conservation Service, Los Lunas Plant Materials Center, Los Lunas, New Mexico. 8p. http://www.nm.nrcs.usda.gov/news/publications/polecutting.pdf

LLPMC. 2007a. Guidelines for planting dormant whip cuttings to revegetate and stabilize streambanks. Deep planting - the ground water connection. U. S. Department of Agriculture, Natural Resources Conservation Service, Los Lunas Plant Materials Center, Los Lunas, New Mexico. 8p. http://www.nm.nrcs.usda.gov/news/publications/dormant-willow-planting.pdf

LLPMC. 2007b. Guidelines for planting long-stem transplants for riparian restoration in the Southwest. Deep planting - the ground water connection. U. S. Department of Agriculture, Natural Resources Conservation Service, Los Lunas Plant Materials Center, Los Lunas, New Mexico. 8p. http://www.nm.nrcs.usda.gov/news/publications/deep-planting.pdf

LLPMC. 2009. Revegetating riparian areas in the Southwest "lessons learned". U. S. Department of Agriculture, Natural Resources Conservation Service, Los Lunas Plant Materials Center, Los Lunas, New Mexico. 6p. http://www.nm.nrcs.usda.gov/technical/tech-notes/pmc/pmc70a1.pdf

Mahoney, J.M. and S.B. Rood. 1991. A device for studying the influence of declining water table on poplar growth and survival. Tree Physiology 8:305-314.

Merritt, D.M., M.L. Scott, N.L. Poff, G.T. Auble, and D.A. Lytle. 2010. Theory, methods, and tools for determining environmental flows for riparian vegetation: riparian vegetation-flow response guilds. Freshwater Biology 55:206-225/

Noy-Meir, I. 1973. Desert ecosystems: environment and producers. Annual Review of Ecology and Systematics 4:25-51.

Reever Morghan, K.J., Sheley, R.L., Denny, M.K., Pokorny, M.L. 2005. Seed islands may promote establishment and expansion of native species in reclaimed mine sites (Montana). Ecological Restoration. 23(3): 214-215.

Roundy, B.A., L.B. Abbott, and M. Livingston. 1997. Surface soil water loss after summer rainfall in a semidesert grassland. Arid Soil Research and Rehabilitation 11:49-62.

Ryel, R.J., A.J. Leffler, M.S. Peek, C.Y. Ivans, and M.M. Caldwell. 2004. Water conservation in Artemisia tridentata through redistribution of precipitation. Oecologia 141:335-345.

Scott, M.L., G.T. Auble, and J.M. Friedman. 1997. Flood dependency of cottonwood establishment along the Missouri River, Montana, USA. Ecological Applications 7(2):677-690.

Seiler, R.L. 1997. Methods to identify areas susceptible to irrigation-induced selenium contamination in the western United States. U.S. Department of Interior, United States Geological Service, Fact Sheet FS-038-97.

Schwinning, S., B.I. Starr, N.J. Wojcik, M.E. Miller, J.E. Ehleringer, and R.L. Sanford Jr. 2005. Effects of nitrogen deposition on an arid grassland in the Colorado Plateau Cold Desert. Rangeland Ecology and Management 58(6):565-574.

Taylor, J.P. and K.C. McDaniel. 1998. Riparian management on the Bosque del Apache National Wildlife Refuge. p 219-232. In Herrera, E.A., T.G. Bahr, C.T. Ortega Klett, and B.J. Creel (Eds). Water Resource Issues in New Mexico, New Mexico Journal of Science Volume 38, November 1998.

VICAIRE 2006 (VIrtual CAmpus In hydrology and water REsources management) Modified : March 9, 2006 Module 3 Groundwater Hydrology, Chapter

2 Porous Media - The Liquid Phase, assessed October 29, 2009, http://echo.epfl.ch/VICAIRE/mod_3/chapt_2/main.htm

Winkel, V.K., B.A. Roundy, and J.R. Cox. 1991. Influence of seedbed microsite characteristics on grass seedling emergence. Journal of Range Management 44(3):210-214.

Wynia, R. 2007. USDA-NRCS Plant Guide - Blue Grama Bouteloua gracilis (Willd. ex Kunth) Lag. ex Griffiths. 3p. http://plants.usda.gov/plantguide/pdf/pg_bogr2.pdf accessed November 23, 2009.

# SYNTHESIS

# REACHING TOWARD THE INTEGRATION OF RESEARCH INTO RESOURCE MANAGEMENT ACTIVITIES: A 20-YEAR EVALUATION OF COLORADO PLATEAU BIENNIAL CONFERENCES

*Martha E. Lee, Carena J. van Riper, Charles van Riper III and Gerald T. Kyle*

## ABSTRACT

The Biennial Conferences of Research on the Colorado Plateau serve a critical purpose of connecting management and science in the southwestern US. The primary goal of the conferences has been to create a forum where managers and scientists could come together to learn about and discuss scientific findings to be incorporated in resource management and planning. The purpose of the present study is to assess whether, after 20 years, the conference is still meeting that goal of scientific integration, and ascertain how conference attendees believe the conference is of use, determine what aspects they would like to see retained and/or removed, and examine ways the conference could be improved. An on-line survey was administered to attendees of the 2009 10th Biennial Conference of Research on the Colorado Plateau. We utilize an importance-performance analysis and assess social networking among conference attendees. Results of our analyses show the conference is performing well and attendees are relatively satisfied with the organization, delivery and content of the meeting. Maintaining a reasonable cost and providing opportunities to network with speakers and other attendees are identified as particularly important elements that need to be maintained or enhanced. We found that interpersonal connections, social cohesion, secondary associations, and perceived utility emerged as dimensions of social networking and shaped the relationships formed among conference attendees. We anticipate that the findings of this conference evaluation will help future planners better meet the needs of scientists, managers, administrators, and students.

## INTRODUCTION

A total of 10 Biennial Conferences of Research on the Colorado Plateau (COPL) have provided a scientific forum for the presentation and discussion of resource management topics related to biological, physical, cultural, and social conditions on the Colorado Plateau. Since the first meeting was held in 1991 at Northern Arizona University (NAU), the COPL conferences have served the critical purpose of facilitating intellectual exchange about salient resource management topics in the southwestern US. This meeting has created a place where managers feel comfortable interfacing with scientists, such that new ideas can be implemented in resource management, and provided opportunities to present and publish innovative research findings in a book series (e.g., van Riper and Cole 2004; van Riper and Mattson 2005; van Riper and Sogge 2008; van Riper et al 2010; http://www.uapress.arizona.edu/BOOKS/bid2255.htm). Over this 20-year time period, the organizers have strived to promote discussion, information sharing and productive communication among participants, with a focus on enhancing social networks among scientists, managers, administrators and student attendees. The COPL conferences have been widely supported across Department of the Interior agencies (e.g., U.S. Geological Survey; National Park Service; Bureau of Land Management; Bureau of Reclamation; U.S. Fish and Wildlife Service), as well as the U.S. Forest Service, and Arizona Game and Fish.

Although the COPL conferences continue to garner support among federal and state agencies, as well as partner universities, the organizers felt it was necessary to take a step back and assess whether the COPL conferences have been meeting the needs and aspirations of attendees. Are the COPL conferences living up to participants' standards? How effective are the meetings for integrating research into management activity? What is the extent to which professional networking occurs at the COPL conferences? How can organizers most efficiently and effectively focus their efforts to improve future meetings? These questions guided the development of this research designed to improve the conference experience and help ensure relevant outcomes for science and resource management on the Colorado Plateau.

### Study Purpose

In this paper, we report the results of a self-evaluation of participant experience at the 2009 10th Biennial Conference of Research on the Colorado Plateau held in Flagstaff, AZ. This evaluation was deemed necessary because of the changes that have occurred over the 10-conference series, and an expanding pool of participants. We determined the perceived benefits of the conference among attendees, examined which aspects could be retained and/or removed without compromising effectiveness, and explored areas for improvement. Our analysis offers suggestions for improving the conference experience so that future hosts and organizers can more efficiently invest their time, energy and programming. The objectives of this study were as follows:

1. Describe the socio-demographic characteristics, professional makeup, conference attendance history, and level of participation among attendees.
2. Determine the relative degrees of importance and satisfaction that attendees associate with the organizational components of the conference.
3. Examine social networking among conference attendees.

## BACKGROUND INFORMATION

Over the past 20 years of COPL conferences, organizers have provided a suite of opportunities to share research findings, facilitate communication and enhance professional skills among attendees. These opportunities are referred to herein as the programmatic components of the COPL conference. For the purposes of this paper, these components were identified by the conference organizers and attendees during two focus group sessions aimed at capturing the most salient aspects of the meeting, such as paper and poster presentations given by academic and agency professionals, panel discussions, Client Day, and a Career Day information session. These events, in addition to optional fieldtrips and social activities, were considered relevant owing to the COPL goals and objectives, as well as their potential to enhance the conference experience.

### Importance-Performance Analysis

We used an Importance-Performance Analysis (IPA) of the COPL programmatic components to determine if the conference was meeting the expectations and needs of conference attendees. This technique was introduced in the field of marketing (Martilla and James 1977) to compare the utility of brands, products and services with consumers' satisfaction of those same elements. The use of IPA has been widely adopted in various service industries for determining the quality or performance of service and education providers. Janes and

| | |
|---|---|
| **Concentrate Here** | **Keep up the Good Work** |
| **Low Priority** | **Possible Overkill** |

(Importance — vertical axis label)

**Performance/Satisfaction**

Figure 16.1    Example of an Importance-Performance Analysis grid.

Wisnom (Janes and Wisnom 2003) reviewed 42 studies that employed IPA in hospitality and tourism industry research. This technique has also been used to evaluate visitor experiences at recreational areas and other theme sites (Hollenhorst et al. 1992; Ellis and Vogelsong 2004; Rivera et al. 2009), service quality at university programs (Kitcharoen 2004; Ford et al. 1999), and professional conferences (Severt et al. 2006; van Riper and Healy 2008).

Our IPA provided a mechanism for assessing COPL conference attendees' preferences for the various programmatic components of the conference, and evaluating how well the COPL conference performed on these same components. IPA involves three steps: 1) selection of relevant attributes, 2) measurement of the importance and performance for each attribute, and 3) plots of the average score given to each component on a two-dimensional grid with importance scores displayed on the vertical axis and performance scores along a horizontal axis (Figure 16.1). The intersection of the importance and performance axes is centered on mean values rather than the center of the grid. This approach emphasizes the relative differences among items being evaluated (Manning 2011).

All programmatic components fall within one of the four grid quadrants. The upper left-hand quadrant includes elements that conference attendees perceive as being important but not offered at the desired performance level (i.e., "Concentrate Here"). The conference organizers need to focus on improving performance related to these components. The lower left quadrant showed programmatic components that are of less importance and average performance (i.e., "Low Priority") because both importance and performance are lower than the average. These elements should receive a low priority in resource allocation decisions. The upper right quadrant shows programmatic components that are important and have high performance (i.e., "Keep Up the Good Work"). Conference organizers are performing well with respect to these components. The lower right quadrant shows components that are of less importance but high performance (i.e., "Possible Overkill"), suggesting that conference organizers should place less emphasis on improving these components (O'Sullivan 1991; Wade and Eagles 2003). IPA helps determine whether the COPL conference is meeting the needs of conference attendees, and provides future conference organizers with insight into the

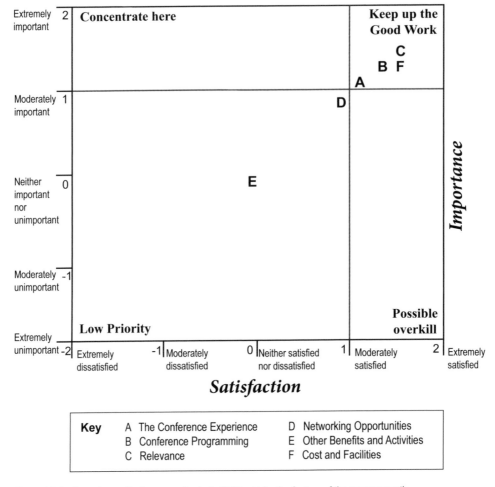

Figure 16.2    Importance-Performance Analysis (IPA) grid for the factors of the programmatic components of the 10th Biennial Conference of Research on the Colorado Plateau.

extent to which attendees believe they should retain, expand or possibly remove elements of the meeting.

Social Networking

This paper offers a perspective on social networking theory derived from the idea that conferences provide an external social structure that allows attendees to generate networks of secondary associations, form high levels of interpersonal trust, perceive mutual aid, and recognize feelings of reciprocity (Arai and Pedlar 2003; Glover and Hemingway 2005; Putnam 2007). Conference participation is assumed to be a voluntary undertaking that fosters collective decisions that further the goals of scientific and management communities tied to the southwestern US. This conceptual framework is guided by part of the COPL conference mission to connect science and management through forming networks and promoting social interaction on formal and informal bases (Glover et al. 2005). Thus, there is a utility for social networking theory to assess the outcomes of conference participation.

Social networking is closely related to the idea of social capital, which is theorized to arise as a consequence of relationships, sociability and informal interaction (Bourdieu 1986; Coleman 1988; Putnam 2007). Individuals collectively build social capital by creating networks, norms and interpersonal trust and by acting to foster a sense of social cohesion (Putnam 1995). A number of underpinning themes emerge in the literature on social networking, including shared norms of reciprocity, interpersonal trust, and social networks. Researchers have explored these themes in the context of the social sciences (Glover and Hemingway 2005), examining how professional networks can be established through shared resources (van Riper and Healy 2008) and how leisure pursuits contribute to a sense of community (Arai and Pedlar 2003). Many of these themes align with the assumed beneficial qualities that are encouraged by the COPL conference organizers and achieved by conference attendees.

There are various forms of social networks that can emerge among members of a community (Ballet et al. 2007). These connections can help to expand and promote social cohesion while also having potential to discourage new relationships outside existing networks (Gargiulo and Benassi 2000). In this sense, both positive and negative outcomes can be produced by social capital. For example, Newman and Dale (Newman and Dale 2007) argue that social capital is comprised of "bridging" and "bonding" ties. Bridging ties refer to intergroup relationships that tend to promote diverse, flexible and adaptive networks. Bonding ties on the other hand, refer to interpersonal relations that exist within a close-knit community. These ties are often homogenous and centralized, in that there's potential for bonding ties to impede the formation of new connections. It is within the concepts of bridging and bonding ties that we frame the emergent benefits of attending the COPL meeting.

## METHODS

### Survey Questionnaire

The 2009 COPL conference was evaluated through an online survey of participants using Survey Monkey™ (Dillman 2007). Electronic survey questionnaires were distributed to all conference attendees with valid e-mail addresses one month after the symposium. We contacted 336 attendees by email, sent the first follow-up reminder a week later, and a second reminder three weeks after the initial e-mail contact. A total of 169 respondents completed the survey (50% response rate). The questionnaire was divided into three sections: 1) background information on respondents and participation in the COPL conference, 2) respondent evaluations of the importance of and satisfaction with the programmatic components of the COPL conference, and 3) measures of social networking and perceived utility associated with the COPL conference. The survey also asked respondents a suite of socio-demographic questions.

### Survey Measures

#### Importance-Performance Analysis

We designed the programmatic components for an IPA using results from two focus group sessions: one conducted during the 2007 Biennial Conference of Research on the Colorado Plateau held in Flagstaff, AZ and a second at the 2008 Colorado River Basin Science and Management Symposium held in Scottsdale, AZ. The topics of the focus group sessions were to answer the following question: "What are the attributes of an excellent conference?" Results of the focus group sessions were used to develop survey questions. We also identified salient themes and gleaned questions from the conference evaluation and social networking literature.

The final list of 37 programmatic components that we utilized addressed the organizational details of the conference, the content and delivery of conference presentations, opportunities to network and interact with speakers and other attendees, conference location and amenities, and additional activities such as field trips. Conference attendees were asked to rank the importance of each of these elements in deciding whether or not to attend. These items were ranked on a 5-point Likert scale ranging from "very important" to "very unimportant." Participants were then asked to report how satisfied they were at the 10th Biennial Conference with each of the conference attributes. This measure of satisfaction was treated as a proxy to measure the "performance" aspect of the IPA. Gap scores between the measures of importance and performance were analyzed for incongruities. These scores were obtained by subtracting the perceived importance scores from the performance scores. If the difference was negative, performance fell short of respondents' importance rating. If the difference was positive, performance exceeded importance. If there was no difference, performance equaled importance. A paired-samples t-test was used to identify significant differences between importance and performance measures for each programmatic component.

We used principal components (with varimax rotation) analysis (PCA) to group the 37 conference elements rated for importance into underlying constructs or themes. The PCA produced six underlying factors that accounted for 51.9 percent of the total variance. Survey respondents were assigned an importance score for each factor based on their averaged responses for the items loading onto each dimension. Mean satisfaction scores for each factor were also calculated. Importance and performance scores for the six factors were plotted on an IPA grid.

### Social Networking Analysis

Social networking was measured using survey items derived from past research (Arai and Pedlar 2003; van Riper and Healy 2008). Data analysis first involved replacing missing data using a multiple imputation procedure in PRELIS 8.70 (Graham et al. 2003; Collins et al. 2001; Schafer and Graham 2002). We then conducted an exploratory factor analysis (EFA) using PASW Statistics version 18.0. All scale items were included in this procedure to determine how items fell into conceptual categories. This step helped us to identify patterns and groupings of items and evaluate overall consistency. We retained and/or rejected survey items according to low loading scores and cross loading using the oblique (Promax) rotation. Eigenvalues greater than 1.0 were accepted. The reliability measures of the proposed dimensions indicated good internal consistency according to Chronbach's Alpha scores. The final EFA solution identified eight latent constructs that accounted for 62.9% of the variance captured by the proposed dimensions. We also conducted a confirmatory factor analysis (CFA) to test the factors structure suggested in the EFA. Factor loadings greater than 0.5 were retained (Hair et al. 1998). This procedure allowed us to further clarify which items conceptually aligned with various identified dimensions.

Three dimensions that reflected the bonding and bridging ties mentioned above (Newman and Dale 2007) were hypothesized to form the construct of social networking, including Interpersonal Connections (e.g., "I connected with individuals outside of my area of expertise at the 10th Biennial Conference"), Social Cohesion (e.g., "The 10th Biennial Conference helps to build a

| Variables | | Percent |
|---|---|---|
| Education | High school graduate | 0.7 |
| | Technical school or Associates' degree | 0.7 |
| | Bachelor's degree | 22.3 |
| | Master's degree | 38.5 |
| | Ph.D., M.D., J.D., or equivalent | 37.8 |
| Gender | Female | 56.8 |
| | Male | 43.2 |
| Current Employer | Federal government | 38.5 |
| | State or county government | 8.8 |
| | Nonprofit organization | 11.5 |
| | Private business | 2.7 |
| | University | 30.4 |
| | Unemployed | 2.7 |
| | Other (e.g., retired, international organization, self-employed) | 5.4 |
| Length of Time Employed by Current Employer | Less than a year | 10.1 |
| | 1 - 5 years | 36.5 |
| | 6-10 years | 21.0 |
| | 11-20 years | 20.3 |
| | More than 20 years | 9.4 |
| Job Title | Agronomist | 0.7 |
| | Attorney | 0.7 |
| | Biostatistician | 0.7 |
| | Economist | 0.7 |
| | Hydrologist | 0.7 |
| | Non-profit Organization Board Member | 0.7 |
| | Forester | 1.4 |
| | Habitat-Resource Specialist | 1.4 |
| | Retired | 1.4 |
| | Archaeologist | 2.1 |
| | Botanist | 2.8 |
| | Consultant | 2.8 |
| | Program Coordinator | 2.8 |
| | Technician | 2.8 |
| | Program/Research Manager | 3.5 |
| | Conservation Biologist | 4.8 |
| | Administration | 9.7 |
| | Researcher | 9.7 |
| | Ecologist | 12.4 |
| | Academic | 12.4 |
| | Biologist | 12.4 |
| | Student | 15.2 |

Table 16.1  Profile of survey respondents at the 10th Biennial Conferences of Research on the Colorado Plateau.

| Variables | Percent |
|---|---|
| Conference Participation | |
| First-time attendee of the Biennial Conference? | |
| Yes | 54.1 |
| No | 45.9 |
| Number of previous conference visits (n=68) | |
| 1 | 32.4 |
| 2-3 | 39.7 |
| 4-5 | 17.6 |
| 5+ | 10.3 |
| Likely to attend the next Biennial Conference? | |
| Very likely | 52.0 |
| Somewhat likely | 27.0 |
| Neither likely nor unlikely | 14.2 |
| Somewhat unlikely | 4.7 |
| Very unlikely | 2.0 |
| Gave a paper or talk at the conference? | |
| Yes | 48.0 |
| No | 52.0 |
| Attended the Colorado River Basin Science and Resource Management Symposium held November 18-20 in Scottsdale, AZ? | 5.4 |
| Satisfaction with the conference overall (n=161)* | |
| Very satisfied | 54.7 |
| Somewhat satisfied | 40.4 |
| Neither satisfied nor dissatisfied | 3.7 |
| Somwhat dissatisfied | 1.2 |
| Very dissatisfied | 0.0 |

*Satisfaction scale, where 2=very satisfied, 1=moderately satisfied, 0=neither satisfied nor dissatisfied, -1=moderately dissatisfied, -2=very dissatisfied.

Table 16.2 Involvement in activities and satisfaction with the 10th Biennial Conference of Research on the Colorado Plateau (n=148)

sense of community in my area of expertise"), and Secondary Associations (e.g., "The 10th Biennial Conference brings together people who come from different areas of expertise"). Respondents were asked to report the extent to which they agreed or disagreed with statements measured on a Likert-type scale ranging from 1 to 5.

Perceived utility of the meeting was also assessed using five survey items that measured the extent to which conference participants associated professional and career-oriented beneficial qualities with conference attendance. In other words, attendees were thought to become satisfied with the meeting according to the extent to which they reaped benefits and perceived the conference to be worthwhile. Respondents rated their agreement or disagreement with the survey items included in this dimension on a five point Likert-type scale.

## RESULTS

### Profile of Survey Respondents

A profile of survey respondents and their level of participation in the COPL conference are reported in Tables 16.1 and 16.2. Almost 100 percent of respondents had a minimum of a bachelor's degree while three-quarters had at least a master's degree. Gender was almost equally divided between men (43%) and women (57%). Close to 40% worked for the federal government, 30% worked for a university, 11% were associated with nonprofit organizations, 9% worked for state or county government, and 5% were students. The most often listed job titles were "ecologist, academic, biologist, student, researcher, administrator, and conservation biologist." Survey respondents were active participants in the COPL conference, in that 48% gave a paper or a talk and 86% had prepared a poster.

Fifty-four percent were first-time attendees at the COPL conference. Most of the 46% of previous attendees had been to one or two previous COPL conferences. Seventy-nine percent said they were likely or very likely to attend the next meeting. In contrast, only 5% had attended the 2008 Colorado River Basin Science and Resource Management Symposium.

### Importance-Performance Analysis

The Importance-Performance portion of the survey was designed to assess COPL conference attendees' opinions regarding the most important elements of the conference and their impressions of how it performed on those same elements. There were 37 programmatic components included in this analysis, which were assessed for "importance" and "performance" (i.e., satisfaction; Table 16.3). Conference attendees rated the majority of the conference elements as being at least somewhat important. The features of the COPL conference rated most important

included "networking opportunities at the conference," "opportunities to talk with other conference attendees," "organization of the conference schedule," "good acoustics and audio-visual equipment at the conference," "conference cost is reasonable," and "keynote speakers are experts in their fields." Rated as least important were "availability of team building activities," "Client's Day as part of the conference," and "Continuing Education Units (CEU) offered for attending the conference."

Conference attendees were very positive in their performance ratings of the conference. Ninety-five percent indicated they were either "somewhat satisfied" or "very satisfied" with the overall conference (40.4% and 54.7%, respectively). Survey participants rated the COPL conference highest on "clean and comfortable facilities," "convenience of the conference location," "food at the conference," "timely program topics," and "a mix of talks with specific and broader themes."

The comparison of the importance and performance ratings on individual programmatic components revealed little difference between these two measures as the majority of programmatic components were rated between moderately and extremely important. However, the comparison revealed components that needed additional emphasis verses those that received too much attention from conference organizers (i.e., overkill). The mean satisfaction rating for the conference cost was rated only slightly less than moderately satisfied and the gap between the two scores was significantly different. Networking opportunities at the conference and having time available to question conference presenters also showed statistically significant differences between the performance scores for these two elements, suggesting that conference cost and ensuring networking opportunities and interaction among presenters were programmatic components that could be

| Component | Importance | Satisfaction | Discrepancy | t-value |
|---|---|---|---|---|
| Organization of the conference schedule | 1.49 | 1.33 | -0.16 | 2.12 |
| Networking opportunities at the conference | 1.53 | 1.34 | -0.19 | 3.37* |
| Conference cost is reasonable | 1.47 | 0.80 | -0.67 | 6.47* |
| Keynote speakers are experts in their fields | 1.47 | 1.38 | -0.08 | 1.24 |
| Applications to resource management are evident | 1.39 | 1.39 | 0.00 | 0.00 |
| Good acoustics and audio-visual equipment at the conference | 1.48 | 1.54 | 0.06 | -0.74 |
| Opportunities to make new contacts | 1.40 | 1.41 | 0.01 | -.022 |
| Conference theme appeals to both managers and scientists | 1.41 | 1.37 | -0.04 | 0.62 |
| Timely program topics | 1.45 | 1.48 | 0.03 | -0.55 |
| A mix of talks with specific and broader themes | 1.42 | 1.47 | 1.33 | -0.85 |
| Mixed audience of managers and scientists | 1.41 | 1.40 | -0.01 | 0.22 |
| Time available to question speakers after their presentations | 1.40 | 0.97 | -0.43 | 5.16* |
| Sessions stay on track | 1.27 | 1.25 | -0.02 | 0.20 |
| Convenience of conference location | 1.21 | 1.72 | 0.51 | 7.94* |
| Online registration | 1.15 | 1.58 | 0.43 | -6.88* |
| Clean, comfortable facilities | 1.08 | 1.81 | 0.73 | -13.8* |
| Conference proceedings are made readily available | 1.03 | 0.97 | -0.06 | 0.61 |
| Conference abstracts available ahead of time | 1.10 | 1.27 | 0.17 | -1.97 |
| Having both oral presentations and a poster session | 1.09 | 1.40 | 0.31 | -3.79* |
| Conference length | 1.01 | 1.45 | 0.44 | -6.16* |
| Adequate time allowed between sessions | 1.01 | 1.06 | 0.05 | -0.65 |
| Panels with interactive speaker discussions | 0.98 | 0.80 | -0.18 | 2.08 |
| Conference organizer contact information is readily available | 0.89 | 1.10 | 0.21 | -2.40 |
| Food at the conference | 0.82 | 1.54 | 1.46 | -9.29* |
| Professional development opportunities for students | 0.97 | 0.70 | -0.27 | 2.84* |
| Built-in downtime at the conference | 0.73 | 1.02 | 0.29 | -3.36* |
| Interactive activities included in the agenda (workshops, hands-on activities) | 0.76 | 0.67 | -0.09 | 0.97 |
| Introduction and wrap-up sessions | 0.64 | 0.80 | 0.16 | -2.01 |
| Social events held during the conference | 0.63 | 0.99 | 0.36 | -4.18* |
| A single theme for the conference (vs. multiple themes) | 0.21 | 0.84 | 0.63 | -8.53* |
| Online evaluations to give post-conference feedback | 0.24 | 0.91 | 0.67 | -8.52* |
| Field trips | 0.27 | 0.14 | -0.13 | 1.26 |
| Client's day as part of the conference | -0.10 | 0.40 | 0.30 | -4.95* |
| Continuing Education Units (CEU) | 0.07 | 0.34 | 0.27 | -2.43 |
| Availability of team-building activities | -0.24 | 0.14 | 0.10 | -3.10* |

*Significant at p < 0.01.*

Table 16.3 Importance and performance scores for the programmatic components of the 10th Biennial Conference of Research on the Colorado Plateau.

improved upon. On the other hand, clean and comfortable facilities were rated as moderately important and close to very satisfied in performance, suggesting that the conference facilities exceeded what was needed. This is particularly important in light of the importance-performance gap found for having a reasonable conference cost.

The factor analysis of programmatic components identified six factor groups or conference dimensions: The Conference Experience, Conference Programming, Networking Opportunities, Other Benefits and Activities, Relevance, and Cost and Facilities (see Table 16.4).

The performance ratings of the programmatic components were put into these same groups and compared in the Importance-Performance Analysis (IPA) grid shown in Figure 16.2. The IPA grid included mean values for the six conference components for importance and performance to emphasize the relative differences among the components. The fact that four of the six factors fall within the "Keep up the Good Work" quadrant of the IPA grid reflects the positive experiences attendees had at the conference and sends a message to future conference organizers to follow the example of the 10th COPL conference. Additionally, future planners should continue to emphasize making the conference relevant to resource management, a well-programmed conference experience at a reasonable cost, and convenient facilities. If needed, less time and effort could be spent on providing other benefits and activities such as field trips and team-building activities. Even though networking opportunities fell within the "Low Priority" portion of the IPA grid, it should be noted that networking opportunities were consistently rated as an important element of the conference.

## Social Networking

Respondents reported high levels of agreement with all items associated with social networking among conference attendees (Table 16.5). Respondents most strongly agreed with the statements, "I think the COPL conference is a beneficial conference" (M = 4.7), "The COPL conference brings together people who come from different areas of expertise" (M = 4.5), "I feel comfortable at the conference" (M = 4.5), and "I value the relationships I established with people who attended the COPL conference" (M = 4.2). These findings support the notion that conference attendance supports and encourages the formation of social networks among attendees.

To further refine our understanding of social networking among COPL attendees, we determined the groupings of survey items and emergent dimensions. Our analysis approach – data cleaning, EFA and CFA – yielded four dimensions. The fit indices for the final model indicated an acceptable model fit ($X^2$ = 1470.98, df = 547, RMSEA= 0.092, NNFI= 0.91, CFI= 0.917) and internal consistency (Cronbach's Alpha ranged from 0.765 to 0.881). First, a dimension of "interpersonal connections" (M = 3.91) emerged, suggesting that social networks were formed through new and existing professional relationships that lasted over time and were considered to be important. Second, a dimension related to "social cohesion" (M = 4.1) contributed to the formation of social networks. This second dimension was formed around the cohesiveness of the larger COPL community and beneficial qualities of the attendance. Third, a dimension of "secondary associations" (M = 3.8) was integral to the formation of social networks at the COPL meeting. According to this dimension, the COPL meeting brought together attendees from within and outside areas of expertise, drawing on the opportunities for informal interaction. Finally, "perceived utility" (M = 3.6) emerged as a dimension, in that attendees united on the basis of shared goals and objectives and feelings of comfort in attending the COPL conference.

| Factor | loadings | mean | α |
|---|---|---|---|
| The Conference Experience | | 1.04 | .832 |
| Built-in downtime at the conference | .477 | | |
| Introduction and wrap-up sessions | .625 | | |
| Adequate time allowed between sessions | .444 | | |
| Mixed audience of managers and scientists | .699 | | |
| Time available to question speakers | .682 | | |
| Panels with interactive speaker discussions | .655 | | |
| Food at the conference | .581 | | |
| Opportunities to talk with other conference participants | .553 | | |
| Conference length | .427 | | |
| Conference Programming | | 1.11 | .796 |
| Organization of the conference schedule | .445 | | |
| Conference proceedings made readily available | .576 | | |
| Conference abstracts available ahead of time | .693 | | |
| Online registration | .654 | | |
| Good acoustics and AV equipment | .601 | | |
| Conference organizer contact information is made readily available | .530 | | |
| Online evaluation | .476 | | |
| Mix of talks with specific and broader themes | .395 | | |
| Keynote speakers are experts in their fields | .542 | | |
| Networking Opportunities | | 0.84 | .816 |
| Networking opportunities | .697 | | |
| Opportunities to make new contacts | .750 | | |
| Having both oral presentations and a poster session | .595 | | |
| Interactive activities included | .534 | | |
| Social events held during the conference | .519 | | |
| Client's Day as part of the conference | .512 | | |
| Professional development opportunities for students | .372 | | |
| Other Benefits and Activities | | -0.96 | .801 |
| Continuing Education Units (CEU) offered | .760 | | |
| Field trips | .697 | | |
| Team-building activities | .793 | | |

Table 16.4 Factor loadings, mean values and internal consistency of the importance scores for the programmatic components of the 10th Biennial Conference of Research on the Colorado Plateau.

| Factor | loadings | mean | α |
|---|---|---|---|
| Relevance | | 1.41 | .715 |
|     Applications to resource management are evident | .723 | | |
|     Theme appeals to managers and scientists | .603 | | |
|     Timely program topics | .511 | | |
|     Conference topics relevant to my job | .581 | | |
| Cost and Facilities | | 1.25 | .540 |
|     Convenience of conference location | .598 | | |
|     Sessions stay on track | .510 | | |
|     Conference cost is reasonable | .571 | | |
|     Clean and comfortable facilities | .567 | | |

*Note. Importance scale ranged from 2 = very important, 1 = moderately important, 0 = neither important nor unimportant, -1 = moderately unimportant, -2 = very unimportant.*

Table 16.4 con't. Factor loadings, mean values and internal consistency of the importance scores for the programmatic components of the 10th Biennial Conference of Research on the Colorado Plateau.

## DISCUSSION

Our evaluation of the 10th Biennial Conference of Research on the Colorado Plateau offer support for the importance of the COPL conference as a venue to enhance communication among scientists, managers, administrators, and student doing work in the southwestern US. We examined the importance and performance of the programmatic components of the COPL meeting and offered insight on how conference participation helped generate forms of social networking among attendees. This information provides conference organizers with insights on how to tailor the conference experience to better suit the needs of the attendees, inform the supporting institutions and/or agencies of their successes, and points out where improvements could occur.

Our findings suggest that the 10th COPL conference was highly successful. The conference attracted a mix of university faculty, federal, state, and county land managers, non-profit organization personnel, and students. The IPA revealed that conference attendees were overall very satisfied with how the conference was designed and delivered. The gap analysis of the importance and performance of individual programmatic components of the conference revealed specific areas that required attention, specifically keeping the cost of the Biennial Conference reasonable, providing opportunities for networking, and providing additional time for talking with presenters and other participants. Keeping the conference theme and topics relevant for both managers and scientists, with a focus on applications for resource management, is another important component rated high in performance, which should be maintained in future COPL conferences.

Another message that emerged from the IPA concerned the difficulties in balancing conference quality and cost. The High Country Conference Center was an outstanding venue for the COPL conference. The facilities, A-V equipment and food were consistently highly rated by survey respondents. The availability of this venue for hosting the meeting is of considerable

| Statement | Factor loadings | Mean | S.D. | α |
|---|---|---|---|---|
| Social Networking Scale Items | | | | |
| Interpersonal Connectsion | | 3.905 | | .877 |
| As a result of the scientific program sessions, I established new professional relationships at the COPL conference | .750 | 3.73 | .93 | |
| Through informal social interactions I established new professional relationships at the COPL conference | .742 | 3.81 | .97 | |
| I connected with individuals outside of my area of expertise at the COPL conference | .728 | 3.95 | .91 | |
| I will continue to associate with the individuals that I met at the COPL conference | .832 | 3.84 | 1.04 | |
| I value the relationships I established with people who attended the COPL conference | .786 | 4.20 | .87 | |
| Social Cohesion | | 4.059 | | |
| The COPL conference promotes a sense of social cohesion on the Colorado Plateau | .758 | 4.02 | .88 | |
| The COPL conference helps to build a sense of community in my area of expertise | .747 | 3.69 | 1.17 | |
| The connections I made at the COPL conference will benefit me in the future | .722 | 4.07 | .91 | |
| Secondary Associations | | 3.794 | | .808 |
| The COPL conference brings together people who come from different areas of expertise | .666 | 4.45 | .67 | |
| The COPL conference brings together people from my discipline | .633 | 3.72 | 1.14 | |
| This conference has helped me connect with other professionals in my area of expertise | .840 | 3.83 | .98 | |
| I formed a social network through informal social interactions at the conference | .831 | 3.33 | 1.16 | |
| Perceived Utility | | 3.580 | | .759 |
| I think the COPL conference is a beneficial Conference | .693 | 4.66 | .58 | |
| My colleagues and I want the same thing from the COPL conference | .557 | 3.31 | 1.00 | |
| It is important to me to attend the COPL conference | .665 | 4.16 | .96 | |
| People who attend the COPL conference conference get along with one another | .587 | 3.98 | .87 | |
| I feel comfortable at the conference | .642 | 4.45 | .67 | |

Table 16.5   Factor loadings, mean values, standard deviations, and internal consistency of conference attendees' agreement with statements measuring social networking (n = 169)

value to conference organizers and attendees. However, the benefits of holding the conference in such facilities must be balanced with the associated costs. Every effort should be made in future conference planning to maintain a reasonable cost for attendees.

Our analysis of social networking suggested that conference attendees built networks and formed social capital on the bases of trust, perceived mutual aid and feelings of reciprocity (Glover and Hemingway 2005). We found that four dimensions contributed to social networks facilitated by participation in the COPL conference: Interpersonal Connections (five survey items), Social Cohesion (three survey items), Secondary Associations (four survey items), and Perceived Utility (five survey items). These dimensions reflected bonding ties rooted in interactions with individuals in similar areas of expertise and bridging ties that facilitated access to resources and ties across groups of attendees (Newman and Dale 2007). The COPL meeting contributed to the formation of bonding and bridging aspects of social capital. These findings illustrated diverse and active networks, which will inevitably further attendees' professional goals and aspirations. Social interaction and networks formed through participation in the COPL conferences helped connect science and management by fostering more science-based decision making about resources management in the southwestern US.

## CONCLUSION

This study draws on IPA and the idea of social networking to assess whether the 10th Biennial Conference of Research on the Colorado Plateau meets the needs of attendees by fostering communication between scientists and managers, providing a place for formal and informal opportunities for intellectual exchange, and facilitating the creation of social networks. Study findings

offer insight on the relative importance and performance of the programmatic components of the meeting, and the extent to which social networking is encouraged by each of these components. Our findings aim to help conference planners better suit the needs of scientists, managers, administrators, and students that attend future COPL conferences.

## LITERATURE CITED

Arai, S., and A. Pedlar. 2003. Moving beyond individualism in leisure theory: A critical analysis of concepts of community and social engagement. Leisure Studies, 22: 185-202.

Ballet, J., N. Sirven, and M. Requiers-Desjardin. 2007. Social capital and natural resource management: A critical perspective. The Journal of Environment & Development, 16(4): 355-374.

Bourdieu, P. 1986. The forms of social capital. In: J. Richardson (Ed.), Handbook of theory and research for the sociology of education (pp.241-258). New York: Greenwood.

Coleman, J. S. 1988. Social capital in the creation of human capital. American Journal of Sociology, 94: S95-S210.

Collins, L. M., J. L. Schafer, and C. M. Kam. 2001. A comparison of inclusive and restrictive strategies in modern missing data procedures. Psychological Methods, 6: 330-351.

Dillman, D. A. 2007. Mail and internet surveys: The tailored design method. Second edition.

Hoboken, NJ: John Wiley and Sons, Inc.

Ellis, C. L. and H. Vogelsong. 2004. Measuring birdwatcher preferences through importance-performance analysis. Pp. 203-210 In, Murdy, J., comp., ed. Proceedings of the 2003 Northeastern Recreation Research Symposium; April 6-8, 2003; Bolton Landing, NY. Gen. Tech. Rep. NE-317. Newton Square, PA: USDA Forest Service Northeastern Research Station.

Ford, J. B., M. Joseph, and B. Joseph. 1999. Importance-performance analysis as a strategic tool for service marketers: the case of service quality perceptions of business students in New Zealand and the USA. Journal of Services Marketing, 13(2):171-186.

Gargiulo, M. and M. Benassi. 2000. Trapped in your own net? Network cohesion, structural holes and the adaptation of social capital. Organization Science, 11(2): 183-196.

Glover, T. D., and J. L. Hemingway. 2005. Locating leisure in the social capital literature.Journal of Leisure Research, 37(4): 387-401.

Graham, J. W., P. E. Cumsille, and E. Elek-Fisk. 2003. Methods for handling missing data. In J. A. Schinkaand and W. F. Velicer (Eds.). Research Methods in Psychology (pp. 87_114). Volume 2 of Handbook of Psychology (I. B. Weiner, Editor-in-Chief). New York: John Wiley and Sons.

Hair, J. F. Jr., R. E. Anderson, R. L. Tatham, and W. C. Black. 1998. Multivariate Data Analysis, (5th Edition). Upper Saddle River, NJ: Prentice Hall.

Hollenhorst, S., D. Olsen, and R. Fortney. 1992. Use of importance-performance analysis to evaluate state park cabins: The case of the West Virginia state park system. Journal of Park and Recreation Administration 10(1): 1-11.

Janes, P. L. and M. S. Wisnom, 2003. The use of importance performance analysis in the hospitality industry: A comparison of practices. Journal of Quality Assurance in Hospitality and Tourism, 4(1/2): 23-45.

Kitcharoen, K. 2004. The importance-performance analysis of service quality in administrative departments of private universities in Thailand. ABAC Journal, 24(3): 20-46.

Manning, R. E. 2011. Studies in outdoor recreation: Search and research for satisfaction Corvallis, OR: Oregon State University Press.

Martilla, J. A. and J. C. James.1977. Importance-performance analysis. Journal of Marketing 41(1):77-79.

Newman, L. and A. Dale, 2007. Homophily and agency: Creating effective sustainable development networks. Environment, Development and Sustainability, 9, 79-90.

O'Sullivan, E. L. 1991. Marketing for parks, recreation, and leisure. State College, PA: Venture Publishing, Inc.

Putnam, R. D. 1995. Bowling alone: America's declining social capital. Journal of Democracy, 6: 65-78.

Putnam, R. D. 2007. Bowling alone: The collapse and revival of American community. New York, NY: Simon and Schuster.

Rivera, M. A., S. Amir, and D. Severt. 2009. Perceptions of service attributes in a religious theme site: an importance-satisfaction analysis. Journal of Heritage Tourism 4(3): 227-243.

Schafer, J. L. and J. W. Graham. 2002. Missing data: Our view of the state of the art. Psychological Methods, 7: 147-177.

Severt, D., D. Wang, P-J Chen, and D. Breiter. 2006. Examining the motivation, perceived performance, and behavioral intentions of convention attendees: Evidence from a regional conference. Tourism Management 28:399-408.

van Riper, C., III., and K. A. Cole. 2004. The Colorado Plateau: Cultural, Biological, and Physical Research. University of Arizona Press, Tucson, AZ. 279 pp.

van Riper, C., III., and D. J. Mattson. 2005. The Colorado Plateau II: Biophysical, Socioeconomic and Cultural Research. University of Arizona Press, Tucson, AZ. 448 pp

van Riper, C., III., and M. K. Sogge. 2008. The Colorado Plateau III: Integrating research and resources management for effective conservation. The University of Arizona Press, Tucson, AZ. i-xiv; 393 pp.

van Riper, C., III, B. W. Wakeling and T. K. Sisk. 2010. The Colorado Plateau IV: Integrating research and resources management for effective conservation. The University of Arizona Press, Tucson, AZ. i-xii; 335 pp.

van Riper, C. J., and N. Healy. 2008. Perceptions of the International Symposium on Society and Resource Management Student Forum. Society and Natural Resources, 21(8): 1-8.

Wade, D. J. and P. F. J. Eagles. 2003. The use of importance-performance analysis and market segmentation for tourism management in parks and protected areas: an application to Tanzania's national parks. Journal of Ecotourism 2(3):196-212.

# SYNTHESIS: INFORMING COLLABORATIVE CONSERVATION AND MANAGEMENT OF COLORADO PLATEAU RESOURCES

*Miguel L. Villarreal, Charles van Riper III, Carena J. van Riper, Matthew J. Johnson and S. Shane Selleck*

The Colorado Plateau Biennial Conference series continues its tradition of bridging cultural, social, and biophysical research interests, and addressing the needs of scientists and land managers working in a complex geographic area. Like its predecessors, this 10th edition of proceedings brings together a diverse collection of previously unpublished research papers, the majority of which were presented at the 10th Biennial Conference of Research on the Colorado Plateau. The theme of the conference was "Collaborative Conservation in Rapidly Changing Landscapes," and the chapters are broadly aligned under this theme. One recurring research thread in this volume is the linkage that exists among historic, recent and/or potential land-management decisions, ecosystem function, and landscape health. Several chapters examine the impact of past management activities, particularly grazing and fire suppression, on current plant community composition and ecosystem structure. Other chapters present assessments of management tools and concepts (e.g. rapid biological inventories, ecosystem valuation, and landscape connectivity) that have been, or can be, utilized by land managers to conserve, maintain, or restore ecosystem health within changing landscapes of the Colorado Plateau. All of the chapters in this volume focus on illuminating the advantages and challenges of collaborative decision-making for biological- and cultural-resource conservation. Diverse topics ranging from the impacts of aspen decline on avian diversity to vegetative "soft capping" of exposed archeological ruins have the common end-goal of presenting land managers with new tools for improving stewardship of resources in a time of rapidly changing social and ecological conditions.

In this synthesis, we present a brief distillation of each chapter, highlighting research approaches and key findings. This chapter is organized by the four thematic sections of the book: 1) Reaching towards integration on the Colorado Plateau, 2) Assessing natural and man-made threats to ecological systems, 3) Synergy between human and environmental systems: Planning and management frameworks, and 4) Tools for conservation and collaborative decision-making. This synthesis guides readers through the overarching themes of the book, with the goal of stimulating new research ideas and collaborative efforts while highlighting the continued vitality of the Colorado Plateau Biennial Conference book series.

## SECTION ONE: REACHING TOWARDS INTEGRATION ON THE COLORADO PLATEAU

In Chapter 1, Holcomb et al. examine how federal land management mandates are manifested on the landscape and how collaborative decision making can improve cross-boundary conditions in multi-jurisdictional landscapes. A number of studies have previously examined the biological and geographical effects of boundaries (Landres et al. 1998; Saunders et al. 1991). There is, however, surprisingly little scientific information describing the effects of administrative boundaries on ecosystems and ecosystem processes on the Colorado Plateau, an area with one of the highest densities of national parks in the United States. In this chapter, titled "Administrative Boundaries and Ecological Divergence: The Divided History and Coordinated Future of Land Management on the Kaibab Plateau, Arizona USA," Holcomb et al. employ a mixed-methods approach to assess the effects of administrative boundaries and management policies on forest structure, composition, and wildfire behavior. In the quantitative portion of the study, the authors collected vegetation plot data on either side of the boundary between Grand Canyon National Park (GCNP) and the Kaibab National Forest (KNF), examining the plot data to identify differences in species composition and stand structure. The authors also modeled fire behavior in the study area to predict whether management decisions would influence the potential for active crown fire. In the qualitative portion of the study, the authors identified key issues and concerns regarding coordinated resource management and collaborative decision making by conducting interviews with resource managers, research scientists, and members of environmental-advocacy groups.

Results from the research presented in Chapter 1 illuminate management-related ecological differences on the Kaibab plateau, with plots located on KNF characterized by greater aspen and less fir dominance than those in GCNP. In addition to compositional differences, fire-behavior models indicated that GCNP had more land area predicted to support active crown fires whereas KNF had more area predicted to burn as surface fires and passive crown fires. The authors suggest that land management, specifically disturbances related to timber harvest and road building, resulted in the higher

dominance of aspen on KNF. Discussing the salient points from interviews with resource managers, researchers, and environmental advocates, the authors note a general consensus that greater interagency coordination on the Kaibab Plateau would improve ecological conditions; however, political limitations to such coordination remain.

Chapter 2, titled "Assessment of Mixed Conifer Forest Conditions, North Kaibab Ranger District, Arizona USA" contributes to the scientific literature examining and establishing long-term changes in forest conditions of the southwestern US (Fulé et al. 1997; Vankat et al. 2005; Vankat 2011). The goal of this chapter was to inform forest restoration and fuel-reduction activities by identifying biophysical and environmental factors that contribute to mixed conifer-stand density. To accomplish this goal, Sesnie et al. compared contemporary mixed-conifer forest conditions with historical data from 1909 and analyzed the relationship between environmental gradients, forest composition, and forest structure. Their results suggest major changes in mixed conifer forest conditions after 1909. Tree density and basal area, particularly of shade tolerant species, increased considerably, and forest conditions now appear more homogenous across the landscape. Analyzing contemporary stand data with abiotic environmental data, the authors found that elevation and solar radiation have a greater effect on determining species composition than did stand density. The information presented in this chapter adds to a rich literature describing long-term forest dynamics on the Colorado Plateau.

In Chapter 3, Mork et al. examine the impacts of post-wildfire livestock grazing on ponderosa pine understory communities. Their study is the first to document concurrent effects of both fire and grazing disturbances on these communities of the Colorado Plateau, and they focus their research on post-fire invasions by non-native species. Disturbances such as grazing and wildfire are known to alter ecological conditions in a way that can encourage exotic invasion, and new invaders can quickly alter system dynamics (D'antonio and Vitousek 1992). This chapter focuses specifically on the effects of grazing and wildfire on the establishment of cheat-grass (Bromus tectorum), a highly flammable invasive annual that, once established, is capable of altering fire regimes and ecosystem processes (Knapp 1996).

Results from a controlled grazing enclosure experiment initiated two years after the Warm Fire on the Kaibab National Forest indicated greater abundance of Bromus tectorum in grazed plots when compared to ungrazed. Although not statistically significant, the enclosure data suggest that long-term grazing following wildfire may encourage the spread of non-native plants, especially introduced grasses. These results, while preliminary, encourage further long-term research into the relationship between post-fire grazing and non-native species in ponderosa pine understory.

Chapter 4, "Challenges and Opportunities for Ecosystem Services and Policy in Arid and Semi-Arid Environments," closes the first section of the book. This chapter by Bagstad et al. presents challenges that face the Colorado Plateau in applying the ecosystem-services concept to decision making. The three central postulates of this chapter are that: 1) temporal and spatial distribution of water drives ecosystem services on the Colorado Plateau; 2) the location of human beneficiaries matters greatly in the perceived value of an ecosystem service; and, 3) low human population densities over the Plateau lower the perceived value of ecosystem services in general.

The chapter focuses on water as a driver of ecosystem services on the Colorado Plateau, as it is the primary limiting resource in arid and semiarid ecosystems and controls the rates and the timing of biological processes

in dryland species. In addition to riparian and spring systems, the authors discuss the Colorado Plateau uplands, which provide a range of important ecosystem services including carbon sequestration and storage, dust and sediment regulation, and forage provision (Miller et al. 2011). The chapter also examines the importance of watershed and airshed position as related to ecosystem services. On the Colorado Plateau, researchers and managers increasingly recognize the importance of accounting for the supply and demand of ecosystem services and their spatial and temporal flow patterns in preparing rigorous assessments of managed landscapes. The flow of surface and groundwater carries benefits toward human recipients far from the Plateau, reaching as far west as California and south into Mexico. The arrangement of water users and supplies in the Southwest has evolved dramatically over the last century and is likely to continue to do so in the future.

This final chapter ends with an in-depth discussion of the implications of ecosystem services for science and policy in the southwest. The authors discuss the difficulty of linking water quantity, quality, location, and timing to ecosystem service provision, watershed position, and long-distance beneficiary flows, which partially explains why an ecosystem services approach to date has failed to impact watershed management decisions on a meaningful level. They also point out that further evolution of science and policy is needed to bring ecosystem services into decision making on the Colorado Plateau. Continued development of the ecosystem services field will allow managers and scientists to improve resource-management tools and policies in the future.

## SECTION 2: ASSESSING NATURAL AND MAN-MADE THREATS TO ECOLOGICAL SYSTEMS

In Chapter 5, Floyd et al. present results of research at six long-term monitoring sites in Chaco Culture National Historic Park, where they compared the consequences of past grazing regimes on plant species richness, cover of biological soil crusts, shrub density, vegetative cover, and plant community composition. Livestock grazing has long been held to have serious ecological consequences in the arid Southwest and may become even more problematic with global climate change projections of warmer and more arid conditions (Garfin et al. 2010). The authors examined the conditions of two grazing exclosures, one that was fenced in the 1930s and another in the 1990s, such that long-term protection (>60 years of exclosure), recent protection (≤12 years), and current grazing treatments were immediately adjacent at six sites. They found that plant community structure reestablished differentially with protection from grazing depending on the inherent biotic potential of each site (e.g. variability of soil quality). The greatest differences between surveys were in forb characteristics, although much of this was related to the amount of precipitation the site received prior to the recent survey.

In areas of long-term protected treatments, signs of recovery included the abundance of biological soil crust and plant biodiversity, and plant species richness and total cover was greater under long-term protection at all six sites. Depending on soil substrate, some sites had higher shrub recovery while others experienced increases in grasses. Information on signs of biological soil crust recovery, important for stabilizing soils and mitigating dust in arid ecosystems, and the variable vegetation trends presented in this chapter can be used to guide management decisions on grazing and restoration on the Colorado Plateau.

In Chapter 6, Bombaci and Korb report the effects of sudden aspen decline (SAD) on avian species composition in southwest Colorado Plateau forests. This is a timely chapter, in that Worrall et al. (2008) found that aspen (Populus tremuloides) forests in southwestern Colorado have experienced the greatest SAD-related declines in the region, losing over 10 percent of total aspen cover. Compounding this loss is the fact that aspen regeneration typically occurs vegetatively after a disturbance via a clonal root system (Schier et al. 1985) and SAD loss is characterized by a lack of suckering growth beneath declining aspen stands.

Aspen habitat is considered one of the most biologically diverse ecosystems in the west (Kay 1997), and it is especially important for birds (Robbins et al. 1986). In this chapter, the authors conducted an assessment of changes in avian species composition and biodiversity associated with SAD on the Colorado Plateau with a goal of providing a measure of preliminary changes within aspen ecosystems to establish a basis for long-term monitoring of SAD-related avian community changes. Given the relationship between habitat quality and species diversity, they hypothesized that there would be lower avian biodiversity in SAD-affected stands when compared to healthy stands.

After measuring vegetation structure at avian count stations on the Dolores Ranger District of the San Juan National Forest, the authors found some unexpected results. Differences were detected in both overstory and understory vegetation structure among low, moderate, and high SAD stands, and many of the observed differences were responses to increased aspen mortality. However, contrary to their hypothesis, avian biodiversity was greater in high SAD areas, as indicated by associated higher mean species richness and diversity values. The authors suggest that significantly greater biodiversity values found in high SAD stands may indicate that certain avian species are responding to the newly-available niches provided by changes in vegetation structure associated with SAD. Hole-nesting species had consistently higher numbers in SAD stands, possibly the result of more dead trees that provide greater numbers of nesting cavities. This chapter provides a solid foundation of information on differences in avian species composition, abundance, and biodiversity associated with SAD in southwest Colorado, providing insight into responses of forest-bird communities to future SAD disturbances. Furthermore, given the future likelihood of prolonged and climate-induced drought in western North America (Garfin et al. 2010), and the relationship between SAD and drought (Johnson et al. 2008), this research provides crucial insight into responses of forest-bird communities to future SAD disturbances under climate change.

Chapter 7, by van Riper and Crow, also addresses avian responses to habitat change, shifting the focus to pinyon-juniper vegetation. The authors investigated relationships between birds and vegetation in fuels-reduction treatment areas within pinyon-juniper woodlands over the Colorado Plateau. Their goal was to document differences in obligate bird communities to chained versus hand-cut pinyon-juniper fuels-reduction treatments, relative to undisturbed control sites. They selected 73 study plots in southern Utah and northern Arizona, of which 33 had been previously thinned by hand cutting or chaining, and 40 control sites in untreated pinyon-juniper woodlands. At the 73 locations, they documented vegetation structure and counted birds during the 2005 and 2006 breeding seasons utilizing the variable circular plot technique (Reynolds et al. 1980). Results indicated that the density of pinyon pines was the most important variable in predicting bird species richness at all sites. Within treatments, chaining resulted in significantly fewer birds and decreased species richness, while there

was no difference between selective cutting and control sites. Initially, study areas with selective hand cutting had reduced numbers of birds, but within several years, these sites supported greater bird numbers than non-treatment control sites.

For the 16 pinyon-juniper obligate bird species examined during this study, the authors found that the abundance of five was positively related to pinyon-pine density and two were positively related to juniper density. The five related to pinyon-pine are the Black-chinned Hummingbird (*Archilochus alexandri*), Blue-gray Gnatcatcher *(Polioptila caerulea)*, Black-throated Gray Warbler *(Dendroica nigrescens)*, House Finch *(Carpodacus mexicanus)* and Western Scrub Jay *(Aphelocoma californica)*; the two related to juniper density are the Bewick's Wren *(Thyomanes bewickii)* and Brewer's Sparrow *(Spizella breweri)*. These responses, along with other bird-vegetation relationships influenced by treatment type, can be considered by land managers when planning fuel reduction treatments in pinyon-juniper woodland habitat on the Colorado Plateau.

In chapter 8, Stumpf reviews the literature related to predation and parasitism of songbirds and explores how habitat characteristics influence these factors. This chapter specifically focuses on songbirds that are subject to both high rates of predation and parasitism and discusses how bird species may respond differently to characteristics at different scales, especially in southwestern riparian habitats where linear patches may alter predator and/or parasite diversity and abundance. In most cases, smaller overall habitat patches, linear patches with very narrow cross-sections, and habitat patches adjacent to disturbed or agricultural landscapes lead to higher predation rates. Parasitism is especially prevalent in isolated habitat patches adjacent to and near agricultural areas that have active pesticide programs. In the conclusions to this chapter, Stumpf finds that landscape scale characteristics, including patch size and surrounding landscape, are most often associated with nest predation, and that all three spatial scales (landscape, territory, and nest site) are equally likely to be associated with parasitism including patch size, habitat type, and surrounding landscape. She concludes by suggesting that managers seeking to design and implement restoration plans should examine individual patch characteristics as well the surrounding landscape in order to minimize the risk of predation and parasitism to songbirds. Habitat restoration and management of landscapes will benefit from more rigorous, long-term studies incorporating these landscape characteristic measurements, especially in areas that experience high rates of nest predation and parasitism.

## SECTION 3: SYNERGY BETWEEN HUMAN AND ENVIRONMENTAL SYSTEMS: PLANNING AND MANAGEMENT FRAMEWORKS

In Chapter 9, titled "Knowing the Cogs and Wheels: The Utility of Bioblitzes for Conservation," Fertig et al. document the results of a bioblitz in Deer Creek, Utah. A Bioblitz is a biological inventory conducted during a short time period (e.g., 24-48 hours) within a specified area. As a citizen science effort, this kind of event helps to connect the public with biologists, promote interest in nature, and serve as a potential first step in acquiring baseline data on species presence for areas of conservation concern. The idea of assembling teams of taxonomic specialists to quickly establish baseline conditions of biological diversity in threatened habitats took hold in the late 1980s with the development of the Rapid Assessment Program (RAP) by several international conservation groups (Wilson 1992). The Deer Creek bioblitz specifically focused on identifying the number and diversity of species in the area, educating

and involving the public, and discovering previously unreported species in the area.

The authors present results from two bioblitzes in July 2007 and May 2008 that occurred in the Deer Creek drainage, an area east of Boulder and adjacent to Grand Staircase-Escalante National Monument and Dixie National Forest, Utah. A team of botanists, bryologists, biologists, entomologists, ecologists, and interested landowners participated in the bioblitzes and documented 588 taxa, including more than two dozen species not previously recorded in the area. Nearly 80% of the vascular plant and vertebrate taxa from the study area were previously documented from other sites in the Deer Creek watershed. Although comparable numbers of species were observed both years, there was nearly 50% turnover in recorded species from 2007 to 2008. The bioblitzes increased communication among biologists from multiple disciplines as well as demonstrated the high biological significance of the Deer Creek drainage, fostering interest in conservation in the local community.

Chapter 10 is a valuation study of the Verde River watershed in central Arizona. West et al. present findings from the first phase of a larger program of research designed to value ecosystem services provided by the Verde River and its watershed. Drawing on mailback questionnaires and 35 semi-structured interviews conducted with stakeholders, the authors identified a range of values ascribed to "ecosystem services", which are characteristics and functional processes of the natural environment that provide benefits to sustain and fulfill human life (Costanza et al. 1997). The study explores how select stakeholders and community members view the functions, processes, and services provided by the Verde River Watershed as well as how they connect to and appreciate this landscape. The findings of West et al. establish a baseline understanding of how and why the area is considered to be important.

The results from this chapter illustrate a diversity of values associated with the Verde River as a place of meaning and value, rather than simply a resource for human use. Nearly 500 ways of valuing the river and its watershed are discussed in the study interviews, the most prominent of which was the existence value or intrinsic worth of the watershed. West et al. referred to this ecosystem service as a "cultural" value that reflected the reasons why local residents believed this area was important (Millennium Ecosystem Assessment 2003). This chapter identifies what people believe contribute to quality of life and elucidates societal connections with important natural resource conditions in the southwest.

Future research can look to this study for guidance on what residents and visitors believe to be the most salient areas of management concern. For example, stakeholders expressed concern over environmental impacts and threats to the future health of the watershed, including water depletion, human development, invasive species, and climate change. In addition to providing a foundation for future research, the issues highlighted in this study will inform educational efforts and help ensure land managers are receptive to public interests.

In Chapter 11, Lim et al. present ongoing laboratory and field research at Mesa Verde National Park, utilizing the archeological ruins at Far View House to address the question of how best to protect exposed masonry walls from environmental stresses. Exposed ruin walls at archeological sites on the Colorado Plateau are typically protected from water (in all its forms) and invasive vegetation by the application of "hard caps" of lime, amended soils and lime, or cement mortars (Ashurst 2007). These hard caps fail when cracks develop and water flows into the walls; this seepage, along with temperature differences, can ultimately lead to masonry collapse. The repeated repairs needed to maintain hard caps have been inadequate in the long-term management of ruins. In order to best conserve and manage archaeological structures, the authors argue that managers need to understand the sources and effects of environmental factors and then identify remedial and preventive conservation methods to minimize their damaging impacts upon features and sites.

The authors explore the usefulness of "soft capping," a procedure introduced in recent years at several archaeological sites in England and Turkey (Wood 2005). Soft capping replaces hard caps with vegetation planted on top of a layer of soil, with optional layers of gravel, and sometimes geo¬synthetics (Matero et al. 2008). The authors present information where they compare methods for the protection of the wall tops and the prevention of water penetration into the wall core, utilizing two types of caps that were installed on top of a test wall: 1) a hard mortar and stone cap based on the Park's current methods of stabilization; and 2) an experimental soft vegetative cap. Temperature and moisture probes were installed in the wall masonry and surrounding ground/fill in order to monitor and evaluate the performance of each cap.

The comparison between the soft- and hard-cap temperature regime demonstrated the soft cap's ability to dampen temperature fluctuation within the wall. The soft caps were more robust and kept in less moisture because they dry faster due to exposure. The authors conclude that a soft cap should perform very well in dry climates with concentrated precipitation events, like that at Mesa Verde. Considered in combination with water-susceptible materials such as soil-based mortars, soft caps appear to be a far better solution to the problem of wall protection in the arid Southwest than hard caps. A second finding from this study shows the need to understand how water moves through the wall system, especially as it is influenced by grade and associated fill. They demonstrate that moisture uptake by the wall from the adjacent soil occurs for an extended period of time, especially as snow melts. This has direct implication on selecting and integrating preservation methods. They conclude this chapter with recommendations for future preservation activities including monitoring and assessment, and close with a discussion about soft capping as an attractive alternative to hard caps for archeological preservation.

Chapter 12 explores two competing subwatershed plans that were developed for an urbanizing area in the Town of Oakville, Ontario. Eagles et al. examine mapping schemes associated with these two plans, one designed by the Town's urban planners and the other by the area's land owners. Both groups developed their plans independently, but were given a special focus on ecological cores and linkages (e.g., Environmentally Sensitive Areas) to be protected from development. The similarities and differences in the two designs point to different methods and philosophies that underpinned the two groups in their planning process. The authors explore the differing interpretations in light of the politics of the planning process specifically related to negotiation and power dynamics within each group. In the Oakville case, more green space was set aside by the Town's team of urban planners and this conservative plan was adopted as a template for development prior to any conflict resolution undertaken by the participating parties. Most of the subsequent negotiations with representatives for the group of land owners occurred "behind closed doors." The authors explore the limitations of applying landscape connectivity literature to land use planning in urban areas and these findings should prove useful to managers on the Colorado Plateau.

Chapter 13, "Recovering endangered fish in the San Juan River using adaptive management" by Mark McKinstry, discusses the implementation of the San Juan River Basin Recovery Implementation Program (SJRIP). The SJRIP was begun in 1992 and authorized by Congress in 2000 to protect and recover endangered fishes (Colorado pikeminnow and razorback sucker) while allowing water development to proceed. The SJRIP is coordinated by the U. S. Fish and Wildlife Service and the Bureau of Reclamation, with the aim of using adaptive management to assist in the recovery of these fish species. The SJRIP program is a nationally recognized effort that has served as a model to address other Endangered Species Act issues throughout the country. McKinstry describes how the overall program works, defines the goals and objectives of the program, and details the responsibilities of the participants. The author explains the aggressive efforts that are being implemented through program elements and comprehensive plans to: construct fish passages, fish screens, and propagation facilities; restore and enhance aquatic habitat; improve water use efficiency; stock native fish and control non-native fish species; and conduct hydrologic evaluations that allow the adoption of natural-flow mimicry. This chapter highlights management activities under this program and how they are formulated in an adaptive management framework whereby specific actions are conducted, with subsequent monitoring of the fish and habitat used to inform and modify future activities. This adaptive management approach has allowed the SJRIP to make progress towards recovery in the face of uncertainty about various aspects of the biology and ecology of these endangered species.

## SECTION 4: TOOLS FOR CONSERVATION AND COLLABORATIVE DECISION-MAKING

In Chapter 14, Poff et al. employ statistical modeling techniques and Geographic Information Systems (GIS) to forecast changes in forest conditions based on multiple (and sometimes conflicting) management preferences and goals. Spatial modeling exercises that provide multiple potential outcomes are becoming indispensable tools for collaborative and comprehensive environmental planning processes (Sheppard and Meitner 2005). The modeling process employed in this study, called Multi-Objective Decision Making (MODM) fosters collaborative decision-making by identifying a compromise point between competing goals and preferences of multiple land managers.

In this study, members of the Mogollon Rim Ranger District Interdisciplinary Team were asked to identify ideal forest conditions based on their areas of expertise and management agendas, which included wildfire, weed control, silviculture, and wildlife habitat management. When coupled with GIS analyses and spatial data, managers' preferences, measured and modeled in terms of stand density, were visualized both over the landscape and into the future. Information garnered from the MODM exercise can help managers with competing agendas identify a compromise position and, when acted upon, allows for a more transparent decision making process.

In Chapter 15, Dreesen et al. present a number of novel planting techniques for stream restoration that aim to increase plant establishment and survival rates. Many riparian areas throughout the southwestern United States have undergone considerable changes over the past century due to human land uses and climatic changes (Webb et al. 2007). The deep-planting methods described in the paper are novel in that they maximize the use of existing soil moisture reservoirs

and minimize or eliminate the need for irrigation. The authors argue that riparian plants favor a deep-planting, root-burial type of planting technique because it mimics the effects of post-flooding sediment dynamics common to riparian systems.

The authors found that by deep-planting a long pole cutting directly into the capillary fringe many replanted phreatophytes can expect a 90% survival rate. For upland species, the entire root crown is deep-planted into the capillary fringe; however the upland plants typically require more irrigation and maintenance than pole-cut phreatophytes. Upland plant survival rates varied between 70-90%, with most of the mortality related to long-term root inundation during wet periods, and failure to properly irrigate following water table decline. The chapter provides sound guidance for ecologists and land managers seeking to enhance or restore degraded riparian zones and river channels, areas of particular importance for maintaining regional biodiversity and habitat connectivity.

## SECTION 5: A 20-YEAR EVALUATION OF THE COLORADO PLATEAU BIENNIAL CONFERENCE

The book closes with a chapter by Lee et al. that offers insight into the perceived value of the Biennial Conference of Research on the Colorado Plateau (COPL) among attendees. The authors conducted a series of evaluations from participants who attended the 2009 10th COPL conference to determine whether attendees believed the organizers' goal of scientific integration was attained through conference participation. Using an on-line survey of participants, Lee et al. examine the programmatic components (e.g., conference programming, networking opportunities, and additional activities) of the meeting and descriptively analyzed how social networks were developed and/ or social capital was built among attendees. This analytic approach allowed the authors to examine the perceived importance and satisfaction of various aspects of the meeting, while documenting the ways in which the conference helped to form relationships among scientists, managers and students.

The study findings suggest that after 20 years, the COPL meetings have brought a suite of professional benefits to conference attendees and can be improved in several important ways. The programmatic components of the Biennial conference are considered to be generally important and satisfactory; however, maintaining a reasonable cost and providing opportunities to network with speakers and other conference participants warrants special attention from future conference organizers. The analysis of social capital survey items suggests that interpersonal connections, social cohesion, secondary associations, and perceived utility were relevant dimensions of social networking that shaped the COPL conference experience. These findings detail how the COPL conferences could continue to provide meaningful experiences among attendees. This final chapter maintains a critical perspective on how the 10th Biennial Conference of Research on the Colorado Plateau can be improved so that future hosts and organizers can more efficiently invest their time, energy, and programming.

# REFERENCES

Ashurst, J. A. 2007. Conservation of Ruins. Burlington, MA, Butterworth-Heinemann.

Costanza, R., d'Arge, R., fe Groot, R., Farber, S., Grasso, M., and B. Hannon. 1997. The value of the world's ecosystem services and natural capital. Nature 387:253-260.

D'Antonio, C. M. and P. M. Vitousek. 1992. Biological invasions by exotic grasses, the grass/fire cycle, and global change. Annual Review of Ecology and Systematics 23:63 – 87.

Fulé, P. Z., W. W. Covington, and M. M. Moore. 1997. Determining Reference Conditions for Ecosystem Management of Southwestern Ponderosa Pine Forests. Ecological Applications 7(3): 895-908.

Garfin, G. M., J. K. Eischeid, M. T. Lenart, K. L. Cole, K. Ironsides, and N. Cobb. 2010. Yellow-billed Cuckoo Distribution and Habitat Associations in Arizona, 1998-1999. Pp. 21-44, In: The Colorado Plateau IV: Integrating research and resources management for effective conservation (van Riper, C., III, B. F. Wakeling, and T. D. Sisk, Eds). University of Arizona Press, Tucson, AZ. 335 pp.

Kay, C. E., 1997. Is aspen doomed? Journal of Forestry. 95(5): 4-11.

Landres, P.B., R.L. Knight, S.A. Pickett, and M.L. Cadenasso. 1998. Ecological Effects of Administration Boundaries, in "Stewardship Across Boundaries", Richard L. Knight and Peter B. Landres, Eds., Island Press, Washington D.C.

Matero, F, K. Wong and S. Stokely. 2008. Gordion 2008: Field Report for Terrace Building 2. Field Report submitted to the University of Pennsylvania, Department of Historic Preservation.

Millennium Ecosystem Assessment. 2003. Ecosystems and Human Well-being: A Framework for Assessment. Washington D.C., Island Press.

Miller, M. E., R. T. Belote, M. A. Bowker, and S.L. Garman. 2011. Alternative states of a semiarid grassland ecosystem: Implications for ecosystem services. Ecosphere 2(5):1-18.

Knapp, P.A., 1996. Cheatgrass (Bromus tectorum L) dominance in the Great Basin Desert: History, persistence, and influences to human activities. Global Environmental Change 6: 37-52.

Reynolds, R.T., J.M. Scott, and R.A. Nussbaum. 1980. A variable circular-plot method for estimating bird numbers. Condor 82:309-313.

Robbins, C. S, D. Bystrak, and P. H. Geissler. 1986. The breeding bird survey: its first fifteen years, 1965-1979. U.S.D.I. Fish and Wildlife Service, Resource Publication 157. Washington, D.C.

Saunders, D.A., R.J. Hobbs, and C.R. Margules. 1991. Biological Consequences of Ecosystem Fragmentation: A Review. Conservation Biology 5(1): 18-32.

Schier, G. A., J. R. Jones, and R. P. Winokur. 1985. Vegetative regeneration. in Aspen: Ecology and management in the western United States. (N. V. DeByle and R. P. Winokur, Eds.). USDA Forest Service General Technical Report RM-119: 29–33. Rocky Mountain Forest and Range Experiment Station, USDA Forest Service, Fort Collins, Colorado.

Sheppard, S. R. J., and M. Meitner. 2005. Using multi-criteria analysis and visualisation for sustainable forest management planning with stakeholder groups. Forest Ecology and Management 207 (1–2):171-187.

Vankat, J.L., D.C. Crocker-Bedford, D.R. Bertolette, P. Leatherbury, T. McKinnon, and C.L. Sipe. 2005. Indications of large changes in mixed conifer forests of Grand Canyon National Park. C. van Riper III, D.J. Mattson (Eds.), The Colorado Plateau II: Biophysical, Socioeconomic, and Cultural Research. Proceedings of the 7th Biennial Conference of Research on the Colorado Plateau, University of Arizona Press, Tucson, Arizona, pp. 121–129.

Vankat, J.L. 2011. Post-1935 changes in forest vegetation of Grand Canyon National Park, Arizona, USA: Part 2—Mixed conifer, spruce-fir, and quaking aspen forests. Forest Ecology and Management 261(3): 326-341.

Webb R.H., S.A. Leake, and R.M. Turner. 2007. The Ribbon of Green: Change in Riparian Vegetation in the Southwestern United States. Tucson, Arizona, University of Arizona Press.

Wilson, E.O. 1992. The Diversity of Life. Cambridge, MA, Harvard University Press.

Worrall, J. J., L. Egeland, T. Eager, R. A. Mask, E.W. Johnson, P. A. Kemp, and W. D. Shepperd. 2008. Rapid mortality of Populus tremuloides in southwestern Colorado, USA. Forest Ecology and Management 255: 686-696.

Wood, C. 2005. Soft Capping: Justifying a Return to Picturesque. IN Context 90: 22-24.

# CONTRIBUTORS

**William Auberle**
Landsward Institute
Northern Arizona University
PO Box 5845
Flagstaff, AZ 86011-5845
william.auberle@nau.edu
"Valuing the Verde River Watershed:
An Assessment"

**Ethan N. Aumack**
Grand Canyon Trust
2601 N. Fort Valley Road
Flagstaff, AZ 86001
eaumack@grandcanyontrust.org
"Administrative Boundaries and Ecological
Divergence: The Divided History and
Coordinated Future of Land Management
on the Kaibab Plateau, Arizona, USA"

**Kenneth J. Bagstad**
Mendenhall Postdoctoral Fellow, Research
Economist
U.S. Geological Survey
Rocky Mountain Geographic Science Center
P.O. Box 25046, MS 516, Denver, CO 80225
Kenneth.Bagstad@uvm.edu
"Challenges and opportunities for ecosystem
services science and policy in arid and semiarid
environments"

**Sara P. Bombaci**
Fort Lewis College, Natural and Behavioral
Sciences Department
1000 Rim Drive, Durango, CO 81301
sbombaci@fortlewis.edu
"The Effects of Sudden Aspen Decline on
Avian Species Composition and Biodiversity
in Southwestern Colorado"

**Claire Crow**
Zion National Park
Springdale, Utah 84767
claire_crow@nps.gov
"Avian Community Responses to Pinyon-Juniper
Woodland Thinning Treatments on the
Colorado Plateau"

**Brett G. Dickson**
School of Earth Sciences and Environmental
Sustainability
Laboratory of Landscape Ecology and
Conservation Biology
Northern Arizona University
Flagstaff, AZ 86001-5694
Brett.Dickson@nau.edu
"Administrative Boundaries and Ecological
Divergence: The Divided History and
Coordinated Future of Land Management
on the Kaibab Plateau, Arizona, USA"
"Assessment of Mixed Conifer Forest
Conditions, North Kaibab Ranger District,
Arizona, USA"

**David R. Dreesen**
Los Lunas Plant Materials Center
USDA-NRCS
1036 Miller St. SW
Los Lunas, NM 87031
david.dreesen@nm.usda.gov
"Innovative Planting Methods for the Arid and
Semi-Arid Southwest"

**Paul F. J. Eagles**
University of Waterloo, School of Planning
200 University Avenue
Waterloo, Ontario, Canada N2L 3G1
eagles@uwaterloo.ca
"Moving from Landscape Connectivity Theory
to Land Use Planning: Urban Planning in
Oakville, Ontario"

**Walter Fertig**
Moenave Botanical Consulting
1117 W Grand Canyon Dr., Kanab
Utah 84741
walt@kanab.net
"Knowing the Cogs and Wheels: Using Bioblitz
Methods to Rapidly Assess Floras and Faunas"

**Gregory A. Fenchel**
Los Lunas Plant Materials Center
USDA-NRCS
1036 Miller St. SW
Los Lunas, NM 87031
gregory.fenchel@nm.usda.gov
"Innovative Planting Methods for the Arid
and Semi-Arid Southwest"

Thomas L. Fleischner
Environmental Studies Program, Prescott
College
220 Grove Ave, Prescott, AZ 86301
tfleischner@prescott.edu
"Revisiting Trends in Vegetation Recovery
Following Protection from Grazing,
Chaco Culture National Historic Park,
New Mexico"

M. Lisa Floyd
Environmental Studies Program
Prescott College
220 Grove Avenue, Prescott, AZ 86301
lfloyd-hanna@prescott.edu
"Revisiting Trends in Vegetation Recovery
Following Protection from Grazing,
Chaco Culture National Historic Park,
New Mexico"

Brian W. Geils
Forest Service, Rocky Mountain Research
Station
2500 South Pine Knoll Drive
Flagstaff, AZ 86001
bgeils@fs.fed.us
"Spatio-Temporal Multi-Objective Decision
Making in Forest Management"

Danny G. Goodson
Los Lunas Plant Materials Center
USDA-NRCS
1036 Miller St. SW
Los Lunas, NM 87031
danny.goodson@nm.usda.gov
"Innovative Planting Methods for the Arid
and Semi-Arid Southwest"

David D. Hanna
Environmental Studies Program, Prescott
College
220 Grove Ave, Prescott, AZ 86301
dhanna@prescott.edu
"Revisiting Trends in Vegetation Recovery
Following Protection from Grazing,
Chaco Culture National Historic Park,
New Mexico"

Michael Henry
The Architectural Conservation Laboratory
University of Pennsylvania
Philadelphia, PA 19104
henrmic@design.upenn.edu
"Vegetative Capping of Archaeological
Masonry Walls"

Christopher M. Holcomb
School of Earth Sciences and Environmental
Sustainability
Laboratory of Landscape Ecology and
Conservation Biology, Northern Arizona
University
P.O. Box 5694, Flagstaff, AZ 86011-5694
christopher.holcomb@nau.edu
"Administrative Boundaries and Ecological
Divergence: The Divided History and
Coordinated Future of Land Management
on the Kaibab Plateau, Arizona, USA"

Matthew J. Johnson
Colorado Plateau Research Station,
Northern Arizona University, Box 5614,
Flagstaff, AZ 86011
Matthew.Johnson@nau.edu
"Synthesis: Informing Collaborative
Conservation and Management of Colorado
Plateau Resources"

Julie Korb
Fort Lewis College, Natural and Behavioral
Sciences Department
1000 Rim Drive, Durango, CO 81301
korb_j@fortlewis.edu
"The Effects of Sudden Aspen Decline on
Avian Species Composition and Biodiversity
in Southwestern Colorado"

Gerard T. Kyle
Human Dimensions of Natural Resources
Laboratory
Department of Recreation, Park and Tourism
Sciences
Texas A&M University
College Station, TX, 77843, USA
gerard@tamu.edu
"Reaching Toward the Integration of Research
into Resource Management Activities:

A 20-Year Evaluation of Colorado Plateau
Biennial Conferences"

## Martha E. Lee

School of Forestry, Box 15018
Earth and Environmental Sciences
Northern Arizona University
Flagstaff, AZ, 86011, USA
e-mail: Martha.lee@nau.edu
"Reaching Toward the Integration of Research
into Resource Management Activities:
A 20-Year Evaluation of Colorado Plateau
Biennial Conferences"

## Alex B. Lim

The Architectural Conservation Laboratory
University of Pennsylvania
Philadelphia, PA 19104
alexlim@design.upenn.edu
"Vegetative Capping of Archaeological
Masonry Walls"

## Frank G. Matero

The Architectural Conservation Laboratory
University of Pennsylvania
Philadelphia, PA 19104
fgmatero@design.upenn.edu
"Vegetative Capping of Archaeological
Masonry Walls"

## Mark C. McKinstry

Bureau of Reclamation, Office of Adaptive
Management
125 South State Street, UC-735, Salt Lake City,
Utah 84138
mmckinstry@usbr.gov
"Recovering Endangered Fish in The San Juan
River Using Adaptive Management"

## Lauren A. Mork

School of Earth Sciences and Environmental
Sustainability
Laboratory of Landscape Ecology and
Conservation Biology
Northern Arizona University, P.O. Box 5694
Flagstaff, AZ 86011
lmork@restorethedeschutes.org
"Livestock Grazing Following Wildfire:
Understory Community Response in an Upland
Ponderosa Pine Forest"

## Daniel G. Neary

Forest Service, Rocky Mountain Research
Station
2500 South Pine Knoll Drive
Flagstaff, AZ 86001
dneary@fs.fed.us
"Spatio-Temporal Multi-Objective Decision
Making in Forest Management"

## Elke Meyfarth O'Hara

University of Waterloo, School of Planning
200 University Avenue
Waterloo, Ontario, Canada N2L 3G1
ermeyfar@uwaterloo.ca
"Moving from Landscape Connectivity Theory
to Land Use Planning: Urban Planning in
Oakville, Ontario"

## Boris Poff

Forest Service, Rocky Mountain Research
Station
2500 South Pine Knoll Drive
Flagstaff, AZ 86001
borispoff@hotmail.com
"Spatio-Temporal Multi-Objective Decision
Making in Forest Management"

## Jill M. Rundall

Laboratory of Landscape Ecology and
Conservation Biology
School of Earth Sciences and Environmental
Sustainability
Northern Arizona University, PO Box 5767.
Flagstaff, AZ 86011-5767
jill.rundall@nau.edu
"Assessment of Mixed Conifer Forest
Conditions, North Kaibab Ranger District,
Arizona, USA"

## S. Shane Selleck

Sonoran Desert Research Station
University of Arizona, SNRE
1110 E. South Campus Drive, #116
Tucson, AZ 85704
sselleck@usgs.gov
"Synthesis: Informing Collaborative
Conservation and Management of Colorado
Plateau Resources"

**Darius Semmens**
U.S. Geological Survey
Rocky Mountain Geographic Science Center
P.O. Box 25046, MS 516
Denver, CO 80225
"Challenges and opportunities for ecosystem
services science and policy in arid and semiarid
environments"

**Steven E. Sesnie**
School of Earth Sciences and Environmental
Sustainability
Laboratory of Landscape Ecology and
Conservation Biology
Northern Arizona University, P.O. Box 5694
Flagstaff, AZ 86011
steven.sesnie@nau.edu
"Administrative Boundaries and Ecological
Divergence: The Divided History and
Coordinated Future of Land Management on the
Kaibab Plateau, Arizona, USA"
"Assessment of Mixed Conifer Forest
Conditions, North Kaibab Ranger District,
Arizona, USA"

**Brad Shattuck**
Canyon De Chelly National Monument
PO Box 588, Chinle, AZ 86503.
brad_shattuck@nps.gov
"Revisiting Trends in Vegetation Recovery
Following Protection from Grazing,
Chaco Culture National Historic Park, New
Mexico"

**Thomas D. Sisk**
School of Earth Sciences and Environmental
Sustainability
Laboratory of Landscape Ecology and
Conservation Biology
Northern Arizona University, P.O. Box 5694
Flagstaff, AZ 86011
Thomas.Sisk@nau.edu
"Administrative Boundaries and Ecological
Divergence: The Divided History and
Coordinated Future of Land Management on the
Kaibab Plateau, Arizona, USA"
"Livestock Grazing Following Wildfire:
Understory Community Response in an Upland
Ponderosa Pine Forest"
"Assessment of Mixed Conifer Forest
Conditions, North Kaibab Ranger District,
Arizona, USA"

**Dean Howard Smith**
P.O. Box 15066
Northern Arizona University
Flagstaff AZ, 86011-5066
Dean.smith@nau.edu
"Valuing the Verde River Watershed: An
Assessment"

**John Spence**
Glen Canyon National Recreation Area
691 Scenic View Drive, Page, AZ 86040-1507
John_Spence@nps.gov
"Knowing the Cogs and Wheels: Using Bioblitz
Methods to Rapidly Assess Floras and Faunas"

**Katie J. Stumpf**
Northern Arizona University
Department of Biology
P.O. Box 5640
Flagstaff, AZ 86011
Katie.stumpf@nau.edu
"Landscape Scale Features Predict Predation
and Parasitism On Passerine Nests: A Literature
Review"

**Aregai Tecle**
School of Forestry
Northern Arizona University
P.O. Box 5018
Flagstaff, AZ 86011
Aregai.Tecle@nau.edu
"Spatio-Temporal Multi-Objective Decision
Making in Forest Management"

**Carena J. van Riper**
Human Dimensions of Natural Resources
Laboratory
Department of Recreation, Park and Tourism
Services
Texas A&M University
College Station, TX 77843
cvanripe@tamu.edu
"Reaching Toward the Integration of Research
into Resource Management Activities:
A 20-Year Evaluation of Colorado Plateau
Biennial Conferences"
"Synthesis: Informing Collaborative
Conservation and Management Of Colorado
Plateau Resources"

## Charles van Riper III

U.S. Geological Survey
Southwest Biological Science Center
Tucson, AZ 85711
charles_van_riper@usgs.gov
"Reaching Toward the Integration of Research
into Resource Management Activities:
A 20-Year Evaluation of Colorado Plateau
Biennial Conferences"
"Challenges and opportunities for ecosystem
services science and policy in arid and semiarid
environments"
"Avian Community Responses to Pinyon-Juniper
Woodland Thinning Treatments on the
Colorado Plateau"
"Synthesis: Informing Collaborative
Conservation and Management Of Colorado
Plateau Resources"

## Miguel Villarreal, Ph.D.

Mendenhall Fellow - Research Geographer
U.S. Geological Survey
Western Geographic Science Center
520 N. Park Avenue, Suite #102G, Tucson, AZ
85719-5035
mvillarreal@usgs.gov
"Synthesis: Informing Collaborative
Conservation and Management Of Colorado
Plateau Resources"

## Patricia West

Landsward Institute
Northern Arizona University
PO Box 5845
Flagstaff, AZ 86011-5845
patty.west@nau.edu
"Valuing the Verde River Watershed: An
Assessment"

## Keith L. White

Los Lunas Plant Materials Center
USDA-NRCS
1036 Miller St. SW
Los Lunas, NM 87031
keith.white@one.usda.gov
"Innovative Planting Methods for the Arid and
Semi-Arid Southwest"

## Graham Whitelaw

Queen's University, School of Planning
99 University Avenue
Kingston, Ontario, Canada K7L 3N6
graham.whitelaw@queensu.ca
"Moving from Landscape Connectivity Theory
to Land Use Planning:  Urban Planning in
Oakville, Ontario"

## Linda Whitham

The Nature Conservancy
PO Box 1329, Moab, Utah 84532
lwhitham@tnc.org
"Knowing the Cogs and Wheels: Using Bioblitz
Methods to Rapidly Assess Floras and Faunas"

# INDEX